Probabilistic Models of Cosmic Backgrounds

Combining research methods from various areas of mathematics and physics, Probabilistic Models of Cosmic Backgrounds describes the isotropic random sections of certain fibre bundles and their applications to creating rigorous mathematical models of both discovered and hypothetical cosmic backgrounds.

Previously scattered and hard-to-find mathematical and physical theories have been assembled from numerous textbooks, monographs, and research papers, and explained from different or even unexpected points of view. This consists of both classical and newly discovered results necessary for understanding a sophisticated problem of modelling cosmic backgrounds.

The book contains a comprehensive description of mathematical and physical aspects of cosmic backgrounds with a clear focus on examples and explicit calculations. Its reader will bridge the gap of misunderstanding between the specialists in various theoretical and applied areas who speak different scientific languages.

The audience of the book consists of scholars, students, and professional researchers. A scholar will find basic material for starting their own research. A student will use the book as supplementary material for various courses and modules. A professional mathematician will find a description of several physical phenomena at the rigorous mathematical level. A professional physicist will discover mathematical foundations for well-known physical theories.

Anatoliy Malyarenko received his PhD degree at Taras Shevchenko National University of Kyiv in 1985. He is a professor in the Division of Mathematics and Physics at Mälardalen University, Sweden. His research interests include random fields on manifolds with physical applications and financial mathematics.

Probabilistic Models of Cosmic Backgrounds

Anatoliy Malyarenko

CRC Press
Taylor & Francis Group
Boca Raton London New York

CRC Press is an imprint of the
Taylor & Francis Group, an **informa** business

Designed cover image: shutterstock_481604845

First edition published 2024
by CRC Press
2385 NW Executive Center Drive, Suite 320, Boca Raton FL 33431

and by CRC Press
4 Park Square, Milton Park, Abingdon, Oxon, OX14 4RN

CRC Press is an imprint of Taylor & Francis Group, LLC

© 2024 Anatoliy Malyarenko

ISBN: 978-1-032-38198-5 (hbk)
ISBN: 978-1-032-38293-7 (pbk)
ISBN: 978-1-003-34435-3 (ebk)

DOI: 10.1201/9781003344353

Typeset in LM Roman
by KnowledgeWorks Global Ltd.

To all freedom fighters

Contents

Foreword

The role of group representation theory in Quantum Mechanics, Quantum Field Theory, and Particle Physics has been the object of an enormous amount of research and a huge number of books, for instance *Group Theory and Physics* by S. Sternberg and the references therein for a classic reference, among the many one may quote. Although it is less discussed, a deep connection exists as well between Group Representation Theory and Cosmology: this book is an attempt to bridge the gap and explore the relations between these areas. In some sense, it is a follow-up of another monograph by the same author, 'Invariant Random Fields on Spaces with a Group Action', which appeared in press about ten years ago.

The book is divided into three main parts: the first part provide a general background on Abstract Algebra, with special emphasis on Lie Groups, Vector and Tensor Fields, Group Representations, Spin Bundles, and related topics. The second part probes more deeply into Cosmology and Physics and, in particular, discusses the five fundamental notions of spacetime which have evolved throughout history: *Aristotelian, Galileian, Newtonian, Minkowskian, and Einsteinian*, as in the famous characterisation by Roger Penrose. Much attention is of course devoted to the latter, discussing cosmological solutions to Einstein's equation, the Robertson-Walker metric and the equations for polarised Cosmic Microwave Background (CMB) emission, and neutrinos and primordial gravitational waves. The last part discusses how representations of SU(2) and SO(3) can be exploited when building spectral expansions for spin random fields, i.e. sections of tensor bundles emerging as the natural model for the previously mentioned radiation fields. Overall the material is technically advanced, highly non-trivial, and outside than what can be usually found in standard mathematical references.

The capacity of Mathematics to build connections between its different areas and to provide explanations for physical reality is an endless source of wonder, as summarised by E. Wigner in the celebrated paper 'The Unreasonable Effectiveness of Mathematics in the Natural Sciences'. This book provides one more such example, by joining together many different fields: Abstract Algebra and Group Representations, Cosmology, General Relativity and Differential Geometry, Special Functions, and Spectral Representations for spin-valued Random Fields with their cosmological applications.

It may take a long while before neutrino and gravitational wave backgrounds are experimentally detected, whereas CMB polarisation is already the object of extremely ambitious satellite or ground-based experiments which will dominate cosmological research in the next decade (*LiteBIRD, Simons Observatory, CMB-S4...*). It is hoped that this book will help many mathematicians to recognise this area as a new source of inspiration for challenging problems with extremely deep motivations.

Rome, October 2023   *Domenico Marinucci*

Preface

At the end of the 1950s, in Yadrenko (1959), my teacher M. Ĭ. Yadrenko discovered that wide-sense isotropic random fields on a sphere can be characterised as follows: the random coefficients of their Fourier expansion with respect to the complete orthonormal basis of real-valued spherical harmonics are uncorrelated, only the monopole coefficient may have nonzero expected value, and the coefficients with different azimuthal numbers inside a given multipole have the same variance.

In a different line of research, in the middle of the 1960s, two radio engineers, Arno Penzias and Robert Wilson, tested a new antenna, found the excess temperature, and serendipitously discovered the Cosmic Microwave Background (CMB), an electromagnetic radiation from the early Universe filling all space. In the abstract to Penzias and Wilson (1965), they described the CMB as

> This excess temperature is, within the limits of our observations, isotropic, unpolarized, and free from seasonal variations (July, 1964–April, 1965).

In the beginning of the 1970s, theoretical models predicted the existence of anisotropy of the CMB at the level of about 10^{-5}. About 20 years later, the anisotropy of the CMB was confirmed. A mathematical model of the CMB temperature in the form of a Gaussian isotropic random field on the sky sphere was developed. However, the random field was expanded into the Fourier series with respect to the complete orthonormal basis of *complex-valued* spherical harmonics instead of real-valued ones. The author's attempt to understand the cause for that modification became the first reason for writing this book.

Meanwhile, the theoretical models continued to develop and predicted the polarisation of the CMB caused by several physical mechanisms, including the Thomson scattering and primordial gravitational waves. Physicists have written correct spectral expansions of the CMB polarisation in the 1990s, see (Kamionkowski, Kosowsky, and Stebbins 1997; Zaldarriaga and Seljak 1997). The polarisation of the CMB was confirmed in the 2000s. A mathematical model of this phenomenon is *not* a random field, but a more complicated object: a random cross-section of a certain non-trivial fibre bundle over the sky sphere.

At the shift of millennia, the development of cosmology culminated in the so-called ΛCDM or Lambda Cold Dark Matter model. The above model predicts the existence of the three cosmic backgrounds:

- the Cosmic Gravitational Background, decoupled from the rest of the universe about 10^{-33} sec after the Big Bang;

- the Cosmic Neutrino Background, decoupled about 1 sec after the Big Bang;

- the CMB decoupled about $380,000$ years after the Big Bang.

The rigourous models of these backgrounds appeared only in 2010s in (Baldi and Rossi 2014; Geller and Marinucci 2010; Malyarenko 2011). They included a variety of other types of 'spherical harmonics'.

At the same time, an interesting idea was formulated in Baez (2012). On the one hand, in the current standard model of particle physics, both six quarks and six leptons are divided into three *generations*: pairs of particles that demonstrate a similar physical behaviour. For example, the first generation of leptons includes electron and electron neutrino, the second one muon and muon neutrino, and the third one tau and tau neutrino. On the other hand, modern quantum physics is based on the theory of complex unitary representations of topological groups. The latter come in three flavours: the representations of *real*, *complex*, and *quaternionic* type. It is supposed that there exists a link between three generations of particles and three types of complex representations.

In 2017, I participated in the workshop 'Isotropic Random Fields in Astrophysics' organised at the Cardiff University, UK. The scientists working in the spectral theory of isotropic random fields, astrophysics, and data analysis have being invited to present the results of their work. After the workshop I realised that it could be good to write a book that introduces the mathematical foundations of the theory of the cosmic backgrounds for both mathematicians and physicists and helps them to understand each other by the explanation of both mathematical and physical foundations of the theory.

The mathematicians who start to apply their knowledge in the theory of cosmic backgrounds, usually come from the probability and statistics community. In Chapter 1, they will find the survey of the mathematical foundations of the above theory.

On the one hand, we explain the elements of linear algebra and representations of algebras and groups over the fields of real numbers $\mathbb{R}$, complex numbers $\mathbb{C}$, and the skew field of quaternions $\mathbb{H}$. In this way, the concept of a *spinor* appears.

On the other hand, the most important representations of a given group G called the *induced* ones, act in the Hilbert spaces of square-integrable cross-sections of a special kind of fibre bundles called the *homogeneous bundles*. In order to explain these concepts, a solid background in modern differential geometry is included in the chapter. In particular, we consider in a unified way homogeneous fibre bundles whose fibres are (right) finite-dimensional linear spaces over the above three (skew) fields. The elements of the variational calculus, important for physical applications, are explained as well.

The structure of an induced representation is determined by two fundamental results: *Schur's Lemma* and the *Frobenius Reciprocity*. While the formulation of the former result for the case of a complex representation is very
simple, the real and quaternionic cases are more sophisticated. In my opinion,
here is the reason why physicists usually write the spectral expansions of cosmic backgrounds in terms of complex-valued bases. The latter result will find
important applications in Chapter 3.

Finally, in our short explanation of the necessary concepts of probability
we emphasise the role of random elements of finite-dimensional linear spaces
over all three (skew) fields $\mathbb{R}$, $\mathbb{C}$, and $\mathbb{H}$.

In Chapter 2, the mathematicians will find the necessary material to understand the physical foundations of the theory, while the physicists will be
introduced into a mathematical language that rigorously describes familiar
physical theories. In particular, we describe in details the three different space-
times that Penrose (1968) associates with Aristotle, Galilei, and Newton, as
well as the spacetimes by Minkowski and Einstein. In the former three ones,
the space and time are separated, while in the latter two, they constitute a
unified structure.

The physical laws acting in space-times and spacetimes are deduced from
variational principles. These include the elements of classical mechanics and
electrodynamics, as well an special and general relativity and cosmology. For
the latter, the non-variational *Copernican principle* is also important: the Universe is homogeneous and isotropic at big scales. This principle is formalised
mathematically into a concept of an *isotropic spacetime manifold*, and the
celebrated Robertson–Walker metric is deduced.

Physical fields appear in such introduced theories as cross-sections of fibre
bundles over spacetimes. Among them, the most important class for applications to cosmic backgrounds appears as the special solutions to relativistic
wave equations called the *plane waves*. A uniform description of neutrino,
electromagnetic, and gravitational plane waves is also included.

An important part of the theory of cosmic backgrounds should be em-
phasised here. Specifically, at least two of the neutrino generations described
above, have a nonzero mass. By this reason, they may propagate through the
spacetime with an arbitrary speed less than that of light. Such a propaga-
tion fails to be isotropic and should be described by a different mathematical
theory lying outside the scope of this book.

On the other hand, the polarisation of an arbitrary cosmic background
appears as a linear operator acting in a special linear space. For the case
of the massless neutrino (resp., electromagnetic waves, resp., gravitational
waves), the above linear space consists of scalars (resp., vectors, resp., skew-
symmetric rank 2 tensors) and has dimension 1 (resp., 2, resp., 2). By this
reason, the Cosmic Neutrino Background cannot be polarised.

In Chapter 3, we introduce isotropic random cross-sections of homogeneous fibre bundles. Some of them serve as a rigourous mathematical model
of a certain cosmic background. Any (non-necessarily isotropic) random

cross-section, can be represented as a Fourier series with respect to an orthonormal basis of the Hilbert space $L \uparrow G$ where an induced representation acts, with random coefficients. The most convenient random Fourier series are those with uncorrelated coefficients. For an isotropic random cross-section, the coefficients are uncorrelated if the vectors of the basis of $L \uparrow G$ are *special*. This means the following. Under the action of an induced representation of G, a given non-zero vector $\mathbf{x} \in L \uparrow G$ draws the orbit $\{ g \cdot \mathbf{x} \colon g \in G \} \subset L \uparrow G$. The minimal closed subspace of $L \uparrow G$ that contains that orbit, carries a representation of G. We call $\mathbf{x}$ *special* if the constructed representation is irreducible.

In other words, a special basis is the union of the bases of finite-dimensional subspaces of $L \uparrow G$, where the irreducible components of the induced representation act. While obtaining a spectral expansion of the random cross-section, one uses Schur's Lemma to prove that the Fourier coefficients with respect to the basis are uncorrelated. That is where we use the fact that the representation is irreducible.

To construct the special bases, we use the Frobenius reciprocity that establishes an isomorphism between the spaces of the isotypic components of the induced representation of G and the spaces where the irreducible representations of G act. The orthonormal bases of the latter spaces are well-known. Their images under the Frobenius isomorphism are special orthogonal bases. A significant part of the zoo of different spherical harmonics and the corresponding spectral expansions are constructed in Chapter 3. We also give physical applications of the spectral expansions in question. Finally, we shortly describe where the reader can use the knowledge obtained while reading this book.

Appendix A is written for those who would like to go further into some beautiful theoretical concepts on which the material of the book is based. These includes commutative diagrams, categories, universality, and spin manifolds.

An important feature of the book is as follows: it does not contain many proofs but does contain many examples which may help to better understand complicated constructions.

Both Master and PhD students can use various parts of the book as a supplementary reading. Early career researchers can use the book as a starting point for future research in this complex subject. Established researchers may use it as a reference book.

The practitioners who process the results of high-precision observations of cosmic backgrounds may obtain a service that is not available elsewhere: a simple explanation how the spectral expansions of the above backgrounds appear and why they are useful in practice.

Västerås, October 2023   *Anatoliy Malyarenko*

Acknowledgements

First and foremost, I am grateful to my teacher Professor M. Ĭ. Yadrenko, who introduced me to the beautiful world of Probability and especially to random fields.

My colleagues from the Division of Mathematics and Physics at the Mälardalen University created a very friendly working and research environment, which helped me a lot while writing this book.

Very useful discussions with my colleagues Victor Abramov, Paolo Baldi, Domenico Marinucci, and Maurizia Rossi, as well as with my coauthors Mykola Leonenko, Andriy Olenko, and Martin Ostoja-Starzewski improved the explanation of several non-trivial concepts.

I am also grateful to the members of the TeX–LaTeX Stack Exchange community at `tex.stackexchange.com` for excellent figures, to Marc van Leeuwen for an exhaustive explanation concerning complexifying of vector spaces in Leeuwen (2011), and to Pablo Carlos Budassi for the nice picture of the CMB sphere.

Finally, many thanks to the organisers and participants of the workshop 'Isotropic Random Fields in Astrophysics' which was held at the Cardiff University, UK, in June 2017. The idea to write this book appeared right there.

Västerås, October 2023

Anatoliy Malyarenko

List of Figures

List of Tables

List of Symbols

A	An algebra
A	An atlas
$\mathsf{A}_{\max}$	The maximal atlas
A^+	The even Clifford algebra
$a_{\ell m}$	The random multipole coefficients
B	A symmetric bilinear form
$\mathbf{B}$	Magnetic field
$\mathfrak{B}(L)$	The Borel σ-field
c	The speed of light in vacuum
C	A category
C_ℓ	The temperature power spectrum
$\mathbb{C}$	The field of complex numbers
$\mathcal{C}$	The causal cone
$\mathcal{C}_+$	The future causal cone
$\mathcal{C}_-$	The past causal cone
$C_{\alpha\beta\gamma\delta}$	The Weyl tensor
$\mathrm{Cl}(L, B)$	A Clifford algebra
Cl_n	A complex Clifford algebra
Cl_n^+	An even complex Clifford algebra
$\mathrm{Cl}_{p,q}$	A real Clifford algebra
$\mathrm{Cl}_{p,q}^+$	An even complex Clifford algebra
$\mathbb{C}P^1$	The complex projective line
$\mathcal{C}_0$	The mantle of the forward light cone
D_A	The angular diameter distance
D_l	The lookback distance
D_L	The luminosity distance
$D^\ell_{mm'}(\varphi, \theta, \psi)$	The Wigner D function
d	The degree of polarisation
$\mathrm{d}g$	A Haar measure
E	The expected value
$\mathbf{E}$	Electric field
(E, π_L, M)	A fibre bundle
$\mathbf{e}_\theta, \mathbf{e}_\varphi$	The canonical basis
$\mathbf{e}^+, \mathbf{e}^-$	The helicity basis
$\mathbb{F}$	A field
$F^{\alpha\beta}$	The electromagnetic tensor
$g_{\mu\nu}$	A Riemannian metric

G	A Lie group
G_N	The Newton gravitational constant
$G \cdot p$	The orbit of a point p
$\mathrm{GL}(n, \mathbb{K})$	The general linear group
g	The determinant of a metric
$g_{\mu\nu}$	The metric
H	The Hubble constant
$\mathbb{H}$	The skew field of quaternions
$\check{H}_0(M; G)$	The non-abelian Čech cohomology group
$\check{H}_1(M; G)$	The non-abelian Čech cohomology set
$h_\times$	The amplitude of the 'cross' polarisation
h_+	The amplitude of the 'plus' polarisation
$\mathrm{Hom}_\mathsf{C}(X, Y)$	The morphisms from X to Y
$\mathrm{Hom}_\mathbb{K}(L_1, L_2)$	The set of all $\mathbb{K}$-linear maps from L_1 to L_2
$\mathrm{Hom}_{\mathbb{K}G}(L_1, L_2)$	The set of all intertwining operators from L_1 to L_2
i	The imaginary unit
I, Q, U, V	The Stokes parameters
j	The basic quaternion
$K_{\mathbf{XY}}$	The mutual correlation operator
$\mathbb{K}$	A skew field
$\mathbb{K}^1$	A one-dimensional linear space over $\mathbb{K}$
k	The basic quaternion
L	A linear space over $\mathbb{K}$
(L, R, φ)	An adjunction
L^*	The linear space dual to L
$L \uparrow G$	The space of square-integrable cross-sections
$L(x^i(t), \dot{x}^i(t), t)$	The Lagrangian of a mechanical system
$\mathcal{L}(\Phi, \dot{\Phi}, \mathbf{x}, t)$	The Lagrangian density
$\mathcal{L}_p^+$	The future light cone
ℓ	The multipole number
M	A manifold
$\mathrm{Mor}\,\mathsf{C}$	The morphisms of the category C
m	The azimuthal number
$\mathrm{O}(n)$	The compact real orthogonal group
$\mathrm{O}(n, \mathbb{C})$	The complex orthogonal group
$\mathrm{O}(p, q)$	The real orthogonal group
$\mathrm{Ob}\,\mathsf{C}$	The objects of the category C
P	The polarisation tensor
P_ℓ	The Legendre polynomial
(P, π, M, G)	A principal bundle
q	A quaternion
qcU_0	The G-space of the trivial irreducible quaternionic representation
R	The Ricci tensor
$\check{R}^\varepsilon{}_{\alpha\beta\gamma}$	The Riemann curvature tensor

$R_{\alpha\beta}$	The Ricci tensor
$\mathbb{R}$	The field of real numbers
r	The comoving distance
S	A linear space of spinors
s_i	The normalised Stokes parameters
S^{d-1}	The centred unit sphere in $\mathbb{R}^d$
$S[\gamma]$	The action of a mechanical system
$S[\Phi]$	The action of the field Φ
$\mathrm{SL}(n,\mathbb{F})$	The special linear group over $\mathbb{F}$
$\mathrm{SO}(L,g)$	Either $\mathrm{SO}_0(p,q)$ or $\mathrm{SO}(n,\mathbb{C})$
$\mathrm{SO}(n)$	The special compact orthogonal group
$\mathrm{SO}(n,\mathbb{C})$	The special complex orthogonal group
$\mathrm{SO}_0(p,q)$	The special real orthogonal group
$\mathrm{Sp}(n)$	The compact symplectic group
$\mathrm{Sp}(p,q)$	The symplectic group
$\mathrm{Spin}(L,g)$	The spin group
$\mathrm{Spin}^0(L,g)$	The reduced spin group
$\mathrm{Spin}(n)$	The compact real spin group
$\mathrm{Spin}(n,\mathbb{C})$	The complex spin group
$\mathrm{Spin}(p,q)$	The real spin group
$\mathrm{Spin}^0(p,q)$	The reduced real spin group
$\mathrm{SU}(n)$	The compact special unitary group
$\mathrm{SU}(p,q)$	The special unitary group
$\mathrm{SU}^*(2n)$	The special linear group over $\mathbb{H}$
T^k_l	The linear space of multi-linear forms
$T_p M$	The tangent space of M at p
$T^{\mu\nu}$	The energy-momentum tensor
$T^\mu_{\alpha\beta}$	The torsion tensor
$\mathsf{T}^{2\ell',\ell m}$	The pure-orbital tensor spherical harmonics
$\mathsf{T}^{E2,\ell m}$, $\mathsf{T}^{B2,\ell m}$	The pure-spin tensor spherical harmonics
t	Cosmic time
tr	The trace
U	A real linear space
U_0	The G-space of the trivial irreducible real representation
$(\mathcal{U},\mathbf{h})$	A chart
$\{\mathcal{U}_{\alpha\beta}, f_{\alpha\beta}\}$	A cocycle
$\mathrm{U}(n)$	The compact unitary group
$\mathrm{U}(p,q)$	The unitary group
V	A complex linear space
V	The variance
W	A quaternionic linear space
w_q	The Stiefel–Whitney characteristic classes
$\mathbf{x},\ldots$	Vectors
X	A vector field
$(X \to F)$	Comma category

$Y_{\ell,m}$	The complex-valued spherical harmonics
${}_sY_{\ell,m}$	The complex-valued spin-weighted spherical harmonics
Y_ℓ^m	The real-valued spherical harmonics
$\mathbf{Y}^{\ell',\ell m}$	The pure-orbital vector spherical harmonics
$\mathbf{Y}^{E,\ell m}$, $\mathbf{Y}^{B,\ell m}$	The pure-spin vector spherical harmonics
$Y^G_{(\ell m)}$	The gradient tensor spherical harmonics
$Y^C_{(\ell m)}$	The curl tensor spherical harmonics
${}_sZ_{\ell m}$	The real-valued spin-weighted spherical harmonics
z	The redshift
$\alpha, \ldots$	Real numbers
γ	A Clifford map
γ_i	Mathematical gamma matrices
γ_{n+1}	Mathematical chirality operator
Γ	The Clifford–Lipschitz group
ΓE	The space of continuous cross-sections
Γ_i	Physical gamma matrices
Γ_{n+1}	Physical chirality operator
$\Gamma^\mu_{\alpha\beta}$	The Christoffel symbols
$\delta J(\gamma)$	The first variation of the functional J
$\frac{\delta S}{\delta \Phi}$	The variational derivative
ε	A skew-symmetric bilinear form
η_{ij}	The metric tensor of the space $\mathrm{R}^{p,q}$
ζ	A complex number
θ	The expansion scalar
θ_g	The operator of a group representation
Λ	The cosmological constant
$\xi[F]$	A fibre bundle with fibre F
ρ	The operator of an algebra representation
ρ_0	The critical density
σ	A Clifford map for Cl_{2m+1} and $\mathrm{Cl}_{p,q}$ with odd $p+q$
$\sigma^{A\dot{A}}{}_i$	The Infeld–van der Waerden symbols
σ_i	The Pauli matrices
τ	The tilt angle
Φ	A scalar field
$\varphi_X(t)$	The characteristic function
φ^*T	The pull back of T
ψ_A	A left-handed Weyl field
$\psi_{\dot{A}}$	A right-handed Weyl field
ψ^*X	The push forward of X
$\chi_L(g)$	The character of a representation L
$(\Omega, \mathfrak{F}, \mathsf{P})$	A probability space
1_X	The identity morphism
$\square$	The d'Alembert operator
$\eth_s$	The spin-raising operator

$\overline{\eth}_s$	The spin-lowering operator
$\nabla_{A\dot{A}}$	The spinor covariant derivative
$\nabla_X Y$	A Koszul connection
$\nabla_\alpha Y_\beta$	The covariant derivative of Y
$\oplus$	The direct sum
$\otimes$	The tensor product
$\smile$	The cup product
$\wedge$	The grade involution

1

Mathematical Preliminaries

In the first few paragraphs of these introductory notes, we give a non-formal and non-rigourous explanation of mathematical tools that are used for the description of cosmic backgrounds. During the explanation, we introduce several mathematical concepts and notions. Afterwards, we describe the content of every subsequent section where the introduced notions are rigorously explained.

A detector of a particular cosmic background observes the values of the so-called polarisation tensor of the background's radiation on a finite subset of the sky sphere. It turns out that the background is described mathematically as a cross-section of a certain *fibre bundle* over the sphere. It is supposed that the detector observes a single realisation of a *random* cross-section, and the randomness was introduced by quantum fluctuations in early Universe.

The fibres of the above bundle are linear spaces of polarisation tensors, and the bundle itself is a *spinor* one. In fact, a *spinor bundle* is the central concept of this chapter.

To explain spinor bundles, we need to explain spinors. Non-formally, a spinor is an element of a certain finite-dimensional linear space. A defining feature of such a space is that it is equipped with a special action of a certain *spin group*. The elements of a spin group constitute a closed subset of the set of invertible elements in a special algebra called a *Clifford algebra*. Such an algebra can be defined over an arbitrary field, but for our purposes only the real and complex cases are essential.

Every real or complex Clifford algebra is isomorphic either to the algebra $\mathrm{Hom}_{\mathbb{K}}(L)$ or to the direct sum of two copies of $\mathrm{Hom}_{\mathbb{K}}(L)$. Here, the symbol $\mathbb{K}$ denotes an element of the set $\{\mathbb{R}, \mathbb{C}, \mathbb{H}\}$ that contains the field $\mathbb{R}$ of real numbers, the field $\mathbb{C}$ of complex numbers, and the skew field $\mathbb{H}$ of quaternions. The symbol L denotes a finite-dimensional $\mathbb{K}$-linear space, and the symbol $\mathrm{Hom}_{\mathbb{K}}(L)$ is the algebra of $\mathbb{K}$-linear maps from L to itself. Therefore, we start our rigourous explanation in Section 1.1, where we introduce the above (skew) fields.

Finite-dimensional $\mathbb{K}$-linear spaces are introduced in Section 1.2. We choose the so-called abstract index notation to denote their elements because this notation is familiar to physicists. The topics explained in this section, include operations over linear spaces and additional structures on *complex* linear ones.

DOI: 10.1201/9781003344353-1

Algebras and their representations appear in Section 1.3. In particular, for all complex Clifford algebras Cl_n with $0 \leq n \leq 4$, and for all real Clifford algebras $Cl_{p,q}$ with $0 \leq p + q \leq 4$, we fix our choice of the basis-dependent Dirac and Pauli matrices. For all complex even Clifford algebras Cl_n^+ with $1 \leq n \leq 4$, and for all real even Clifford algebras $Cl_{p,q}^+$ with $1 \leq p + q \leq 4$, we describe their irreducible representations in Section 1.4. The elements of the $\mathbb{K}$-linear spaces where the above representations act, are called *algebraic spinors*, but most of them also have special names which are listed.

On the one hand, the set of invertible elements of every *even* Clifford algebra Cl_n^+ (resp., $Cl_{p,q}^+$) contains a closed subset $\mathrm{Spin}(n, \mathbb{C})$ (resp., $\mathrm{Spin}^0(p, q)$) which is an analytic manifold. On the other hand, in modern physical theories the space and time constitute a smooth manifold. We explain the concept of a manifold in Section 1.5.

The sets $\mathrm{Spin}(n, \mathbb{C})$ and $\mathrm{Spin}^0(p, q)$ are topological groups. Each of them contains a normal subgroup, say G, which has two elements. The projections $\pi \colon \mathrm{Spin}(n, \mathbb{C}) \to \mathrm{Spin}(n, \mathbb{C})/G$ and $\pi \colon \mathrm{Spin}^0(p, q) \to \mathrm{Spin}^0(p, q)/G$ are *two-fold coverings*. The concept of covering is explained in Section 1.5 as well.

The concept of *Lie group*, which has both the structure of an analytic manifold and a group, is introduced in Section 1.6. In particular, all the so-called classical groups are Lie groups. We give the explicit description for all of them, except the groups $\mathrm{Spin}(n, \mathbb{C})$ with $n \geq 5$ and $\mathrm{Spin}^0(p, q)$ with $p + q \geq 5$. For the spinor groups, we give the explicit constructions of the two-fold coverings $\mathrm{Spin}(n, \mathbb{C}) \to \mathrm{SO}(n, \mathbb{C})$ and $\mathrm{Spin}^0(p, q) \to \mathrm{SO}_0(p, q)$.

There exist different approaches to *bundles*. We review three of them in Section 1.7. In the *group action approach*, a Lie group G acts smoothly from the right on a smooth manifold P called the *total space* in such a way that the set of orbits of that action is again a smooth manifold M and the *bundle projection* $\pi \colon P \to M$ that maps each point $x \in P$ to its orbit is a smooth map, for every point $p \in M$ the *fibre* $\pi^{-1}(p)$ is a copy of the underlying manifold of G, and the bundle is *locally trivial*: over a certain neighbourhood $\mathcal{U}$ of p, the inverse image $\pi^{-1}(\mathcal{U})$ may be realised as the Cartesian product $\mathcal{U} \times G$. It turns out that the cross-sections of a principal bundle are known to physicists as *gauges*. If, in addition, G acts smoothly from the left on a smooth manifold F, then one may construct a *fibre bundle* $\xi[F]$ over M whose fibres are copies of F.

In the *classical approach*, one first constructs a special fibre bundle over a smooth manifold M called a *tangent bundle*. Its fibres $\pi^{-1}(p)$ are copies of a linear space whose dimension is equal to that of M, and the constructed bundle is locally trivial over any chart $(\mathcal{U}, \mathbf{x})$ of M. Afterwards, any operation over linear spaces described in Subsection 1.2.2 leads to *vector and tensor bundles*.

In the *cocycle approach*, a vector bundle whose fibre is a linear space L is assembled from local 'pieces' $\mathcal{U} \times L$, where $\mathcal{U}$ runs over an atlas of M.

A certain choice of a cross-section in a special tensor bundle leads to *Riemannian manifolds* considered in Subsection 1.7.4.

In order to define differentiation of vector and tensor fields, we choose an axiomatic approach through Koszul connections. At this moment, we meet for the first time a tensor subject to *sign conventions*. We follow Guidry (2019), where four different tensors are defined with sign conventions taken into account. Potentially, there exists eight different conventions accordingly to the choice of three numbers S_1, S_2, and S_3, each of which belongs to the set $\{-1, 1\}$. We make our first choice $S_2 = 1$ in Subsection 1.7.6.

A special case of a group action, when a Lie group G acts linearly and continuously on a linear space L over a skew field $\mathbb{K}$, is called a representation and is considered in Section 1.8. Some special features of our approach to the subject are as follows.

First, we consider the representations over real numbers, complex numbers, and quaternions simultaneously. Second, we use the idea by Adams (1969) for a convenient system of notation. Third, we pay much attention to a detailed description of classical spinors and spin-tensors.

On some manifolds, one can create an additional *spin structure* and construct spin and spin-tensor bundles. In Section 1.9 we construct such bundles using all the three approaches mentioned earlier. We also consider the question on how to differentiate the smooth sections of these bundles.

A special class of fibre bundles called *homogeneous fibre bundles* is studied in Section 1.10. It turns out that the Hilbert space of square-integrable cross-sections of such a bundle carries a special group representation called the *induced representation*. The fine structure of such a representation is given by an important result called the *Frobenius Reciprocity*, which is proved there.

Finally, our treatment of *probability* has the following special feature: we consider random elements in finite-dimensional linear spaces over the three skew fields $\mathbb{R}$, $\mathbb{C}$, and $\mathbb{H}$ simultaneously.

1.1 Skew Fields

It is supposed that the definition of a skew field and the field $\mathbb{R}$ are familiar to the reader. The elements of $\mathbb{R}$ are called *real numbers*. We denote them by lowercase Greek letters α, β, $\ldots$.

The *imaginary unit* is denoted by the symbol i. The field of *complex numbers* $\zeta = \alpha + \beta$i is denoted by the symbol $\mathbb{C}$.

The *basic quaternions* are denoted by the symbols i, j, and k. The skew field of *quaternions* $q = \alpha + \beta$i$ + \gamma$j$ + \delta$k is denoted by the symbol $\mathbb{H}$.

A generic *skew field*, that is, an element of the set $\{\mathbb{R}, \mathbb{C}, \mathbb{H}\}$, is denoted by the symbol $\mathbb{K}$. The *conjugation* on $\mathbb{K}$ is denoted by the over-line and is defined by $\overline{\alpha} = \alpha$, $\overline{\alpha + \beta \text{i}} = \alpha - \beta$i, and $\overline{\alpha + \beta \text{i} + \gamma \text{j} + \delta \text{k}} = \alpha - \beta$i$ - \gamma$j$ - \delta$k.

On the other hand, a generic *field*, that is, an element of the set $\{\mathbb{R}, \mathbb{C}\}$, is denoted by the symbol $\mathbb{F}$.

1.2   Linear Spaces

We suppose that the definition of a linear space over a skew field is familiar to the reader. Note that for the case of $\mathbb{K} = \mathbb{H}$ we consider both *right* $\mathbb{H}$-*linear spaces*, when a quaternion acts from the right of the elements of the space, and *left* $\mathbb{H}$-*linear spaces*, when it acts from the left of the elements of the space.

1.2.1   Notation

We denote a finite-dimensional linear space over $\mathbb{R}$ by the symbol U, over $\mathbb{C}$ by the symbol V, over $\mathbb{H}$ by the symbol W. A generic finite-dimensional linear space over $\mathbb{K}$ or $\mathbb{F}$ is denoted by the symbol L. The elements of L may have various different names, like 'vectors', 'tensors', 'spinors', etc. Because of this reason, we avoid to use the term 'vector space'.

In the *coordinate-free notation*, the elements of L are denoted by lowercase boldface Latin letters $\mathbf{x}$, $\mathbf{y}$, ... We use this notation occasionally for the reasons explained below.

The *abstract index notation* was introduced by Penrose (1968). His idea is as follows. It is important to distinguish the following two types of equations.

- The equations for elements of a linear space L that *always* hold true.

- The equations for the *components* of the above elements that hold true *only in a particular basis* of L.

One does not introduce a basis but rather uses a notation for the elements of a linear space that mirrors the expressions for their basis components. The indices are placeholders and are not related to any basis.

In particular, a vector is denoted by a Latin letter followed by an upper lowercase *Latin* index, e.g., x^i. This notation does not require any particular basis in L. The *components* of a vector x^i in a *particular basis* are denoted by the same letter followed by an upper lowercase *Greek* index, e.g., x^α. This convention is due to Wald (1984).

The definition of a left $\mathbb{H}$-linear space does not differ from that of a linear space over a field. In the case of the *right* $\mathbb{H}$-linear space W, the compatibility of scalar-vector multiplication with multiplication in $\mathbb{H}$ takes the form

$$(x^i q_1) q_2 = x^i (q_1 q_2),$$

the identity element of scalar-vector multiplication acts by $x^i 1 = x^i$, the distributivity of scalar-vector multiplication with respect to addition in W becomes $(x^i + y^i) q = x^i q + y^i q$, and the distributivity of scalar-vector multiplication with respect to addition in $\mathbb{H}$ becomes $x^i(q_1 + q_2) = x^i q_1 + x^i q_2$.

Example 1. The scalar-vector multiplication $\mathbb{R} \times \mathbb{C} \to \mathbb{C}$, $(\alpha, \zeta) \mapsto \alpha\zeta$, turns $\mathbb{C}$ into a two-dimensional real linear space.

Similarly, the scalar-vector multiplication $\mathbb{R} \times \mathbb{H} \to \mathbb{H}$, $(\alpha, q) \mapsto \alpha q$ turns $\mathbb{H}$ into a four-dimensional real linear space.

The scalar-vector multiplication $\mathbb{C} \times \mathbb{H} \to \mathbb{H}$, $(\zeta, q) \mapsto \zeta q$ turns $\mathbb{H}$ into a two-dimensional complex linear space.

1.2.2 Operations over Linear Spaces

Let L_1 and L_2 be linear spaces over $\mathbb{K}$.

Definition 1. The *direct sum* L_1 and L_2 is the set

$$L_1 \oplus L_2 = \{\, (x^i, y^j) \colon x^i \in L_1, y^j \in L_2 \,\}$$

with component-wise addition and scalar multiplication.

For two linear spaces L_1 and L_2 over the *same* skew field $\mathbb{K}$, let $\mathrm{Hom}_{\mathbb{K}}(L_1, L_2)$ be the set of all $\mathbb{K}$-linear maps $f \colon L_1 \to L_2$. We abbreviate this notation to $\mathrm{Hom}_{\mathbb{K}}(L)$ if $L_1 = L_2 = L$. The set $\mathrm{Hom}_{\mathbb{F}}(U)$ is a $\mathbb{F}$-linear space with respect to the point-wise addition of maps and the scalar-vector multiplication $(\alpha f)(x^i) = f(\alpha x^i)$.

We wish to write the $\mathbb{K}$-linear maps on the left. By this reason, in the case of $\mathbb{K} = \mathbb{H}$, we define the set $\mathrm{Hom}_{\mathbb{H}}(W_1, W_2)$ only in the case when both W_1 and W_2 are *right* finite-dimensional quaternionic linear spaces. The map $f \colon W_1 \to W_2$ is $\mathbb{H}$-linear if and only if $f(x^i q_1 + y^i q_2) = f(x^i)q_1 + f(y^i)q_2$ for all x^i, $y^i \in W$ and q_1, $q_2 \in \mathbb{H}$.

Observe!

The map $f \colon W \to W$, $f(x^i) = x^i \mathrm{i}$ is *not* $\mathbb{H}$-linear! Indeed, for any $q \in \mathbb{H}$ we must have $f(x^i q) = f(x^i)q$, which becomes $(x^i q)\mathrm{i} = (x^i \mathrm{i})q$. By the compatibility of scalar multiplication with multiplication in $\mathbb{H}$, we obtain $x^i(q\mathrm{i}) = x^i(\mathrm{i}q)$. For $q = \mathrm{j}$, this gives $x^i(-\mathrm{k}) = x^i \mathrm{k}$ which is true if and only if $x^i = 0$.

However, the map $f(x^i) = x^i q$ is linear if $q \in \mathbb{R} \subset \mathbb{H}$. That is, the set $\mathrm{Hom}_{\mathbb{H}}(W)$ is a *real* linear space. By this reason, we introduce the following notation.

Denote by $\mathbb{K}'$ the field given by $\mathbb{R}' = \mathbb{H}' = \mathbb{R}$, $\mathbb{C}' = \mathbb{C}$. In general, the set $\mathrm{Hom}_{\mathbb{F}}(L)$ is a $\mathbb{F}$-linear space, while the set $\mathrm{Hom}_{\mathbb{K}}(L)$ is a $\mathbb{K}'$-linear space.

Remark 1. The elements of the set $\mathrm{Hom}_{\mathbb{K}}(L_1, L_2)$ are morphisms of a category, see Example 92.

The space $\mathrm{Hom}_{\mathbb{K}}(L, \mathbb{K}^1)$ is called the space *dual* to L and is denoted by L^*. The elements of L^* are called *linear forms*. They are denoted by a Latin letter followed by a *lower* lowercase Latin letter, e.g., x_i. The components of a linear form *in a particular basis* are denoted by the same letter followed by a lower Greek index, e.g., x_α.

For a right quaternionic linear space W, the *real* linear space W^* can be turned into a *left quaternionic* one by defining

$$(qf)(x^i) = qf(x^i), \qquad f \in W^*, \quad q \in \mathbb{H}, \quad x^i \in W.$$

Observe!

The finite-dimensional linear spaces U and U^* have the same dimension. However, an isomorphism between U and U^* cannot be constructed without using additional structures on U. In other words, any such construction is not 'natural'. For complete explanation, see Section A.2.

Let r be a positive integer, and let U_1, ..., U_r be finite-dimensional *real* linear spaces. A map $B \colon U_1 \times U_2 \times \cdots \times U_r \to \mathbb{R}^1$ is called a *multi-linear form* if for all i, $1 \le i \le r$, and for all $\mathbf{x}_1 \in U_1$, ..., $\mathbf{x}_{i-1} \in U_{i-1}$, $\mathbf{x}_{i+1} \in U_{i+1}$, ..., $\mathbf{x}_r \in U_r$, the map

$$U_i \to \mathbb{R}^1, \qquad \mathbf{x} \mapsto B(\mathbf{x}_1, \ldots, \mathbf{x}_{i-1}, \mathbf{x}, \mathbf{x}_{i+1}, \ldots, \mathbf{x}_r)$$

is a linear form.

Observe!

The nonzero right bilinear forms over finite-dimensional quaternionic linear spaces do not exist. Indeed, let $B \colon W_1 \times W_2 \to \mathbb{H}^1$ be such a form. Then, for any $(\mathbf{x}, \mathbf{y}) \in W_1 \times W_2$ we have

$$B(\mathbf{x}, \mathbf{y})\mathrm{k} = B(\mathbf{x}, \mathbf{y})\mathrm{ij} = B(\mathbf{x}, \mathbf{y}\mathrm{i})\mathrm{j} = B(\mathbf{x}\mathrm{j}, \mathbf{y}\mathrm{i})$$
$$= B(\mathbf{x}\mathrm{i}, \mathbf{y})\mathrm{i} = B(\mathbf{x}, \mathbf{y})\mathrm{ji} = -B(\mathbf{x}, \mathbf{y})\mathrm{k}.$$

We obtain $B(\mathbf{x}, \mathbf{y}) = 0$, because $\mathrm{k} \neq 0$.

By this reason, we should be more flexible. Call a function $B \colon L \times L \to \mathbb{K}$ *bi-additive form* if

$$B(\mathbf{x}_1 + \mathbf{x}_2, \mathbf{y}) = B(\mathbf{x}_1, \mathbf{y}) + B(\mathbf{x}_2, \mathbf{y}), \qquad B(\mathbf{x}, \mathbf{y}_1 + \mathbf{y}_2) = B(\mathbf{x}, \mathbf{y}_1) + B(\mathbf{x}, \mathbf{y}_2).$$

Such a form is called *bilinear* if

$$B(\alpha\mathbf{x}, \mathbf{y}) = \alpha B(\mathbf{x}, \mathbf{y}), \qquad B(\mathbf{x}, \alpha\mathbf{y}) = \alpha B(\mathbf{x}, \mathbf{y}), \quad \alpha \in \mathbb{K},$$

Hermitian bilinear if

$$B(\mathbf{x}\alpha, \mathbf{y}) = \overline{\alpha} B(\mathbf{x}, \mathbf{y}), \qquad B(\mathbf{x}, \mathbf{y}\alpha) = B(\mathbf{x}, \mathbf{y})\alpha, \quad \alpha \in \mathbb{K},$$

symmetric if

$$B(\mathbf{y}, \mathbf{x}) = B(\mathbf{x}, \mathbf{y}),$$

skew-symmetric if
$$B(\mathbf{y}, \mathbf{x}) = -B(\mathbf{x}, \mathbf{y}),$$

Hermitian symmetric if
$$B(\mathbf{y}, \mathbf{x}) = \overline{B(\mathbf{x}, \mathbf{y})},$$

and *Hermitian skew-symmetric* if
$$B(\mathbf{y}, \mathbf{x}) = -\overline{B(\mathbf{x}, \mathbf{y})}.$$

A bilinear or Hermitian bilinear form is called *non-degenerate* if $B(\mathbf{x}, \mathbf{y}) = 0$ for all $\mathbf{y} \in L$ implies $\mathbf{x} = \mathbf{0}$ and $B(\mathbf{x}, \mathbf{y}) = 0$ for all $\mathbf{x} \in L$ implies $\mathbf{y} = \mathbf{0}$.

In what follows, we reserve the symbol ε to denote *skew-symmetric* forms.

There exists eight types of non-degenerate bi-additive forms. For each type, there exist a basis of the space L in which the form is calculated as follows.

- Real symmetric bilinear form $B\colon U \times U \to \mathbb{R}$:

$$B(\mathbf{x}, \mathbf{y}) = \sum_{\alpha=1}^{p} x_\alpha y_\alpha - \sum_{\beta=1}^{q} x_{p+\beta} y_{p+\beta}. \tag{1.1}$$

- Real skew-symmetric bilinear form $\varepsilon\colon U \times U \to \mathbb{R}$:

$$\varepsilon(\mathbf{x}, \mathbf{y}) = \sum_{\alpha=1}^{n} (x_{2\alpha-1} y_{2\alpha} - x_{2\alpha} y_{2\alpha-1}). \tag{1.2}$$

- Complex symmetric bilinear form $B\colon V \times V \to \mathbb{C}$:

$$B(\mathbf{x}, \mathbf{y}) = \sum_{\alpha=1}^{n} x_\alpha y_\alpha. \tag{1.3}$$

- Complex skew-symmetric bilinear form $\varepsilon\colon V \times V \to \mathbb{C}$ given by Equation (1.2).

- Complex Hermitian symmetric bi-additive form $B\colon V \times V \to \mathbb{C}$:

$$B(\mathbf{x}, \mathbf{y}) = \sum_{\alpha=1}^{p} \overline{x_\alpha} y_\alpha - \sum_{\beta=1}^{q} \overline{x_{p+\beta}} y_{p+\beta}. \tag{1.4}$$

- Complex Hermitian skew-symmetric bi-additive form $\varepsilon\colon V \times V \to \mathbb{C}$:

$$\varepsilon(\mathbf{x}, \mathbf{y}) = \sum_{\alpha=1}^{p} \mathrm{i}\overline{x_\alpha} y_\alpha - \sum_{\beta=1}^{q} \mathrm{i}\overline{x_{p+\beta}} y_{p+\beta}.$$

- Quaternionic Hermitian symmetric bi-additive form $B\colon W \times W \to \mathbb{H}$:

$$B(\mathbf{x}, \mathbf{y}) = \sum_{\alpha=1}^{p} \overline{x_\alpha} y_\alpha - \sum_{\beta=1}^{q} \overline{x_{p+\beta}} y_{p+\beta}. \tag{1.5}$$

- Quaternionic Hermitian skew-symmetric bi-additive form $\varepsilon\colon W \times W \to \mathbb{H}$:

$$\varepsilon(\mathbf{x}, \mathbf{y}) = \sum_{\alpha=1}^{n} \overline{x_\alpha} \mathbf{i} y_\alpha.$$

Definition 2. The *tensor product* of the empty family of linear spaces over the field $\mathbb{R}$ (when $r = 0$) is the one-dimensional linear space $\mathbb{R}^1$. When $r \geq 1$, the tensor product $U_1 \otimes \cdots \otimes U_r$ is the linear space of all r-linear forms over $U_1^*, \ldots, U_r^*$. An element of the above product is called a *rank r tensor*. When $r \geq 1$, the *tensor product* of the vectors $\mathbf{x}_i \in U_i$, $1 \leq i \leq r$, is the r-linear form given by

$$\mathbf{x}_1 \otimes \cdots \otimes \mathbf{x}_r(\mathbf{y}_1, \ldots, \mathbf{y}_r) = \prod_{i=1}^{r} \mathbf{y}_i(\mathbf{x}_i),$$

where $\mathbf{y}_i \in U_i^*$, $1 \leq i \leq r$.

For an important particular case, when $U_1 = U_2 = \cdots = U_k = U^*$ and $U_{k+1} = U_{k+2} = \cdots = U_{k+l} = U$, we denote by $T_l^k(U)$, or just T_l^k, the linear space of multi-linear forms. In the abstract index notation, an element of the above space is denoted by a letter followed by k upper lowercase Latin and l lower lowercase Latin indices: $T^{i_1 \cdots i_k}{}_{j_1 \cdots j_l}$ and is called *covariant* of order k and *contravariant* of order l, or just of type (k, l). The abstract indices determine the number and type of variables the tensor acts on. In any equation, the same slot on both hand sides must be denoted by the same letter.

We see that a tensor of type $(1, 0)$ is a linear form on U^*, an element of the linear space U^{**}. For a vector $x^i \in U$, denote by $(x^i)^{**}$ the element of U^{**} which acts on a linear form $y_i \in U^*$ by $(x^i)^{**}(y_i) = y_i(x^i)$. It is well-known that the map $U \to U^{**}$, $x^i \mapsto (x^i)^{**}$ is an isomorphism between U and U^{**}. With the help of this isomorphism, we identify U and U^{**}. That is, a tensor of type $(1, 0)$ is an element of U, or a *contravariant vector*.

Similarly, a tensor of type $(1, 1)$ is a bilinear map $U^* \times U \to \mathbb{R}$. For a fixed $x^i \in U$, $T(\cdot, x^i)$ is an element of U, and the map $U \to U$, $x^i \mapsto T(\cdot, x^i)$ is a linear operator in U. As an easy exercise, the reader can identify $T_1^1(U)$ with the linear space of linear operators in U^*.

Assume that $k > 0$ and $l > 0$, and let $1 \leq i \leq k$ and $1 \leq j \leq l$. The *contraction* is the map $C_{ij}\colon T_l^k \to T_{l-1}^{k-1}$ given by

$$C_{ij} T^{a_1 \cdots a_k}{}_{b_1 \cdots b_l} = T^{a_1 \cdots a_{i-1} m a_{i+1} \cdots a_k}{}_{b_1 \cdots b_{j-1} m b_{j+1} \cdots b_l},$$

where we used the *Einstein summation convention*: when an index variable appears twice in a single term, that term is summed up over the values of the index.

Example 2. The contraction $T^i{}_i \in \mathbb{R}^1$ of a tensor $T^i{}_j \in T_1^1$ is the *trace* of the linear operator $T^i{}_j$.

Let T (resp., T') be a tensor of type (k, l) (resp., (k, l')).

Definition 3. The *tensor product* $T \otimes T'$ is the element of the linear space $T_{l+l'}^{k+k'}$ given by

$$(T \otimes T')(\mathbf{x}_1, \dots, \mathbf{x}_{k+k'}, \mathbf{x}_1^*, \dots, \mathbf{x}_{l+l'}^*) = T(\mathbf{x}_1, \dots, \mathbf{x}_k, \mathbf{x}_1^*, \dots, \mathbf{x}_l^*)$$
$$\times T(\mathbf{x}_{k+1}, \dots, \mathbf{x}_{k+k'}, \mathbf{x}_{l+1}^*, \dots, \mathbf{x}_{l+l'}^*).$$

In the abstract index notation, the $\otimes$ sign is omited, then the tensor product of the tensors $S^{i_1 \cdots i_k}{}_{j_1 \cdots j_l}$ and $T^{i'_1 \cdots i'_{k'}}{}_{j'_1 \cdots j'_{l'}}$ is just $S^{i_1 \cdots i_k}{}_{j_1 \cdots j_l} T^{i'_1 \cdots i'_{k'}}{}_{j'_1 \cdots j'_{l'}}$.

Let Σ_l be the set of all permutations of l symbols, and let $\mathrm{sgn}(\sigma)$ be $+1$ (resp., -1) if σ is even (resp., odd)

Definition 4. The *totally symmetric part* of a tensor $T_{j_1 \cdots j_l} \in T_l^0$ is given by

$$T_{(j_1 \cdots j_l)} = \frac{1}{l!} \sum_{\sigma \in \Sigma_l} T_{\sigma(j_1) \cdots \sigma(j_l)},$$

while its *totally skew-symmetric part* is

$$T_{[j_1 \cdots j_l]} = \frac{1}{l!} \sum_{\sigma \in \Sigma_l} \mathrm{sgn}(\sigma) T_{\sigma(j_1) \cdots \sigma(j_l)}.$$

This notation can also be applied to a subset of covariant or contravariant indices, e.g.,

$$T_{(ij)k} = \frac{1}{2}(T_{ijk} + T_{jik}).$$

A tensor $g \in T_2^0$ is called *symmetric* if it is equal to its totally symmetric part, and *metric* if it is symmetric and non-degenerate, that is, the only case in which we have $g_{ij}x^i y^j = 0$ for all $x_i \in U$ is the case $y^j = 0$. In that case, there exists an inverse tensor $g^{ij} \in T_0^2$ with $g_{ij}g^{jk} = \delta_i{}^k$.

We introduce a special notation η_{ij} for the metric tensor of the linear space $\mathbb{R}^{p,q}$, where the real symmetric bilinear form (1.1) acts. That is, the above bilinear form becomes

$$B(x^i, y^j) = \eta_{ij}x^i y^j.$$

The operation of *index lowering* for contravariant vectors x^i is given by $x_i = g_{ij}x^j$. Similarly, the operation of *index raising* for covariant vectors x_i is given by $x^i = g^{ij}x_j$. In other words, additional structures g_{ij} and g^{ij} are used to establish an isomorphism between U and U^*. The above isomorphism depends on the choice of the metric tensor.

Now, we introduce coordinates. If $\mathbf{e}^1, \dots, \mathbf{e}^n$ is a basis of U and $\mathbf{e}_1^*, \dots, \mathbf{e}_n^*$ is the *dual basis* of U^* with

$$\mathbf{e}_\mu^*(\mathbf{e}^\nu) = \delta^\nu{}_\mu = \begin{cases} 1, & \text{if } \mu = \nu, \\ 0, & \text{otherwise}, \end{cases}$$

then the tensor products $\mathbf{e}^{\alpha_1} \otimes \cdots \otimes \mathbf{e}^{\alpha_k} \otimes \mathbf{e}^*_{\beta_1} \otimes \cdots \otimes \mathbf{e}^*_{\beta_l}$ form a basis of the space $T^k_l(U)$, and any tensor has the form

$$T = T^{\nu_1 \cdots \nu_l}_{\mu_1 \cdots \mu_k} \mathbf{e}^{\mu_1} \otimes \cdots \otimes \mathbf{e}^{\mu_k} \otimes \mathbf{e}^*_{\nu_1} \otimes \cdots \otimes \mathbf{e}^*_{\nu_l}.$$

The tensor product of two quaternionic linear spaces requires another definition, see below.

Finally, we introduce the *Adams operations* after Adams (1969).

Consider the field $\mathbb{C}$ as a real vector space, see Example 1. Define the space cU by $cU = U \otimes_{\mathbb{R}} \mathbb{C}$, where the symbol $\otimes_{\mathbb{R}}$ denotes the tensor product of real linear spaces. The scalar-vector multiplication $\zeta'(x^i \otimes \zeta) = x^i \otimes (\zeta\zeta')$ turns cU into a *complex* linear space.

Define the space qV by $qV = V \otimes_{\mathbb{C}} \mathbb{H}$, where the symbol $\otimes_{\mathbb{C}}$ denotes the tensor product of complex linear spaces and where the set $\mathbb{H}$ is considered as a complex linear space as in Example 1. The scalar-vector multiplication $(x^i \otimes q)q' = x^i \otimes (qq')$ turns qV into a *right quaternionic* linear space.

If W is a right quaternionic linear space, let $c'W$ has the same underlying set as W but regard it as a complex linear space.

Similarly, if V is a complex linear space, let rV has the same underlying set as V but regard it as a real linear space.

Finally, if V is a complex linear space, let tV has the same underlying set as V, but make $\zeta \in \mathbb{C}$ act on tV as $\overline{\zeta}$ used to act on V. If $x^i \in V$, then we follow Budinich and Trautman (1988) and denote by tx^i the element x^i when it is considered as an element of tV. With this notation, the scalar-vector multiplication in tV takes the form $\alpha(tx^i) = t(\overline{\alpha}x^i)$.

Remark 2. The operation c can be regarded as an object function, that can be extended to a covariant functor from the category of finite-dimensional real linear spaces to the category of complex ones, see Section A.2. Similarly for q, c', r, and t.

1.2.3 Additional Structures on Complex Linear Spaces

It is often convenient to consider a real linear space U as a complex linear space V together with a real structure.

Definition 5. A map $j\colon V \to V$ is called a *real structure* if for all x^i, $y^i \in V$ and α, $\beta \in \mathbb{C}$ we have

$$j(\alpha x^i + \beta y^i) = \overline{\alpha}j(x^i) + \overline{\beta}j(y^i), \qquad j^2(x^i) = x^i.$$

Given U, define $V = cU$ with the structure map $j(x^i \otimes \zeta) = x^i \otimes \overline{\zeta}$. Conversely, given V with a real structure j, let $U = \{\, x^i \in V \colon jx^i = x^i \,\}$.

Similarly, it is often convenient to consider a quaternionic linear space W as a complex linear space V together with a quaternionic structure.

Definition 6. A map $j\colon V \to V$ is called a *quaternionic structure* if for all x^i, $y^i \in V$ and α, $\beta \in \mathbb{C}$ we have

$$j(\alpha x^i + \beta y^i) = \overline{\alpha}j(x^i) + \overline{\beta}j(y^i), \qquad j^2(x^i) = -x^i.$$

Given W, define $V = c'W$ with the structure map $j(x^i) = x^i\mathrm{j}$. Conversely, given V with a quaternionic structure j, let $W = V$ as a set, and define the scalar multiplication by

$$x^i\mathrm{i} = \mathrm{i}x^i, \qquad x^i\mathrm{j} = j(x^i), \qquad x^i\mathrm{k} = j(\mathrm{i}x^i). \tag{1.6}$$

1.3 Algebras

In this section, we introduce algebras and their representations. An important class of algebras called the *Clifford algebras*, their representations, and involutions are studied in details.

1.3.1 Definitions

Definition 7. A linear space A over a field $\mathbb{F}$ is called an *algebra*, if there is a map $A \times A \to A$, $(\mathbf{x}, \mathbf{y}) \mapsto \mathbf{xy}$, such that

$$(\mathbf{x} + \mathbf{y})\mathbf{z} = \mathbf{xz} + \mathbf{yz}, \qquad \mathbf{z}(\mathbf{x} + \mathbf{y}) = \mathbf{zx} + \mathbf{zy}, \qquad (\alpha\mathbf{x})(\beta\mathbf{y}) = (\alpha\beta)(\mathbf{xy})$$

for all $\mathbf{x}$, $\mathbf{y}$, $\mathbf{z} \in A$, and for all α, $\beta \in \mathbb{F}$.

Definition 8. An algebra A is called *associative* if

$$(\mathbf{xy})\mathbf{z} = \mathbf{x}(\mathbf{yz})$$

for $\mathbf{x}$, $\mathbf{y}$, $\mathbf{z} \in A$.

Definition 9. An algebra A is called *unital* if there exists an *unit element* $1_A \in A$ with $1_A\mathbf{x} = \mathbf{x}1_A = \mathbf{x}$ for all $\mathbf{x} \in A$.

In what follows we consider only associative unital algebras.

Definition 10. An algebra A is called *division* if it is associative, unital, and every $\mathbf{x} \in A \setminus \{\mathbf{0}\}$ has a multiplicative inverse $\mathbf{x}^{-1}$ with $\mathbf{x}^{-1}\mathbf{x} = \mathbf{xx}^{-1} = 1$.

Example 3. The linear space $\mathbb{R}^1$ is a division algebra over $\mathbb{R}$. The multiplication of complex numbers turns the real linear space $\mathbb{C}$ of Example 1 into another real division algebra. Similarly, the multiplication of quaternions turns the real linear space $\mathbb{H}$ of Example 1 into a real division algebra. The linear space $\mathbb{C}^1$ is a division algebra over $\mathbb{C}$.

Theorem 1 (Frobenius). *There exists neither real nor complex finite-dimensional associative division algebras except those described in Example 3.*

Example 4. The $\mathbb{K}'$-linear space $\mathrm{Hom}_{\mathbb{K}}(L)$ is an algebra over the field $\mathbb{K}'$ with respect to the composition of maps. Denote $n = \dim_{\mathbb{K}} L \geq 1$. By introducing a basis in L, we may identify the algebra $\mathrm{Hom}_{\mathbb{K}}(L)$ with the algebra $\mathbb{K}(n)$ of $n \times n$ matrices with entries in the (skew) field $\mathbb{K}$.

Definition 11. The *centre* of an algebra A is the set of elements of A that commute with all elements of A.

Definition 12. An algebra A is called *central* if its centre is equal to

$$\{\, \alpha \cdot 1_A : \alpha \in \mathbb{F} \,\}.$$

Let A_1 and A_2 be two algebras over the same field $\mathbb{F}$. The multiplication

$$(\mathbf{x}_1, \mathbf{y}_1)(\mathbf{x}_2, \mathbf{y}_2) = (\mathbf{x}_1 \mathbf{x}_2, \mathbf{y}_1 \mathbf{y}_2)$$

turns the *direct sum* $A_1 \oplus A_2$ into an algebra.

Similarly, the tensor product $A_1 \otimes A_2$ of two algebras may be turned into an algebra, the *tensor product* of the above algebras. There exists a unique multiplication on $A_1 \otimes A_2$ that satisfies the condition

$$(\mathbf{x}_1 \otimes \mathbf{y}_1)(\mathbf{x}_2 \otimes \mathbf{y}_2) = \mathbf{x}_1 \mathbf{x}_2 \otimes \mathbf{y}_1 \mathbf{y}_2$$

for all $\mathbf{x}_1$, $\mathbf{x}_2 \in A_1$ and $\mathbf{y}_1$, $\mathbf{y}_2 \in A_2$.

1.3.2 Representations of Algebras

Let A be a unital algebra over a field $\mathbb{F}$, let L be a finite-dimensional linear space over a (skew) field $\mathbb{K}$, and let 1_L be the identity map in L: $1_L \mathbf{x} = \mathbf{x}$ for all $\mathbf{x} \in L$.

Definition 13. A linear map $\rho \colon A \to \operatorname{Hom}_{\mathbb{K}}(L)$ is called a *representation* of A if $\rho(1_A) = 1_L$ and $\rho(\mathbf{xy}) = \rho(\mathbf{x})\rho(\mathbf{y})$ for all $\mathbf{x}$, $\mathbf{y} \in A$.

By introducing a basis in L, we may consider a representation ρ as the map $\rho \colon A \to \mathbb{K}(n)$ and speak about a *matrix representation*.

Definition 14. A linear operator $f \colon L_1 \to L_2$ is called an *intertwining operator* for the representations $\rho_1 \colon A \to \operatorname{Hom}_{\mathbb{K}}(L_1)$ and $\rho_2 \colon A \to \operatorname{Hom}_{\mathbb{K}}(L_2)$ if $\rho_2(f(\mathbf{x})) = f(\rho_1(\mathbf{x}))$ for all $\mathbf{x} \in A$.

Definition 15. Two representations $\rho_1 \colon A \to \operatorname{Hom}_{\mathbb{K}}(L_1)$ and $\rho_2 \colon A \to \operatorname{Hom}_{\mathbb{K}}(L_2)$ are called *equivalent*, if there exists an intertwining isomorphism $f \colon L_1 \to L_2$.

Definition 16. A representation ρ is called *irreducible* if the only invariant subspaces of the family $\{\, \rho(\mathbf{x}) : \mathbf{x} \in A \,\}$ are L and $\{\mathbf{0}\}$.

It is well-known that a finite-dimensional representation of a unital algebra is a direct sum of irreducible components. In what follows, we give special names to some representations.

1.3.3 Clifford Algebras

Let B be a symmetric bilinear form on a finite-dimensional space L over a field $\mathbb{F}$, and let A be an algebra over $\mathbb{F}$ with unit element 1_A.

Definition 17. A pair (A, γ) is called a *Clifford algebra* for the pair (L, B) if

- $\gamma \colon L \to A$ is a linear map;

- $\gamma(\mathbf{x})\gamma(\mathbf{y}) + \gamma(\mathbf{y})\gamma(\mathbf{x}) = -2B(\mathbf{x}, \mathbf{y})1_A$ for all $\mathbf{x}, \mathbf{y} \in L$;

- A is the minimal algebra that contains the sets $\{\gamma(\mathbf{x}) \colon \mathbf{x} \in L\}$ and $\{\alpha 1_A \colon \alpha \in \mathbb{F}\}$.

The linear map γ is called a *Clifford map*.

Remark 3. The Clifford map is a "square root" of the quadratic form $Q(\mathbf{x}) = -B(\mathbf{x}, \mathbf{x})$, because $(\gamma(\mathbf{x}))^2 = -Q(\mathbf{x})1_A$.

Once and forever, we fix our choice of $\mathbb{F}$, L, and B. When $\mathbb{F} = \mathbb{R}$, we choose two non-negative integers p and q. Let $\mathbf{e}_1$, ..., $\mathbf{e}_{p+q}$ be a basis of a real linear space $L = \mathbb{R}^{p,q}$ of dimension $p + q$, and let x_i, $1 \leq i \leq p + q$, be the coordinates of a vector $\mathbf{x} \in L$ in the above basis. The symmetric bilinear form B is given by Equation (1.1). When $\mathbb{F} = \mathbb{C}$, we choose $L = \mathbb{C}^n$, and B given by Equation (1.3).

It turns out that there exists a unique Clifford algebra of dimension $2^{\dim L}$. We denote it by $\mathrm{Cl}(L, B)$, or by $\mathrm{Cl}_{p,q}$ if $L = \mathbb{R}^{p,q}$ and B is given by Equation (1.1), or by Cl_n if $L = \mathbb{C}^n$ and B is given by Equation (1.3).

Let ρ be a matrix representation of a Clifford algebra $\mathrm{Cl}(L, B)$. It follows that for any $\mathbf{x} \in L$, $(\rho \circ \gamma)(\mathbf{x})$ is a matrix. Let $L = \mathbb{R}^{p,q}$ and B is given by Equation (1.1).

Definition 18. A *mathematical gamma matrix* is the matrix

$$\gamma_i = (\rho \circ \gamma)(\mathbf{e}_i), \qquad 1 \leq i \leq \dim L,$$

while a *physical gamma matrix* is the matrix

$$\Gamma_i = -\mathrm{i}\gamma_i, \qquad 1 \leq i \leq \dim L.$$

Observe!

The mathematical gamma matrices for the case of $p = 0$, $q = 3$ are traditionally denoted by σ_i instead of γ_i. In the case of $n = 4$, it is customary to start enumeration of indices from 0 instead of 1. See Example 5 below.

The gamma matrices depend on the choice of a matrix representation ρ. We raise the indices of gamma matrices with the help of the metric tensor η^{ij}:

$$\gamma^i = \eta^{ij}\gamma_j, \qquad \Gamma^i = \eta^{ij}\Gamma_j.$$

Assume that $n = p + q = 2k$ is even, and $\mathbb{K} = \mathbb{C}$.

Definition 19. The *mathematical chirality operator* is given by

$$\gamma_{n+1} = -i^{k+q}\gamma_1 \cdots \gamma_n,$$

while the *physical chirality operator* is given by

$$\Gamma_{n+1} = -i^{k+q}\Gamma_1 \cdots \Gamma_n.$$

1.3.4 Examples of Clifford Algebras in Low Dimensions

Example 5. The linear map $\gamma\colon \mathbb{R}^{0,0} = \{\mathbf{0}\} \to \mathbb{R}^1$ with $\gamma(\mathbf{0}) = 0$ is a Clifford map for the space $\mathbb{R}^{0,0}$. That is, $\mathrm{Cl}_{0,0} = \mathbb{R}$. There are no gamma matrices here.

The linear map $\gamma\colon \mathbb{R}^{1,0} \to \mathbb{C}^1$ with $\gamma(\mathbf{e}_1) = i$ is a Clifford map for the space $\mathbb{R}^{1,0}$. That is, $\mathrm{Cl}_{1,0} = \mathbb{C}$, where the field $\mathbb{C}$ is considered as a real algebra, see Example 3. Let $\rho\colon \mathbb{C} \to \mathbb{C}(1)$ be the representation of $\mathrm{Cl}_{1,0}$ given by $\rho(\zeta) = \zeta$. Then

$$\gamma_1 = i. \tag{1.7}$$

The physical gamma matrix is $\Gamma_1 = 1$.

The linear map $\gamma\colon \mathbb{R}^{2,0} \to \mathbb{H}^1$ with $\gamma(\mathbf{e}_1) = \gamma_1 = i$ and $\gamma(\mathbf{e}_2) = \gamma_2 = j$ is a Clifford map for the space $\mathbb{R}^{2,0}$. That is, $\mathrm{Cl}_{2,0} = \mathbb{H}$, where the skew field $\mathbb{H}$ is considered as a real algebra, see Example 3.

Another choice of mathematical gamma matrices is given by combining the map γ with the matrix representation $\rho\colon \mathrm{Cl}_{2,0} \to \mathbb{C}(2)$ with $\rho(i) = \gamma_1$ and $\rho(j) = \gamma_2$, where

$$\gamma_1 = \begin{pmatrix} 0 & i \\ i & 0 \end{pmatrix}, \qquad \gamma_2 = \begin{pmatrix} 0 & 1 \\ -1 & 0 \end{pmatrix}. \tag{1.8}$$

In this case, the physical gamma matrices are

$$\Gamma_1 = \begin{pmatrix} 0 & 1 \\ 1 & 0 \end{pmatrix}, \qquad \Gamma_2 = \begin{pmatrix} 0 & -i \\ i & 0 \end{pmatrix}, \tag{1.9}$$

and the chirality operators are

$$\gamma_3 = \begin{pmatrix} -1 & 0 \\ 0 & -1 \end{pmatrix}, \qquad \Gamma_3 = \begin{pmatrix} 1 & 0 \\ 0 & -1 \end{pmatrix}.$$

The linear map $\gamma\colon \mathbb{R}^{1,1} \to \mathbb{R}(2)$ with

$$\gamma_1 = \begin{pmatrix} 0 & -1 \\ 1 & 0 \end{pmatrix}, \qquad \gamma_2 = \begin{pmatrix} 0 & 1 \\ 1 & 0 \end{pmatrix} \tag{1.10}$$

is a Clifford map for the space $\mathbb{R}^{1,1}$. That is, $\mathrm{Cl}_{1,1} = \mathbb{R}(2)$. The matrix representation ρ is the identity map.

The linear map $\gamma\colon \mathbb{R}^{0,2} \to \mathbb{R}(2)$ with

$$\gamma(\mathbf{e}_1) = \gamma_1 = \begin{pmatrix} -1 & 0 \\ 0 & 1 \end{pmatrix}, \qquad \gamma(\mathbf{e}_2) = \gamma_2 = \begin{pmatrix} 0 & 1 \\ 1 & 0 \end{pmatrix} \tag{1.11}$$

is a Clifford map for the space $\mathbb{R}^{0,2}$. That is, $\mathrm{Cl}_{0,2} = \mathbb{R}(2)$.

Another choice of mathematical gamma matrices is given by combining the map γ with the matrix representation $\rho\colon \mathrm{Cl}_{0,2} \to \mathbb{C}(2)$ with

$$\rho(\gamma(\mathbf{e}_1)) = \gamma_1, \qquad \rho(\gamma(\mathbf{e}_2)) = \gamma_2,$$

where

$$\gamma_1 = \begin{pmatrix} 0 & i \\ i & 0 \end{pmatrix}, \qquad \gamma_2 = \begin{pmatrix} 0 & -i \\ i & 0 \end{pmatrix}.$$

In this case, the physical gamma matrices are

$$\Gamma_1 = \begin{pmatrix} 0 & 1 \\ 1 & 0 \end{pmatrix}, \qquad \Gamma_2 = \begin{pmatrix} 0 & -1 \\ 1 & 0 \end{pmatrix}, \tag{1.12}$$

and the chirality operators are

$$\gamma_3 = \begin{pmatrix} -1 & 0 \\ 0 & 1 \end{pmatrix}, \qquad \Gamma_3 = \begin{pmatrix} 1 & 0 \\ 0 & -1 \end{pmatrix}. \tag{1.13}$$

The linear map $\sigma\colon \mathbb{R}^{0,3} \to \mathbb{C}(2)$ with

$$\sigma_1 = \begin{pmatrix} 0 & 1 \\ 1 & 0 \end{pmatrix}, \qquad \sigma_2 = \begin{pmatrix} 0 & -i \\ i & 0 \end{pmatrix}, \qquad \sigma_3 = \begin{pmatrix} 1 & 0 \\ 0 & -1 \end{pmatrix} \tag{1.14}$$

is a Clifford map for the space $\mathbb{R}^{0,3}$. That is, $\mathrm{Cl}_{0,3} = \mathbb{C}(2)$. The above matrices are classical Pauli matrices.

The linear map $\gamma\colon \mathbb{R}^{4,0} \to \mathbb{H}(2)$ with

$$\gamma(\mathbf{e}_0) = \gamma_0 = \begin{pmatrix} i & 0 \\ 0 & -i \end{pmatrix}, \qquad \gamma(\mathbf{e}_1) = \gamma_1 = \begin{pmatrix} j & 0 \\ 0 & -j \end{pmatrix},$$

$$\gamma(\mathbf{e}_2) = \gamma_2 = \begin{pmatrix} k & 0 \\ 0 & -k \end{pmatrix}, \qquad \gamma(\mathbf{e}_3) = \gamma_3 = \begin{pmatrix} 0 & -1 \\ 1 & 0 \end{pmatrix} \tag{1.15}$$

is a Clifford map for the space $\mathbb{R}^{4,0}$. That is, $\mathrm{Cl}_{4,0} = \mathbb{H}(2)$. There are two other popular choices. The first one is given by combining the map γ with the matrix representation $\rho\colon \mathrm{Cl}_{4,0} \to \mathbb{C}(4)$ with

$$\rho(\gamma(\mathbf{e}_i)) = \gamma_i, \qquad 0 \le i \le 3,$$

where

$$\gamma_0 = \begin{pmatrix} 0 & i\sigma_0 \\ i\sigma_0 & 0 \end{pmatrix}, \qquad \gamma_k = \begin{pmatrix} 0 & \sigma_k \\ -\sigma_k & 0 \end{pmatrix},$$

and where σ_0 is the 2×2 identity matrix and $1 \leq k \leq 3$. The second choice is given by combining the map γ with the matrix representation $\rho \colon \text{Cl}_{4,0} \to \mathbb{C}(4)$ with

$$\rho(\gamma(\mathbf{e}_i)) = \gamma_i, \qquad 0 \leq i \leq 3,$$

where

$$\gamma_0 = \begin{pmatrix} 0 & \sigma_0 \\ \sigma_0 & 0 \end{pmatrix}, \qquad \gamma_k = \begin{pmatrix} 0 & i\sigma_k \\ i\sigma_k & 0 \end{pmatrix}.$$

The calculation of physical gamma matrices and chirality operators may be left to the reader. For a future reference, we write down the physical chirality operator for the second choice:

$$\Gamma_4 = \begin{pmatrix} 1 & 0 & 0 & 0 \\ 0 & 1 & 0 & 0 \\ 0 & 0 & -1 & 0 \\ 0 & 0 & 0 & -1 \end{pmatrix}. \tag{1.16}$$

Example 6. Consider the direct sum $\mathbb{R} \oplus \mathbb{R}$ of two copies of the Clifford algebra $\text{Cl}_{0,0} = \mathbb{R}$. We write down the elements of $\mathbb{R} \oplus \mathbb{R}$ as matrices $\begin{pmatrix} \alpha_1 & 0 \\ 0 & \alpha_2 \end{pmatrix}$. The linear map $\gamma \colon \mathbb{R}^{0,1} \to \mathbb{R} \oplus \mathbb{R}$ with $\gamma_1 = \begin{pmatrix} 1 & 0 \\ 0 & -1 \end{pmatrix}$ is a Clifford map for the space $\mathbb{R}^{0,1}$. That is, $\text{Cl}_{0,1} = \mathbb{R} \oplus \mathbb{R}$.

Similarly, consider the direct sum $\mathbb{H} \oplus \mathbb{H}$ of two copies of the Clifford algebra $\text{Cl}_{2,0} = \mathbb{H}$. The linear map $\gamma \colon \mathbb{R}^{3,0} \to \mathbb{H} \oplus \mathbb{H}$ with

$$\gamma_1 = \begin{pmatrix} i & 0 \\ 0 & -i \end{pmatrix}, \qquad \gamma_2 = \begin{pmatrix} j & 0 \\ 0 & -j \end{pmatrix}, \qquad \gamma_3 = \begin{pmatrix} k & 0 \\ 0 & -k \end{pmatrix} \tag{1.17}$$

is a Clifford map for the space $\mathbb{R}^{3,0}$. That is, $\text{Cl}_{3,0} = \mathbb{H} \oplus \mathbb{H}$.

Example 7. Let $\rho_1, \ldots, \rho_{p+q}$ be Clifford matrices of the Clifford algebra $\text{Cl}_{p,q}$. The reader can check that the matrices

$$\begin{pmatrix} \rho_1 & 0 \\ 0 & -\rho_1 \end{pmatrix}, \ldots, \begin{pmatrix} \rho_p & 0 \\ 0 & -\rho_p \end{pmatrix}, \begin{pmatrix} 0 & -1_{\text{Cl}_{p,q}} \\ 1_{\text{Cl}_{p,q}} & 0 \end{pmatrix},$$

$$\begin{pmatrix} \rho_{p+1} & 0 \\ 0 & -\rho_{p+1} \end{pmatrix}, \ldots, \begin{pmatrix} \rho_{p+q} & 0 \\ 0 & -\rho_{p+q} \end{pmatrix}, \begin{pmatrix} 0 & 1_{\text{Cl}_{p,q}} \\ 1_{\text{Cl}_{p,q}} & 0 \end{pmatrix}$$

are gamma matrices of the tensor product $\text{Cl}_{p,q} \otimes \text{Cl}_{1,1} = \text{Cl}_{p+1,q+1}$.

In particular, the matrices

$$\gamma_1 = \begin{pmatrix} 1 & 0 & 0 & 0 \\ 0 & -1 & 0 & 0 \\ 0 & 0 & -1 & 0 \\ 0 & 0 & 0 & 1 \end{pmatrix}, \qquad \gamma_2 = \begin{pmatrix} 0 & 0 & -1 & 0 \\ 0 & 0 & 0 & -1 \\ 1 & 0 & 0 & 0 \\ 0 & 1 & 0 & 0 \end{pmatrix},$$

$$\gamma_3 = \begin{pmatrix} 0 & 0 & 1 & 0 \\ 0 & 0 & 0 & 1 \\ 1 & 0 & 0 & 0 \\ 0 & 1 & 0 & 0 \end{pmatrix} \tag{1.18}$$

are gamma matrices for the Clifford algebra $\text{Cl}_{1,2} = \mathbb{R}(2) \oplus \mathbb{R}(2)$.

The matrices

$$\gamma_1 = \begin{pmatrix} i & 0 \\ 0 & -i \end{pmatrix}, \qquad \gamma_2 = \begin{pmatrix} 0 & -1 \\ 1 & 0 \end{pmatrix}, \qquad \gamma_3 = \begin{pmatrix} 0 & 1 \\ 1 & 0 \end{pmatrix} \tag{1.19}$$

are gamma matrices for the Clifford algebra $Cl_{2,1} = \mathbb{C}(2)$.

The matrices

$$\gamma_0 = \begin{pmatrix} i & 0 \\ 0 & -i \end{pmatrix}, \quad \gamma_1 = \begin{pmatrix} j & 0 \\ 0 & -j \end{pmatrix}, \quad \gamma_2 = \begin{pmatrix} 0 & -1 \\ 1 & 0 \end{pmatrix}, \quad \gamma_3 = \begin{pmatrix} 0 & 1 \\ 1 & 0 \end{pmatrix} \tag{1.20}$$

are gamma matrices for the Clifford algebra $Cl_{3,1} = \mathbb{H}(2)$. Another popular choice is given by combining the map γ with the matrix representation $\rho \colon Cl_{3,1} \to \mathbb{C}(4)$ with

$$\rho(\gamma(\mathbf{e}_i)) = \gamma_i, \qquad 0 \le i \le 3,$$

where

$$\gamma_0 = \begin{pmatrix} 0 & -\sigma_0 \\ -\sigma_0 & 0 \end{pmatrix}, \qquad \gamma_k = \begin{pmatrix} 0 & -\sigma_k \\ \sigma_k & 0 \end{pmatrix}.$$

The matrices

$$\gamma_1 = \begin{pmatrix} 0 & 0 & -1 & 0 \\ 0 & 0 & 0 & -1 \\ 1 & 0 & 0 & 0 \\ 0 & 1 & 0 & 0 \end{pmatrix}, \quad \gamma_2 = \begin{pmatrix} -1 & 0 & 0 & 0 \\ 0 & 1 & 0 & 0 \\ 0 & 0 & 1 & 0 \\ 0 & 0 & 0 & -1 \end{pmatrix},$$

$$\gamma_3 = \begin{pmatrix} 0 & 1 & 0 & 0 \\ 1 & 0 & 0 & 0 \\ 0 & 0 & 0 & -1 \\ 0 & 0 & -1 & 0 \end{pmatrix}, \quad \gamma_4 = \begin{pmatrix} 0 & 0 & 1 & 0 \\ 0 & 0 & 0 & 1 \\ 1 & 0 & 0 & 0 \\ 0 & 1 & 0 & 0 \end{pmatrix}$$

are gamma matrices for the Clifford algebra $Cl_{1,3} = \mathbb{R}(4)$. Another popular choice is given by combining the map γ with the matrix representation $\rho \colon Cl_{1,3} \to \mathbb{C}(4)$ with

$$\rho(\gamma(\mathbf{e}_i)) = \gamma_i, \qquad 0 \le i \le 3,$$

where

$$\gamma_0 = \begin{pmatrix} 0 & i\sigma_0 \\ i\sigma_0 & 0 \end{pmatrix}, \qquad \gamma_k = \begin{pmatrix} 0 & i\sigma_k \\ -i\sigma_k & 0 \end{pmatrix}. \tag{1.21}$$

The matrices

$$\gamma_0 = \begin{pmatrix} 0 & -1 & 0 & 0 \\ 1 & 0 & 0 & 0 \\ 0 & 0 & 0 & 1 \\ 0 & 0 & -1 & 0 \end{pmatrix}, \quad \gamma_1 = \begin{pmatrix} 0 & 0 & -1 & 0 \\ 0 & 0 & 0 & -1 \\ 1 & 0 & 0 & 0 \\ 0 & 1 & 0 & 0 \end{pmatrix},$$

$$\gamma_2 = \begin{pmatrix} 0 & 1 & 0 & 0 \\ 1 & 0 & 0 & 0 \\ 0 & 0 & 0 & -1 \\ 0 & 0 & -1 & 0 \end{pmatrix}, \quad \gamma_3 = \begin{pmatrix} 0 & 0 & 1 & 0 \\ 0 & 0 & 0 & 1 \\ 1 & 0 & 0 & 0 \\ 0 & 1 & 0 & 0 \end{pmatrix} \tag{1.22}$$

are Dirac matrices for the Clifford algebra $Cl_{2,2} = \mathbb{R}(4)$.

TABLE 1.1
Complex Clifford Algebras

$n \mod 2$	Cl_n	N
0	$\mathbb{C}(N)$	$2^{n/2}$
1	$\mathbb{C}(N) \oplus \mathbb{C}(N)$	$2^{(n-1)/2}$

TABLE 1.2
Real Clifford Algebras

$p - q \mod 8$	$\mathrm{Cl}_{p,q}$	N
0, 6	$\mathbb{R}(N)$	$2^{n/2}$
1, 5	$\mathbb{C}(N)$	$2^{(n-1)/2}$
2, 4	$\mathbb{H}(N)$	$2^{(n-2)/2}$
3	$\mathbb{H}(N) \oplus \mathbb{H}(N)$	$2^{(n-3)/2}$
7	$\mathbb{R}(N) \oplus \mathbb{R}(N)$	$2^{(n-1)/2}$

By Porteous (1995, Proposition 15.29), Cl_n is isomorphic to $c\,\mathrm{Cl}_{p,q}$ for any three non-negative integers n, p, and q with $p + q = n$.

Example 8. We have $\mathrm{Cl}_0 = c\,\mathrm{Cl}_{0,0} = \mathbb{C}$. Similarly, $\mathrm{Cl}_1 = c\,\mathrm{Cl}_{1,0} = \mathbb{C} \oplus \mathbb{C}$.

The Clifford algebra Cl_2 may be constructed either as $c\,\mathrm{Cl}_{1,1} = \mathbb{C}(2)$ or $c\,\mathrm{Cl}_{2,0}$.

We have $\mathrm{Cl}_3 = c\,\mathrm{Cl}_{2,1} = \mathbb{C}(2) \oplus \mathbb{C}(2)$.

The Clifford algebra Cl_4 may be constructed either as $c\,\mathrm{Cl}_{2,2} = \mathbb{C}(4)$ or $c\,\mathrm{Cl}_{3,1}$.

All complex Clifford algebras are shown in Table 1.1. All real Clifford algebras are shown in Table 1.2.

Theorem 2. *For any complex or real Clifford algebra Cl_n (resp., $\mathrm{Cl}_{p,q}$) there is an isomorphism ρ that maps that algebra to the matrix algebra shown in the second column of Table 1.1 (resp., Table 1.2).*

1.3.5 Clifford Involutions

It turns out that there exists a unique automorphism $A \mapsto \tilde{A}$ of a Clifford algebra $\mathrm{Cl}(L, g)$ such that for all $A \in \gamma(L) \subset \mathrm{Cl}(L, g)$ we have $\tilde{A} = -A$. This map is called the *Clifford involution*.

Definition 20. The *even Clifford algebra* of the Clifford algebra $\mathrm{Cl}(L, g)$ is

$$\mathrm{Cl}^+(L, g) = \{\, A \in \mathrm{Cl}(L, g) \colon \tilde{A} = A \,\}.$$

By Vaz and Rocha (2019, Theorem 4.4), the Clifford algebras $\mathrm{Cl}^+_{p,q}$, $\mathrm{Cl}_{q,p-1}$, $\mathrm{Cl}_{p,q-1}$, and $\mathrm{Cl}^+_{q,p}$ are isomorphic. For $n \geq 1$, we have $\mathrm{Cl}^+_n = \mathrm{Cl}_{n-1}$.

TABLE 1.3
Even Real Clifford Algebras

$p - q \mod 8$	$\mathrm{Cl}_{p,q}^+$	N
0	$\mathbb{R}(N) \oplus \mathbb{R}(N)$	$2^{(n-2)/2}$
1, 7	$\mathbb{R}(N)$	$2^{(n-1)/2}$
2, 6	$\mathbb{C}(N)$	$2^{(n-2)/2}$
3, 5	$\mathbb{H}(N)$	$2^{(n-3)/2}$
4	$\mathbb{H}(N) \oplus \mathbb{H}(N)$	$2^{(n-4)/2}$

Example 9. The previous examples give $\mathrm{Cl}_{0,1}^+ = \mathrm{Cl}_{1,0}^+ = \mathbb{R}$, $\mathrm{Cl}_{0,2}^+ = \mathrm{Cl}_{2,0}^+ = \mathbb{C}$, $\mathrm{Cl}_{1,1}^+ = \mathbb{R} \oplus \mathbb{R}$, $\mathrm{Cl}_1^+ = \mathbb{C}$, $\mathrm{Cl}_2^+ = \mathbb{C} \oplus \mathbb{C}$, $\mathrm{Cl}_3^+ = \mathbb{C}(2)$, and $\mathrm{Cl}_4^+ = \mathbb{C}(2) \oplus \mathbb{C}(2)$. Moreover,

$$\mathrm{Cl}_{0,3}^+ = \mathrm{Cl}_{3,0}^+ = \mathbb{H}, \qquad \mathrm{Cl}_{1,2}^+ = \mathrm{Cl}_{2,1}^+ = \mathbb{R}(2),$$
$$\mathrm{Cl}_{0,4}^+ = \mathrm{Cl}_{4,0}^+ = \mathbb{H} \oplus \mathbb{H}, \quad \mathrm{Cl}_{1,3}^+ = \mathrm{Cl}_{3,1}^+ = \mathbb{C}(2).$$

Finally, $\mathrm{Cl}_{2,2}^+ = \mathbb{R}(2) \oplus \mathbb{R}(2)$.

All even real Clifford algebras are shown in Table 1.3.

It turns out that the set of $2^{\dim_{\mathbb{F}} L} - 1$ products $\gamma_{i_1} \cdots \gamma_{i_p}$ with $1 \leq p \leq \dim_{\mathbb{F}} L$ and $i_1 < \cdots < i_p$ together with the empty product $1_{\mathrm{Cl}(L,g)}$ constitute a basis for $\mathrm{Cl}(L, g)$. We say that the elements of the linear span of the above products of p basis vectors are of *degree p* and denote the above linear span by Λ^p. If $A \in \mathrm{Cl}(V, g)$ is of degree p, then $\tilde{A} = A$ if and only if p is even and $\tilde{A} = -A$ otherwise.

The *hat involution* $A \mapsto \hat{A}$ is a linear operator in $\mathrm{Cl}(V, g)$ such that for an element $A \in \mathrm{Cl}(V, g)$ of degree p we have $\hat{A} = A$ if and only if $p \in \{0, 3\}$ (mod 4) and $\hat{A} = -A$ otherwise.

1.4 Algebraic Spinors

Let n be an odd positive integer and $N = (n - 1)/2$. Define the map $\mathrm{pr}_1 \colon \mathbb{C}(N) \oplus \mathbb{C}(N) \to \mathbb{C}(N)$ given by $\mathrm{pr}_1(A \oplus B) = A$.

Definition 21. The *Dirac spinor representation* of the Clifford algebra Cl_{2m+1}^+ is the isomorphism ρ of Theorem 2. The elements of the complex linear space $S = \mathbb{C}^N$, where N is given by Table 1.1, and where the Dirac spinor representation acts, are called *algebraic Dirac spinors*.

We denote Dirac and all subsequent kinds of algebraic spinors by lowercase Greek letters φ, ψ, θ, ...

Example 10 (Algebraic Dirac spinors). The Dirac spinor representation of the Clifford algebra $\mathrm{Cl}_1^+ = \mathbb{C}$ (resp., $\mathrm{Cl}_3^+ = \mathbb{C}(2)$) acts in the linear space

$S = \mathbb{C}^1$ (resp., $S = \mathbb{C}^2$) of Dirac spinors by

$$\rho(A)\varphi = A\varphi, \qquad \varphi \in S. \tag{1.23}$$

Theorem 3. *For odd n, the restriction of the Dirac spinor representation* $\mathrm{pr}_1 \circ \rho$ *of the Clifford algebra* Cl_n *to the subalgebra* Cl_n^+ *is irreducible. For even n, the restriction of ρ to Cl_n^+ splits into two irreducible components, the ± 1-eigenspaces of the physical chirality operator Γ_{n+1}.*

Definition 22. The elements of the complex linear space S, where the irreducible *Weyl spinor representation* $\mathrm{pr}_1 \circ \rho$ of the Clifford algebra Cl_n^+ with n odd acts, are called the *algebraic Weyl spinors*.

The elements of the $+1$-eigenspace S_+ (resp., -1-eigenspace S_-), where the irreducible *positive* or *left-handed* Weyl spinor representation ρ_+ (resp., irreducible *negative* or *right-handed* Weyl spinor representation ρ_-) of the Clifford algebra Cl_n^+ with n even acts, are called the *positive* or *left-handed* (resp., *negative* or *right-handed*) algebraic Weyl spinors.

Example 11 (Algebraic Weyl spinors). The Dirac spinor representation of the Clifford algebra $\mathrm{Cl}_2 = \mathbb{C}(2)$ (resp., $\mathrm{Cl}_4 = \mathbb{C}(4)$) acts in the linear space $S = \mathbb{C}^2$ (resp., $S = \mathbb{C}^4$) of Dirac spinors by Equation (1.23).

The restriction of the above representation to the Clifford algebra $\mathrm{Cl}_2^+ = \mathbb{C} \oplus \mathbb{C}$ (resp., $\mathrm{Cl}_4^+ = \mathbb{C}(2) \oplus \mathbb{C}(2)$) splits into two irreducible components. The physical chirality operator Γ_3 (resp., Γ_4) of the Clifford algebra $\mathrm{Cl}_2^+ = \mathbb{C} \oplus \mathbb{C}$ (resp., $\mathrm{Cl}_4^+ = \mathbb{C}(2) \oplus \mathbb{C}(2)$) is given by Equation (1.13) (resp., (1.16)). For Cl_2^+, the left-handed Weyl spinor representation ρ_+ (resp., the right-handed Weyl spinor representation ρ_-) acts in the $+1$-eigenspace $S_+ = \mathbb{C}^1$ (resp., -1-eigenspace $S_- = \mathbb{C}^1$) of the operator Γ_3. For Cl_4^+, the left-handed Weyl spinor representation ρ_+ (resp., the right-handed Weyl spinor representation ρ_-) acts in the $+1$-eigenspace $S_+ = \mathbb{C}^2$ (resp., -1-eigenspace $S_- = \mathbb{C}^2$) of the operator Γ_4.

We describe a similar construction that involves *real* Clifford algebras $\mathrm{Cl}_{p,q}$. Let N be the number given in the third column of Table 1.2. Define the map $\mathrm{pr}_1 \colon \mathbb{K}(N) \oplus \mathbb{K}(N) \to \mathbb{K}(N)$ given by $\mathrm{pr}_1(A \oplus B) = A$, where $\mathbb{K} \in \{\mathbb{R}, \mathbb{H}\}$.

Definition 23. The *spinor representation* of the Clifford algebra $\mathrm{Cl}_{p,q}$ is the isomorphism ρ of Theorem 2 when $p - q \mod 8 \notin \{3, 7\}$, and the composition $\mathrm{pr}_1 \circ \rho$ otherwise.

Theorem 4. *For odd n, the restriction of the spinor representation of the Clifford algebra $\mathrm{Cl}_{p,q}$ to the subalgebra $\mathrm{Cl}_{p,q}^+$ is irreducible. For even n, the restriction of the spinor representation of $\mathrm{Cl}_{p,q}$ to $\mathrm{Cl}_{p,q}^+$ splits into two irreducible components.*

Definition 24. For odd n, the irreducible representation of the Clifford algebra $\mathrm{Cl}_{p,q}^+$ constructed in Theorem 4 is called *Majorana* (resp., *symplectic Majorana*) if the space of the representation is real (resp., quaternionic). The elements of the real (resp., quaternionic) representation space S are called *Majorana spinors* (resp., *symplectic Majorana spinors*).

Example 12 (Algebraic Majorana spinors). Let $\mathrm{pr}_1 \circ \rho$ be the spinor representation of the Clifford algebra $\mathrm{Cl}_{0,1} = \mathbb{R} \oplus \mathbb{R}$ in the linear space $\mathbb{R}^1$. Its restriction to the subalgebra $\mathrm{Cl}_{0,1}^+ = \mathbb{R}$ is the Majorana representation of $\mathrm{Cl}_{0,1}^+$ in the real linear space $S = \mathbb{R}^1$ of Majorana spinors.

The same representation can be constructed starting from the Clifford algebra $\mathrm{Cl}_{1,0} = \mathbb{C}$. Its spinor representation ρ acts in the linear space $\mathbb{C}^1$. The restriction of ρ to the subalgebra $\mathrm{Cl}_{0,1}^+ = \mathbb{R}$ clearly commutes with the real structure $j(\varphi) = \overline{\varphi}$, $\varphi \in \mathbb{C}^1$. Therefore, it acts in the linear space $S = \mathbb{R}^1$ of Majorana spinors.

Similarly, let $\mathrm{pr}_1 \circ \rho$ be the spinor representation of the Clifford algebra $\mathrm{Cl}_{1,2} = \mathbb{R}(2) \oplus \mathbb{R}(2)$ in the linear space $\mathbb{R}^2$. Its restriction to the subalgebra $\mathrm{Cl}_{1,2}^+ = \mathbb{R}(2)$ is the Majorana representation of $\mathrm{Cl}_{1,2}^+$ in the real linear space $S = \mathbb{R}^2$ of Majorana spinors.

The same representation can be constructed starting from the Clifford algebra $\mathrm{Cl}_{2,1} = \mathbb{C}(2)$. Its spinor representation ρ acts in the linear space $\mathbb{C}^2$. The restriction of ρ to the subalgebra $\mathrm{Cl}_{1,2}^+ = \mathbb{R}(2)$ clearly commutes with the real structure

$$j \begin{pmatrix} \varphi_1 \\ \varphi_2 \end{pmatrix} = \begin{pmatrix} \overline{\varphi_1} \\ \overline{\varphi_2} \end{pmatrix}. \tag{1.24}$$

Therefore, it acts in the linear space $S = \mathbb{R}^2$ of Majorana spinors.

Example 13 (Algebraic symplectic Majorana spinors). Consider the Clifford algebras $\mathrm{Cl}_{0,3}^+ = \mathrm{Cl}_{3,0}^+ = \mathbb{H}$. Its Dirac spinor representation given by the mathematical Dirac matrices (1.8) commutes with the quaternionic structure $j(\varphi_1, \varphi_2)^\top = (i\overline{\varphi_2}, i\overline{\varphi_1})^\top$. Indeed, on the one hand,

$$(\gamma_1 \circ j) \begin{pmatrix} \varphi_1 \\ \varphi_2 \end{pmatrix} = \gamma_1 \begin{pmatrix} i\overline{\varphi_2} \\ i\overline{\varphi_1} \end{pmatrix} = \begin{pmatrix} 0 & i \\ i & 0 \end{pmatrix} \begin{pmatrix} i\overline{\varphi_2} \\ i\overline{\varphi_1} \end{pmatrix} = -\begin{pmatrix} \overline{\varphi_1} \\ \overline{\varphi_2} \end{pmatrix}.$$

On the other hand,

$$(j \circ \gamma_1) \begin{pmatrix} \varphi_1 \\ \varphi_2 \end{pmatrix} = j \begin{pmatrix} 0 & i \\ i & 0 \end{pmatrix} \begin{pmatrix} \varphi_1 \\ \varphi_2 \end{pmatrix} = j \begin{pmatrix} i\overline{\varphi_2} \\ i\overline{\varphi_1} \end{pmatrix} = -\begin{pmatrix} \overline{\varphi_1} \\ \overline{\varphi_2} \end{pmatrix},$$

and similarly for γ_2. The elements of the quaternionic linear space $S = \mathbb{H}^1$ with scalar multiplication given by (1.6) are symplectic Majorana spinors.

For even n, the picture is more sophisticated. The inspection of Table 1.2 shows that the representation ρ acts in a real linear space, say U, if $p - q$ mod $8 \in \{0, 6\}$ and in a quaternionic linear space, say W, otherwise. In the first case, consider the complex representation $c\rho$ of the Clifford algebra $\mathrm{Cl}_{p,q}$ acting in $S = cU$ by

$$(c\rho(A))(\zeta \otimes \boldsymbol{\xi}) = \zeta \otimes (\rho(A)(\boldsymbol{\xi})), \qquad A \in \mathrm{Cl}_{p,q}, \quad \zeta \in \mathbb{C}, \quad \boldsymbol{\xi} \in U.$$

In the second case, let $c'\rho$ be the complex representation of the Clifford algebra $\mathrm{Cl}_{p,q}$ acting in $S = c'W$ in the same way as in W. Upon the restriction to $\mathrm{Cl}_{p,q}^+$ the above complex representations split into the irreducible components acting in the spaces $\pi_\pm S$, where

$$\pi_\pm = \frac{1}{2}(I \pm \Gamma_{n+1}). \tag{1.25}$$

As we will see below, the following three cases are realised.

- There exists a real structure j on S that commutes with the operators $\pi_\pm$. In this case, the restrictions $j_\pm$ of j to the spaces $\pi_\pm S$ are real structures. The real representation that acts in the $+1$-eigenspace S_+ (resp., S_-) of the real structure j_+ (resp., j_-), is called the *positive, or left-handed,* (resp., the *negative, or right-handed*) *Majorana–Weyl spinor representation*. The elements of the space S_+ (resp., S_-) are called the *positive, or left-handed,* (resp., the *negative, or right-handed*) *Majorana–Weyl spinors*.

- There exists a quaternionic structure j on S that commutes with the operators $\pi_\pm$. In this case, the restrictions $j_\pm$ of j to the spaces $\pi_\pm S$ are quaternionic structures. Denote by S_+ (resp., S_-) the space π_+S (resp., π_-S) considered as a quaternionic linear space. The quaternionic representation that acts in S_+ (resp., S_-), is called the *positive, or left-handed,* (resp., the *negative, or right-handed*) *Majorana–Weyl spinor representation*. The elements of the space S_+ (resp., S_-) are called the *positive, or left-handed,* (resp., the *negative, or right-handed*) *symplectic Majorana–Weyl spinors*.

- None of the above structure exist. The irreducible component that acts in the linear space $S_+ = \pi_+S$ (resp. $S_- = \pi_-S$) of the positive, of left-handed (resp. the negative, or right-handed) Weyl spinors, is just the positive, of left-handed (resp. the negative, or right-handed) Weyl spinor representation.

Example 14 (Algebraic Majorana–Weyl spinors). The representation ρ of the Clifford algebra $\mathrm{Cl}_{1,1} = \mathbb{R}(2)$ acts in $\mathbb{R}^2$. The representation $c\rho$ acts in $\mathbb{C}^2$ and commutes with the real structure (1.24). The restrictions $j_\pm$ act by $j_\pm\varphi = \overline{\varphi}$. The left-handed (resp., right-handed) Majorana–Weyl representation of the Clifford algebra $\mathrm{Cl}_{1,1}^+ = \mathbb{R} \oplus \mathbb{R}$ acts in the $+1$-eigenspace $S_+ = \mathbb{R}^1$ (resp., $S_- = \mathbb{R}^1$) of left-handed (resp., right-handed) Majorana–Weyl spinors by $(\alpha,\beta) \mapsto \alpha$ (resp., $(\alpha,\beta) \mapsto \beta$).

The similar construction for the case of the Clifford algebra $\mathrm{Cl}_{2,2}^+ = \mathbb{R}(2) \oplus \mathbb{R}(2)$ may be left to the reader.

Example 15 (Algebraic symplectic Majorana–Weyl spinors). The representation ρ of the Clifford algebra $\mathrm{Cl}_{0,4} = \mathrm{Cl}_{4,0} = \mathbb{H}(2)$ acts in $\mathbb{H}^2$. The representation $c\rho$ acts in $\mathbb{C}^4$. A routine calculation shows that the operators $\pi_\pm$ given by Equation (1.25) with the matrix Γ_4 given by Equation (1.16), commute with the quaternionic structure

$$
j \begin{pmatrix} \alpha \\ \beta \\ \gamma \\ \delta \end{pmatrix} = \begin{pmatrix} \overline{\beta} \\ -\overline{\alpha} \\ \overline{\delta} \\ -\overline{\gamma} \end{pmatrix}.
$$

The restrictions $j_\pm$ act by

$$
j_\pm \begin{pmatrix} \alpha \\ \beta \end{pmatrix} = \begin{pmatrix} \overline{\beta} \\ -\overline{\alpha} \end{pmatrix}.
$$

The symplectic left-handed (resp., right-handed) Majorana–Weyl representation of the Clifford algebra $\mathrm{Cl}_{0,4}^+ = \mathrm{Cl}_{4,0}^+ = \mathbb{H} \oplus \mathbb{H}$ acts in the $+1$-eigenspace $S_+ = \mathbb{H}^1$ (resp., $S_- = \mathbb{H}^1$) of left-handed (resp., right-handed) symplectic Majorana–Weyl spinors by $(\alpha, \beta) \mapsto \alpha$ (resp., $(\alpha, \beta) \mapsto \beta$).

Example 16. The representation ρ of the Clifford algebra $\mathrm{Cl}_{0,2} = \mathbb{R}(2)$ is Majorana. Upon restriction to the subalgebra $\mathrm{Cl}_{0,2}^+ = \mathbb{C}$, the representation ρ is still Majorana, while $c\rho$ splits into the irreducible components acting in one-dimensional complex linear spaces S and tS. They are complex left-handed and right-handed Weyl representations.

The same representations can be constructed starting from the Clifford algebra $\mathrm{Cl}_{2,0} = \mathbb{H}$. Its representation ρ is symplectic Majorana. Upon restriction to the subalgebra $\mathrm{Cl}_{2,0}^+ = \mathbb{C}$, the representation ρ is still symplectic Majorana, while $c'\rho$ splits into the irreducible left-handed and right-handed Weyl representations acting in one-dimensional complex linear spaces S and tS.

Similarly, the representation ρ of the Clifford algebra $\mathrm{Cl}_{1,3} = \mathbb{R}(4)$ is Majorana. Upon restriction to the subalgebra $\mathrm{Cl}_{1,3}^+ = \mathbb{C}(2)$, the representation ρ is still Majorana, while $c\rho$ splits into the irreducible components acting in two-dimensional complex linear spaces S and tS. They are complex left-handed and right-handed Weyl representations.

The same representations can be constructed starting from the Clifford algebra $\mathrm{Cl}_{3,1} = \mathbb{H}(2)$. Its representation ρ is symplectic Majorana. Upon restriction to the subalgebra $\mathrm{Cl}_{3,1}^+ = \mathbb{C}(2)$, the representation ρ is still symplectic Majorana, while $c'\rho$ splits into the irreducible left-handed and right-handed Weyl representations acting in two-dimensional complex linear spaces S and tS.

Information about algebraic spinors is collected in Table 1.4.

1.5 Manifolds and Coverings

The basic idea of this section is as follows. For a non-negative integer n, consider the space $\mathbb{F}^n$ as a simple one, and glue simple spaces together.

TABLE 1.4
Algebraic Spinors of Small Dimension

A	S	Name
$\mathrm{Cl}_1^+ = \mathbb{C}$	$S = \mathbb{C}^1$	Dirac
$\mathrm{Cl}_2^+ = \mathbb{C} \oplus \mathbb{C}$	$S_+ = \mathbb{C}^1$	left-handed Weyl
$\mathrm{Cl}_2^+ = \mathbb{C} \oplus \mathbb{C}$	$S_- = \mathbb{C}^1$	right-handed Weyl
$\mathrm{Cl}_3^+ = \mathbb{C}(2)$	$S = \mathbb{C}^2$	Dirac
$\mathrm{Cl}_4^+ = \mathbb{C}(2) \oplus \mathbb{C}(2)$	$S_+ = \mathbb{C}^2$	left-handed Weyl
$\mathrm{Cl}_4^+ = \mathbb{C}(2) \oplus \mathbb{C}(2)$	$S_- = \mathbb{C}^2$	right-handed Weyl
$\mathrm{Cl}_{0,1}^+ = \mathbb{R}$	$S = \mathbb{R}^1$	Majorana
$\mathrm{Cl}_{0,2}^+ = \mathbb{C}$	$S = \mathbb{R}^2$	Majorana
$\mathrm{Cl}_{0,2}^+ = \mathbb{C}$	$S = \mathbb{C}^1$	left-handed Weyl
$\mathrm{Cl}_{0,2}^+ = \mathbb{C}$	$tS = t\mathbb{C}^1$	right-handed Weyl
$\mathrm{Cl}_{0,3}^+ = \mathbb{H}$	$S = \mathbb{H}^1$	symplectic Majorana
$\mathrm{Cl}_{0,4}^+ = \mathbb{H} \oplus \mathbb{H}$	$S_+ = \mathbb{H}^1$	left-handed symplectic Majorana–Weyl
$\mathrm{Cl}_{0,4}^+ = \mathbb{H} \oplus \mathbb{H}$	$S_- = \mathbb{H}^1$	right-handed symplectic Majorana–Weyl
$\mathrm{Cl}_{1,1}^+ = \mathbb{R} \oplus \mathbb{R}$	$S_+ = \mathbb{R}^1$	left-handed Majorana–Weyl
$\mathrm{Cl}_{1,1}^+ = \mathbb{R} \oplus \mathbb{R}$	$S_- = \mathbb{R}^1$	right-handed Majorana–Weyl
$\mathrm{Cl}_{1,2}^+ = \mathbb{R}(2)$	$S = \mathbb{R}^2$	Majorana
$\mathrm{Cl}_{1,3}^+ = \mathbb{C}(2)$	$S = \mathbb{R}^4$	Majorana
$\mathrm{Cl}_{1,3}^+ = \mathbb{C}(2)$	$S = \mathbb{C}^2$	left-handed Weyl
$\mathrm{Cl}_{1,3}^+ = \mathbb{C}(2)$	$tS = t\mathbb{C}^2$	right-handed Weyl
$\mathrm{Cl}_{2,2}^+ = \mathbb{R}(2) \oplus \mathbb{R}(2)$	$S_+ = \mathbb{R}^2$	left-handed Majorana–Weyl
$\mathrm{Cl}_{2,2}^+ = \mathbb{R}(2) \oplus \mathbb{R}(2)$	$S_- = \mathbb{R}^2$	right-handed Majorana–Weyl

1.5.1   Manifolds

Let M be a set and n be a non-negative integer.

Definition 25. A *chart* in M is a pair $(\mathcal{U}, \mathbf{x})$, where $\mathcal{U}$ is a subset of M and where $\mathbf{x}$ is a one-to-one map of $\mathcal{U}$ to an open subset $\mathbf{x}(\mathcal{U})$ of the space $\mathbb{F}^n$.

Let $p \in \mathcal{U}$. The point $\mathbf{x}(p) \in \mathbb{F}^n$ has the form $\mathbf{x}(p) = (x^1(p), \ldots, x^n(p))^\top$.

Definition 26. The n $\mathbb{F}$-valued functions $x^1(p), \ldots, x^n(p)$ are called the *local coordinates* on $\mathcal{U}$ defined by the chart $(\mathcal{U}, \mathbf{x})$.

Example 17. Let M be a finite-dimensional F-linear space. An arbitrary choice of a coordinate system in M determines a chart supported on M.

For a quaternionic linear space $\mathbb{H}^n$, the map

$$\mathbb{H}^m \to \mathbb{R}^{4n}, \qquad (q_1, \ldots, q_n) \mapsto (\alpha_1, \ldots, \delta_1, \ldots, \alpha_m, \ldots, \delta_m),$$

where $q_k = \alpha_k + \beta_k \mathrm{i} + \gamma_k \mathrm{j} + \delta_k \mathrm{k}$, is a chart on $\mathbb{H}^n$.

Let $S^1 \subset \mathbb{R}^2$ be the centred circle of radius 1:

$$S^1 = \{\,(x,y)^\top \in \mathbb{R}^2 \colon x^2 + y^2 = 1\,\}.$$

The angular polar coordinate φ of a point $p \in \mathcal{U} = S^1 \setminus \{(1,0)^\top\}$ with

$$x = \cos\varphi, \qquad y = \sin\varphi$$

defines a chart $(\mathcal{U},\varphi)$ in S^1 with $\varphi(\mathcal{U}) = (0, 2\pi) \subset \mathbb{R}^1$.

Let $S^2 \subset \mathbb{R}^3$ be the centred sphere of radius 1:

$$S^2 = \{\,(x,y,z)^\top \in \mathbb{R}^3 \colon x^2 + y^2 + z^2 = 1\,\}.$$

The angular spherical coordinates θ and φ of a point

$$p \in \mathcal{U} = S^2 \setminus \{\,(x,0,z)^\top \colon x^2 + z^2 = 1, x > 0\,\}$$

with

$$x = \cos\varphi\sin\theta, \qquad y = \sin\varphi\sin\theta, \qquad z = \cos\theta$$

define a chart $(\mathcal{U}, \mathbf{h})$ in S^2 with $\mathbf{h}(\mathcal{U}) = (0, 2\pi) \times (0, \pi) \subset \mathbb{R}^2$.

The *Euler angles*

$$\begin{pmatrix} \cos\varphi\cos\theta\cos\psi - \sin\varphi\sin\psi & -\cos\varphi\cos\theta\sin\psi - \sin\varphi\cos\psi & \cos\varphi\sin\theta \\ \sin\varphi\cos\theta\cos\psi + \cos\varphi\sin\psi & -\sin\varphi\cos\theta\sin\psi + \cos\varphi\cos\psi & \sin\varphi\sin\theta \\ -\sin\theta\cos\psi & \sin\theta\sin\psi & \cos\theta \end{pmatrix} \tag{1.26}$$

define a chart $(\mathcal{U}, \mathbf{h})$ in the rotation group $\mathrm{SO}(3)$ with $\mathbf{h}(\mathcal{U}) = (0, 2\pi) \times (0, \pi) \times (0, 2\pi) \subset \mathbb{R}^3$.

Two charts $(\mathcal{U}, \mathbf{x})$ and $(\mathcal{V}, \mathbf{y})$ are called *intersecting*, if $\mathcal{U} \cap \mathcal{V} \neq \varnothing$. For two intersecting charts $(\mathcal{U}, \mathbf{x})$ and $(\mathcal{V}, \mathbf{y})$, denote $\mathcal{W} = \mathcal{U} \cap \mathcal{V}$. The sets $\mathbf{x}(\mathcal{W})$ and $\mathbf{y}(\mathcal{W})$ are nonempty subsets of the space $\mathbb{F}^n$. For simplicity, we denote the *transition map* $(\mathbf{y}|_W) \circ (\mathbf{x}|_W)^{-1} \colon \mathbf{x}(\mathcal{W}) \to \mathbf{y}(\mathcal{W})$ by $\mathbf{y} \circ \mathbf{x}^{-1}$. The transition map $\mathbf{x} \circ \mathbf{y}^{-1}$ is called *smooth*, (resp., *real-analytic*, resp.,*holomorphic*,) if the sets $\mathbf{x}(\mathcal{W})$ and $\mathbf{y}(\mathcal{W})$ are open in $\mathbb{F}^n$, $\mathbb{F} = \mathbb{R}$ (resp., $\mathbb{F} = \mathbb{R}$, resp., $\mathbb{F} = \mathbb{C}$) and $\mathbf{y} \circ \mathbf{x}^{-1}$ is infinitely differentiable, (resp., real-analytic, resp., holomorphic).

Definition 27. A set $\mathsf{A} = \{\,(\mathcal{U}_\alpha, \mathbf{x}_\alpha) \colon \alpha \in A\,\}$ of charts in M is called a *smooth atlas* (resp., a *real-analytic atlas*, resp., a *holomorphic atlas*) for M if the supports $\mathcal{U}_\alpha$ of the charts $(\mathcal{U}_\alpha, \mathbf{h}_\alpha)$ cover all of M and the transition map of any two intersecting charts is smooth (resp., real-analytic, resp., holomorphic).

For an arbitrary atlas A for M, let the *maximal atlas* $\mathsf{A}_{\max}$ be the set of all charts $(\mathcal{U}, \mathbf{x})$ in M such that $\mathsf{A} \cup (\mathcal{U}, \mathbf{x})$ is again an atlas. Then, the set $\mathsf{A}_{\max}$ is itself an atlas for M.

Definition 28. A *smooth structure* (resp., a *real-analytic structure*, resp., *holomorphic structure*) on M is a maximal atlas. A *smooth manifold* (resp., *real-analytic manifold*, resp., *holomorphic manifold*) of dimension n is a pair $(M, \mathsf{A}_{\max})$.

In particular, a finite-dimensional linear space over $\mathbb{R}$, the sets S^1, S^2, and $SO(3)$ are real-analytic manifolds.

Introduce a topology on a manifold $(M, \mathsf{A}_{\max})$.

Definition 29. A subset $O \subseteq M$ is called *open* if for any chart $(\mathcal{U}, \mathbf{x}) \in \mathsf{A}_{\max}$ the set $\mathbf{x}(O \cap \mathcal{U}) \subseteq \mathbb{F}^n$ is open in $\mathbb{F}^n$.

To avoid pathologies, we assume that the introduced topology is Hausdorff and has a countable base.

Consider the following examples.

Example 18. Let (ζ_0, ζ_1) be the coordinates in a two-dimensional complex linear space V. Put $P = V \setminus \{\mathbf{0}\}$. Introduce the following equivalence relation in P: a point (ζ_0, ζ_1) is equivalent to all points $(\lambda\zeta_0, \lambda\zeta_1)$ with $\lambda \in \mathbb{C} \setminus \{0\}$. Denote by $[\zeta_0 : \zeta_1]$ the equivalence class of a point $(\zeta_0, \zeta_1) \in P$. It is well-known that the set of equivalence classes is the *complex projective line*. We denote it by $\mathbb{C}P^1$.

Define the domains of two charts by

$$\mathcal{U}_0 = \{\, [\zeta_0 : \zeta_1] \in \mathbb{C}P^1 : \zeta_0 \neq 0 \,\}, \qquad \mathcal{U}_1 = \{\, [\zeta_0 : \zeta_1] \in \mathbb{C}P^1 : \zeta_1 \neq 0 \,\},$$

and the chart maps $h_0 : \mathcal{U}_0 \to \mathbb{C}^1$ and $h_1 : \mathcal{U}_1 \to \mathbb{C}^1$ by

$$h_0([\zeta_0 : \zeta_1]) = \frac{\zeta_1}{\zeta_0}, \qquad h_1([\zeta_0 : \zeta_1]) = \frac{\zeta_0}{\zeta_1}.$$

The overlap map $h_0 \circ h_1^{-1} : \mathbb{C}^1 \setminus \{0\} \to \mathbb{C}^1 \setminus \{0\}$ is $(h_0 \circ h_1^{-1})(\zeta) = \zeta^{-1}$, an analytic function. The constructed atlas may be extended to a holomorphic structure on $\mathbb{C}P^1$.

Example 19. Let the sphere S^2 be embedded into $\mathbb{R}^3$ as in Example 17. Put $\mathcal{U}_\pm = S^2 \setminus \{(0, 0, \pm 1)^\top\}$ and define the charts $\mathbf{h}_\pm : \mathcal{U}_\pm \to \mathbb{C}^1$ by

$$\mathbf{h}_\pm(x_1, x_2, x_3) = \zeta_\pm = \frac{x_1 \pm i x_2}{1 \mp x_3}.$$

In fact, the chart $\mathcal{U}_+$ (resp., $\mathcal{U}_-$) is nothing but the *stereographic projection* from the north (resp., south) pole to the equatorial plane. We have

$$\mathcal{W} = \mathcal{U}_+ \cap \mathcal{U}_- = S^2 \setminus \{(0, 0, 1)^\top, (0, 0, -1)^\top\}$$

and $\mathbf{h}_\pm(\mathcal{W}) = \mathbb{C}^1 \setminus \{0\}$. The transition map $\mathbf{h}_- \circ \mathbf{h}_+^{-1} : \mathbb{C}^1 \setminus \{0\} \to \mathbb{C}^1 \setminus \{0\}$ is given by $\zeta \mapsto \zeta^{-1}$. It is holomorphic, and the above described atlas can be extended to the holomorphic structure on S^2.

Let $p \in M$, and let

$$\mathsf{A}_{\max}(p) = \{\, (\mathcal{U}, \mathbf{h}) \in \mathsf{A}_{\max} : p \in \mathcal{U} \,\}.$$

Definition 30. A map $f\colon M \to N$ is called *smooth* (resp., *real-analytic*, resp., *holomorphic*) if both M and N are smooth (resp., real-analytic, resp., holomorphic) and for any point $p \in M$ and charts $(\mathcal{U}, \mathbf{x}) \in \mathsf{A}_{\max}(p)$, $(\mathcal{V}, \mathbf{y}) \in \mathsf{B}_{\max}(f(p))$ the map $\mathbf{y} \circ f \circ \mathbf{x}^{-1}\colon \mathbf{x}(\mathcal{U}) \to \mathbf{y}(\mathcal{V})$ is smooth (resp., real-analytic, resp., holomorphic).

It is possible to construct a one-to-one analytic map $f\colon \mathbb{C}P^1 \to S^2$ with analytic inverse. For example, put

$$f([\zeta_0 : \zeta_1]) = \begin{cases} (\mathbf{h}_+^{-1} \circ h_0)([\zeta_0 : \zeta_1]), & \text{if } ([\zeta_0 : \zeta_1]) \in \mathcal{U}_0 \\ (0,0,1)^\top, & \text{otherwise} \end{cases}$$

with inverse

$$f^{-1}(x_1, x_2, x_3) = \begin{cases} (h_0^{-1} \circ \mathbf{h}_+)(x_1, x_2, x_3), & \text{if } (x_1, x_2, x_3) \in \mathcal{U}_+ \\ [0 : 1], & \text{otherwise.} \end{cases}$$

In what follows, the symbol $\mathbb{C}P^1$ denotes the *holomorphic* manifold of Examples 18 and 19, while the symbol S^2 denotes the smooth manifold whose smooth structure includes the chart given in Example 17.

We call two manifolds $(M, \mathsf{A}_{\max})$ and $(N, \mathsf{B}_{\max})$ similar if both are either smooth or real-analytic or holomorphic.

For any open set $O \subseteq M$, all charts $(\mathcal{U}, \mathbf{x})$ with $\mathcal{U} \subset \mathcal{O}$ constitute a maximal atlas on O. The set O together with the above atlas is again a manifold called an *open submanifold* of the manifold M. For example, an open interval (a, b) is an open submanifold of the manifold $\mathbb{R}^1$.

Definition 31. The Cartesian product $M \times N$ of a m-dimensional manifold M with maximal atlas $\mathsf{A}_{\max}$ by a similar n-dimensional manifold N with maximal atlas $\mathsf{B}_{\max}$ is the manifold $M_1 \times M_2$ with the maximal atlas build upon the set of charts

$$\{ (\mathcal{U} \times \mathcal{V}, \mathbf{x} \times \mathbf{y})\colon (\mathcal{U}, \mathbf{x}) \in \mathsf{A}_{\max}, (\mathcal{V}, \mathbf{y}) \in \mathsf{B}_{\max} \},$$

where

$$(\mathbf{x} \times \mathbf{y})(p, q) = (\mathbf{h}(p), \mathbf{k}(q)) \in \mathbb{F}^{m+n}, \qquad p \in M, \quad q \in N.$$

1.5.2 Coverings

Definition 32. A *covering* is a triple (P, π, M), where the *projection* π maps the *total space* P to the *base* M and $\pi(P) = M$.

Definition 33. The *fibre* of the covering over a point $p \in M$ is the set $\pi^{-1}(p)$.

Example 20. Let M and F be two sets. The covering $(M \times F, \pi, M)$ with $\pi(p, f) = p$ is called *trivial*.

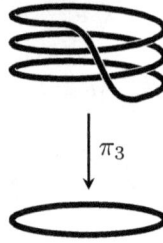

FIGURE 1.1
A non-trivial covering of the circle S^1.

In what follows, we consider only the case when P and B are smooth manifolds and the map π is smooth.

Not all coverings are different.

Definition 34. Two coverings (P_1, π_1, M) and (P_2, π_2, M) of M are called *isomorphic* if there is one-to one smooth map $u \colon P_1 \to P_2$ with smooth inverse that commutes with projections: $\pi_1 = \pi_2 \circ u$ and $\pi_2 = \pi_1 \circ u^{-1}$.

Definition 35. A covering (P, π, M) is called *locally trivial* with fibre F, if for each $p \in M$ there is a chart $(\mathcal{U}, \mathbf{x})$ of M with $p \in \mathcal{U}$ such that the covering $(\pi^{-1}(\mathcal{U}), \pi, \mathcal{U})$ is isomorphic to the trivial covering $(\mathcal{U} \times F, \pi, \mathcal{U})$.

Example 21. Let P and M be two copies of the circle $S^1 = \{\, \zeta \in \mathbb{C} \colon |\zeta| = 1 \,\}$. For a positive integer n, put $\pi_n(\zeta) = \zeta^n$. The covering (S^1, π_n, S^1) is trivial if and only if $n = 1$. For if $n \geq 2$, then any number $z \in S^1$ has n different nth roots, and the coverings (S^1, π_n, S^1) and $(S^1 \times \{1, 2, \dots, n\}, \pi, S^1)$ with $\pi(\zeta, k) = \zeta$ for $1 \leq k \leq n$ are not isomorphic. However, for a small enough open interval around ζ, the sets of different nth roots of numbers in that interval do not intersect, and therefore the covering (S^1, π_n, S^1) is locally trivial. See Fig. 1.1, where $n = 3$.[1]

Consider the cylinder $S^1 \times [-1, 1]$, and introduce an equivalence relation: $(\zeta_1, \alpha_1) \sim (\zeta_2, \alpha_2)$ if and only if $(\zeta_2, \alpha_2) = (\pm\zeta_1, \pm\alpha_1)$. The set of equivalence classes $S^1 \times [-1, 1]/ \sim$ is the Möbius band, and the projection $\pi(\zeta, \alpha) = \zeta^2$ determines the nontrivial Möbius covering of the circle with fibre $[-1, 1]$, see Fig. 1.2.[2]

---

1. Source: https://tex.stackexchange.com/questions/495708/tikz-image-of-a-triple-cover -of-a-circle.

2. Source: https://www.theoreticalphysics.info/index.php/Example: Möbius_Strip_is_ Total_Space_of_a_Fibre_Bundle.

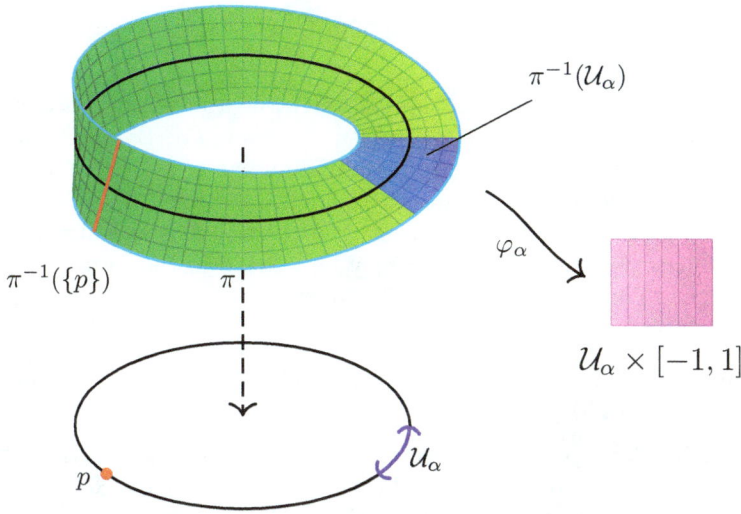

FIGURE 1.2
The Möbius band.

1.6 Lie Groups

The definition and elementary properties of groups are supposed to be known.

Definition 36. A group G is called a real (resp., complex) *Lie group* if G is a real-analytic (resp., holomorphic) manifold, and the map $G \times G \to G$, $(g, h) \mapsto g^{-1}h$ is real-analytic (resp., holomorphic).

In particular, all the so-called classical groups are Lie groups. We give a non-formal definition of a classical group and list all of them.

A *classical group* is a group that belongs to one of the three classes below.

- A group of invertible linear transformations of a finite-dimensional linear space over $\mathbb{R}$, $\mathbb{C}$, or $\mathbb{H}$.

- A closed subgroup of one of the above groups that preserve certain non-degenerate form.

- A certain non-trivial two-fold cover of a subgroup that preserves certain non-degenerate form.

For each classical group, we give a coordinate-free notation, a coordinate notation, or both. Introducing coordinates is equivalent to the identifying the linear space L with the (right) linear space $\mathbb{K}^n$, where n is a positive integer. The classical groups are as follows.

Definition 37. The *general linear group* $\mathrm{GL}(L)$ is the group of invertible linear transformations of L.

When $\mathbb{K} = \mathbb{R}$ (resp., $\mathbb{K} = \mathbb{C}$, resp., $\mathbb{K} = \mathbb{H}$), the above group is denoted by $\mathrm{GL}(U) = \mathrm{GL}(n, \mathbb{R})$ (resp., $\mathrm{GL}(V) = \mathrm{GL}(n, \mathbb{C})$, resp., $\mathrm{GL}(W) = \mathrm{GL}(n, \mathbb{H})$).

We would like to define a determinant of a matrix $C \in \mathbb{H}(n)$. The standard definition of the determinant cannot be applied because the scalars do not commute. One of possible variant of such a definition is as follows. Observe that an arbitrary element $C \in \mathbb{H}(n)$ may be written in the form $A + jB$ with $A, B \in \mathbb{C}(n)$. The map $C \mapsto \left(\begin{smallmatrix} A & -tB \\ B & tA \end{smallmatrix} \right)$ identifies C with a subset of $\mathbb{C}(2n)$. Define $\det C = \det \left(\begin{smallmatrix} A & -tB \\ B & tA \end{smallmatrix} \right)$.

Definition 38. The *special linear group* $\mathrm{SL}(L)$ is the set of elements of determinant 1 in $\mathrm{GL}(L)$.

When $\mathbb{K} = \mathbb{R}$ (resp., $\mathbb{K} = \mathbb{C}$, resp., $\mathbb{K} = \mathbb{H}$), the above group is denoted by $\mathrm{SL}(U) = \mathrm{SL}(n, \mathbb{R})$ (resp., $\mathrm{SL}(V) = \mathrm{SL}(n, \mathbb{C})$, resp., $\mathrm{SL}(W) = \mathrm{SU}^*(2n)$).

Let B be a bilinear form over U or V.

Definition 39. The *isometry group* $\mathrm{O}(U, B)$ (resp., $\mathrm{O}(V, B)$) is the set of all $g \in \mathrm{GL}(U)$ (resp., $g \in \mathrm{GL}(V)$) such that $B(g\mathbf{x}, g\mathbf{y}) = B(\mathbf{x}, \mathbf{y})$ for all $\mathbf{x}$, $\mathbf{y}$ in U (resp., in V).

In particular, when $\mathbb{F} = \mathbb{R}$, B has the form (1.1), and $\min\{p, q\} \geq 1$, the isometry group $\mathrm{O}(U, B)$ is denoted by $\mathrm{O}(p, q)$ and is called the *indefinite orthogonal group*. Otherwise, if $\min\{p, q\} = 0$, the isometry group $\mathrm{O}(U, B)$ is denoted by $\mathrm{O}(n)$ and is called the *real orthogonal group*. When $\mathbb{F} = \mathbb{C}$ and B has the form (1.3), the isometry group $\mathrm{O}(V, B)$ is denoted by $\mathrm{O}(n, \mathbb{C})$ and is called the *complex orthogonal group*. When ε is *skew-symmetric*, then it has the form (1.2). The isometry group $\mathrm{O}(U, B)$ (resp., $\mathrm{O}(V, B)$) is denoted by $\mathrm{Sp}(n, \mathbb{R})$ (resp., $\mathrm{Sp}(n, \mathbb{C})$) and is called the *real symplectic group* (resp., the *complex symplectic group*).

Definition 40. The *special orthogonal group* $\mathrm{SO}(n, \mathbb{F})$ is given by

$$\mathrm{SO}(n, \mathbb{F}) = \mathrm{O}(n, \mathbb{F}) \cap \mathrm{SL}(n, \mathbb{F}).$$

The *indefinite special orthogonal group* $\mathrm{SO}(p, q)$ is given by

$$\mathrm{SO}(p, q) = \mathrm{O}(p, q) \cap \mathrm{SL}(p + q, \mathbb{R}).$$

When $\min\{p, q\} = 0$, we replace notation $\mathrm{SO}_0(p, q)$ with $\mathrm{SO}(n)$. Occasionally, we use notation $\mathrm{O}(L, B)$ to denote either $\mathrm{O}(p, q)$ or $\mathrm{O}(n, \mathbb{C})$, and $\mathrm{SO}(L, B)$ to denote either $\mathrm{SO}_0(p, q)$ or $\mathrm{SO}(n, \mathbb{C})$.

Denote by $\mathbb{C}^{p,q}$ the linear space $\mathbb{C}^{p+q}$ with the Hermitian form (1.4).

Definition 41. The *indefinite unitary group* is the group

$$\mathrm{U}(B) = \mathrm{U}(p, q) = \{ g \in \mathrm{GL}(p + q, \mathbb{C}) \colon B(g(\mathbf{x}), g(\mathbf{y})) = B(\mathbf{x}, \mathbf{y}) \}$$

for all $\mathbf{x}, \mathbf{y} \in \mathbb{C}^{p+q}$, where B is the Hermitian form (1.4), and where $p + q = n \geq 1$.

When $\min\{p, q\} = 0$, we replace notation $\mathrm{U}(p, q)$ with $\mathrm{U}(n)$ and call this group the *unitary group*.

Definition 42. The *indefinite special unitary group* $\mathrm{SU}(p, q)$ is given by

$$\mathrm{SU}(p, q) = \mathrm{U}(p, q) \cap \mathrm{SL}(n, \mathbb{C}).$$

When $\min\{p, q\} = 0$, we replace notation $\mathrm{SU}(p, q)$ with $\mathrm{SU}(n)$ and call this group the *special unitary group*.

Denote by $\mathbb{H}^{p,q}$ the linear space $\mathbb{H}^{p+q}$ with the quaternionic Hermitian form (1.5).

Definition 43. The *indefinite quaternionic unitary group* is the group

$$\mathrm{Sp}(B) = \mathrm{Sp}(p, q) = \{\, g \in \mathrm{GL}(p + q, \mathbb{H}) \colon B(g(\mathbf{x}), g(\mathbf{y})) = B(\mathbf{x}, \mathbf{y}) \,\}$$

for all $\mathbf{x}$, $\mathbf{y} \in \mathbb{H}^{p+q}$, where B is the Hermitian form (1.5), and where $p + q = n \geq 1$.

When $\min\{p, q\} = 0$, we replace notation $\mathrm{Sp}(p, q)$ with $\mathrm{Sp}(n)$ and call this group the *quaternionic unitary group*.

Definition 44. Finally, the group $\mathrm{SO}^*(2n)$ is given by

$$\mathrm{SO}^*(2n) = \{\, g \in \mathrm{GL}(n, \mathbb{H}) \colon C(g(\mathbf{x}), g(\mathbf{y})) = C(\mathbf{x}, \mathbf{y}) \,\}$$

for all $\mathbf{x}$, $\mathbf{y} \in \mathbb{H}^n$, where

$$C(\mathbf{x}, \mathbf{y}) = \sum_{i=1}^{n} x_i \mathrm{j} \overline{y_i}.$$

Definition 45. The *pin group* $\mathrm{Pin}(L, B)$ is the minimal subgroup of the group of all invertible elements of the Clifford algebra $(\mathrm{Cl}(L, B), \gamma)$ that contains the elements $\gamma(\mathbf{l})$ for all $\mathbf{l} \in L$ of length 1.

Definition 46. The *spin group* $\mathrm{Spin}(L, B)$ is given by

$$\mathrm{Spin}(L, B) = \mathrm{Pin}(L, B) \cap (\mathrm{Cl}^+(L, B), \gamma).$$

Definition 47. The *reduced spin group* $\mathrm{Spin}^0(L, B)$ is defined by

$$\mathrm{Spin}^0(L, B) = \{\, \mathbf{a} \in \mathrm{Spin}(L, B) \colon \mathbf{a}\hat{\mathbf{a}} = 1 \,\}.$$

If $L = \mathbb{R}^{p+q}$ and B has the form (1.1), then we use notation $\mathrm{Spin}(p, q)$ (resp., $\mathrm{Spin}^0(p, q)$) instead of $\mathrm{Spin}(L, B)$ (resp., $\mathrm{Spin}^0(L, B)$). Similarly, if $L = \mathbb{C}^n$ and B has the form (1.3), then we use notation $\mathrm{Spin}(n, \mathbb{C})$ (resp., $\mathrm{Spin}^0(n, \mathbb{C})$) instead of $\mathrm{Spin}(L, B)$ (resp., $\mathrm{Spin}^0(L, B)$). When $\min\{p, q\} = 0$, we replace notation $\mathrm{Spin}(p, q)$ (resp., $\mathrm{Spin}^0(p, q)$) with $\mathrm{Spin}(n)$ (resp., $\mathrm{Spin}^0(n)$).

Observe!

Observe that $\mathrm{Spin}^0(1,\mathbb{C}) \neq \mathrm{Spin}(n,\mathbb{C})$ and $\mathrm{Spin}^0(1) \neq \mathrm{Spin}(1)$. However, if $n > 1$, then $\mathrm{Spin}^0(n,\mathbb{C}) = \mathrm{Spin}(n,\mathbb{C})$ and $\mathrm{Spin}^0(n) = \mathrm{Spin}(n)$.

Define the map $\pi\colon \mathrm{Pin}(L,B) \to \mathrm{GL}(\mathrm{Cl}(L,B))$ by

$$\pi(\mathbf{a})\mathbf{x} = \tilde{\mathbf{a}}\mathbf{x}\mathbf{a}^{-1}, \qquad \mathbf{a} \in \mathrm{Spin}_+(L,B), \quad \mathbf{x} \in \mathrm{GL}(\mathrm{Cl}(L,B)). \qquad (1.27)$$

It turns out that

$$\pi(\mathrm{Pin}(L,B)) = \mathrm{O}(L,B) \subset \mathrm{GL}(\mathrm{Cl}(L,B)),$$
$$\pi(\mathrm{Spin}(L,B)) = \mathrm{SO}(L,B) \subset \mathrm{GL}(\mathrm{Cl}(L,B)),$$
$$\pi(\mathrm{Spin}^0(L,B)) = \mathrm{SO}_0(L,B) \subset \mathrm{GL}(\mathrm{Cl}(L,B)),$$

and the map π is two-to-one. That is, the triples $(\mathrm{Pin}(L,B),\pi,\mathrm{O}(L,B))$, $(\mathrm{Spin}(L,B),\pi,\mathrm{SO}(L,B))$, and $(\mathrm{Spin}^0(L,B),\pi,\mathrm{SO}_0(L,B))$ are two-fold coverings.

Remark 4. In other words, the sequences

$$1 \to \mathrm{O}(\mathbb{F}^1,B) \to \mathrm{Pin}(L,B) \to \mathrm{O}(L,B) \to 1,$$
$$1 \to \mathrm{SO}(\mathbb{F}^1,B) \to \mathrm{Spin}(L,B) \to \mathrm{SO}(L,B) \to 1,$$
$$1 \to \mathrm{O}(\mathbb{F}^1,B) \to \mathrm{Spin}^0(L,B) \to \mathrm{SO}_0(L,B) \to 1$$

are exact.

In the following examples, we will use Vaz and Rocha (2019, Proposition 16.15): if $p + q \leq 5$, then

$$\mathrm{Spin}(p,q) = \{\, \mathbf{a} \in \mathrm{Cl}^+_{p,q} \colon \mathbf{a}\hat{\mathbf{a}} = \pm 1_{\mathrm{Cl}_{p,q}} \,\}. \qquad (1.28)$$

Example 22. By Table 1.1, $\mathrm{Cl}_1 = \mathbb{C} \oplus \mathbb{C}$. The subalgebra Cl^+_1 has the form $\{\, \zeta \oplus \zeta \colon \zeta \in \mathbb{C} \,\}$. Its invertible elements form the group $\mathrm{GL}(1,\mathbb{C})$. Both the Clifford and the hat involutions in Cl_1 act on Cl^+_1 trivially. By Equation (1.28), the spin group $\mathrm{Spin}(1,\mathbb{C})$ is its subgroup $\{\pm 1, \pm\mathrm{i}\}$.

The group $\mathrm{Spin}^0(1,\mathbb{C})$ becomes

$$\mathrm{Spin}^0(1,\mathbb{C}) = \{\, \zeta \in \mathrm{Spin}(1,\mathbb{C}) \colon \zeta^2 = 1 \,\} = \mathrm{O}(1).$$

The trivial covering (1.27) $\pi\colon \mathrm{Spin}^0(1,\mathbb{C}) \to \mathrm{SO}(1,\mathbb{C})$ takes the form

$$\pi(\zeta) = \zeta^2. \qquad (1.29)$$

By this reason, spinors are sometimes coined as the 'square roots of vectors'.

By Table 1.1, $\mathrm{Cl}_3 = \mathbb{C}(2) \oplus \mathbb{C}(2)$. The subalgebra Cl^+_3 consists of block-diagonal matrices of the form $\left(\begin{smallmatrix} \mathbf{a} & 0 \\ 0 & \mathbf{a} \end{smallmatrix}\right)$ with $\mathbf{a} \in \mathbb{C}(2)$. Its invertible elements form the group $\mathrm{GL}(2,\mathbb{C})$. The spin group $\mathrm{Spin}(3,\mathbb{C})$ is its subgroup $\mathrm{SL}(2,\mathbb{C})$. The space $\sigma(\mathbb{C}^3)$ is the set of 2×2 complex traceless matrices. We choose the classical Pauli matrices (1.14) times $\frac{1}{\sqrt{2}}$ as an orthonormal basis of the above space. The spin group $SL(2, \mathbb{C})$ acts in $\sigma(\mathbb{C}^3)$ by

$$\pi(\mathbf{a})\mathbf{x} = \mathbf{axa}^{-1}, \qquad \mathbf{a} \in SL(2, \mathbb{C}), \quad \mathbf{x} \in \sigma(\mathbb{C}^3).$$

The simple calculations show that the nontrivial covering $\pi \colon SL(2, \mathbb{C}) \to SO(3, \mathbb{C})$ takes the form

$$\pi\begin{pmatrix} \alpha & \beta \\ \gamma & \delta \end{pmatrix} = \frac{1}{2}\begin{pmatrix} \alpha^2-\beta^2-\gamma^2+\delta^2 & -\alpha^2-\beta^2+\gamma^2+\delta^2 & 2(-\alpha\beta+\gamma\delta) \\ i(\alpha^2-\beta^2+\gamma^2-\delta^2) & -i(\alpha^2+\beta^2+\gamma^2+\delta^2) & -2i(\alpha\beta+\gamma\delta) \\ 2(-\alpha\gamma+\beta\delta) & 2(\alpha\gamma+\beta\delta) & 2(\alpha\delta+\beta\gamma) \end{pmatrix}. \tag{1.30}$$

Example 23. By Table 1.1, $Cl_2 = \mathbb{C}(2)$. The subalgebra Cl_2^+ consists of diagonal matrices. Its invertible elements form the group $GL(1, \mathbb{C}) \times GL(1, \mathbb{C})$. The spin group $Spin(2, \mathbb{C})$ is its subgroup $GL(1, \mathbb{C})$ of the matrices of the form $\mathbf{a} = \begin{pmatrix} z & 0 \\ 0 & z^{-1} \end{pmatrix}$ with $z \in \mathbb{C} \setminus \{0\}$.

The Clifford involution in Cl_2 maps the matrix $\begin{pmatrix} \alpha & \beta \\ \gamma & \delta \end{pmatrix}$ to the matrix $\begin{pmatrix} \alpha & -\beta \\ -\gamma & \delta \end{pmatrix}$. Its restriction to $Spin(2, \mathbb{C})$ is the identical map. The nontrivial covering (1.27) $\pi \colon Spin(2, \mathbb{C}) \to SO(2, \mathbb{C})$ takes the form

$$\pi(\mathbf{a})\mathbf{x} = \begin{pmatrix} z & 0 \\ 0 & z^{-1} \end{pmatrix}\begin{pmatrix} 0 & \beta \\ \gamma & 0 \end{pmatrix}\begin{pmatrix} z & 0 \\ 0 & z^{-1} \end{pmatrix} = \begin{pmatrix} 0 & z^2\beta \\ z^{-2}\gamma & 0 \end{pmatrix}, \tag{1.31}$$

that is, $\pi(\mathbf{a}) = \mathbf{a}^2$.

By Table 1.1, $Cl_4 = \mathbb{C}(4)$. The subalgebra $Cl_4^+ = \mathbb{C}(2) \oplus \mathbb{C}(2)$ consists of block-diagonal matrices. Its invertible elements form the group $GL(2, \mathbb{C}) \times GL(2, \mathbb{C})$. The spin group $Spin(4, \mathbb{C})$ is its subgroup $SL(2, \mathbb{C}) \times SL(2, \mathbb{C})$. The space $\gamma(\mathbb{C}^4)$ is the set of 2×2 complex matrices. We choose the following basis of $\gamma(\mathbb{C}^4)$: the identity matrix, the matrices γ_1 and γ_2 of Equation (1.11), and $\gamma_1\gamma_2$ times $\frac{1}{\sqrt{2}}$.

The group $SL(2, \mathbb{C}) \times SL(2, \mathbb{C})$ acts on $\gamma(\mathbb{C}^4)$ by $\gamma \mapsto g\gamma h^{-1}$, $\gamma \in \gamma(\mathbb{C}^4)$, $(g, h) \in SL(2, \mathbb{C}) \times SL(2, \mathbb{C})$. The nontrivial covering $\pi \colon Spin(4, \mathbb{C}) \to SO(4, \mathbb{C})$ becomes

$$\pi\left(\begin{pmatrix} \alpha_1 & \beta_1 \\ \gamma_1 & \delta_1 \end{pmatrix}, \begin{pmatrix} \alpha_2 & \beta_2 \\ \gamma_2 & \delta_2 \end{pmatrix}\right) = \frac{1}{2}\begin{pmatrix} \alpha_1\delta_2-\beta_1\gamma_2-\gamma_1\beta_2+\delta_1\alpha_2 & -\alpha_1\delta_2-\beta_1\gamma_2+\gamma_1\beta_2+\delta_1\alpha_2 \\ -\alpha_1\delta_2+\beta_1\gamma_2-\gamma_1\beta_2+\delta_1\alpha_2 & \alpha_1\delta_2-\beta_1\gamma_2+\gamma_1\beta_2+\delta_1\alpha_2 \\ -\alpha_1\beta_2+\beta_1\alpha_2+\gamma_1\delta_2-\delta_1\gamma_2 & \alpha_1\beta_2+\beta_1\alpha_2-\gamma_1\delta_2-\delta_1\gamma_2 \\ \alpha_1\beta_2-\beta_1\alpha_2+\gamma_1\delta_2-\delta_1\gamma_2 & -\alpha_1\beta_2-\beta_1\alpha_2-\gamma_1\delta_2-\delta_1\gamma_2 \end{pmatrix}$$

$$\begin{pmatrix} -\alpha_1\gamma_2+\beta_1\delta_2+\gamma_1\alpha_2-\delta_1\beta_2 & \alpha_1\gamma_2+\beta_1\delta_2-\gamma_1\alpha_2-\delta_1\beta_2 \\ \alpha_1\gamma_2+\beta_1\delta_2+\gamma_1\alpha_2-\delta_1\beta_2 & -\alpha_1\gamma_2+\beta_1\delta_2-\gamma_1\alpha_2-\delta_1\beta_2 \\ \alpha_1\alpha_2-\beta_1\beta_2-\gamma_1\gamma_2+\delta_1\delta_2 & -\alpha_1\alpha_2-\beta_1\beta_2+\gamma_1\gamma_2+\delta_1\delta_2 \\ -\alpha_1\alpha_2+\beta_1\beta_2-\gamma_1\gamma_2+\delta_1\delta_2 & \alpha_1\alpha_2+\beta_1\beta_2+\gamma_1\gamma_2+\delta_1\delta_2 \end{pmatrix}. \tag{1.32}$$

Example 24. By Table 1.3, $Cl_{0,1} = \mathbb{R} \oplus \mathbb{R}$. The subalgebra $Cl_{0,1}^+$ has the form $\{x \oplus x \colon x \in \mathbb{R}\}$. Its invertible elements form the group $GL(1, \mathbb{R})$. By Equation (1.28), the spin group $Spin(1)$ is its subgroup $O(1)$. The trivial covering (1.27) $\pi \colon Spin(1) \to SO(1)$ takes the form

$$\pi(x) = x^2. \tag{1.33}$$

The same result can be obtained starting from $\mathrm{Cl}_{1,0} = \mathbb{C}$ and $\mathrm{Cl}_{1,0}^+ = \mathbb{R}$. The invertible elements in $\mathbb{R}$ form the group $\mathrm{GL}(1, \mathbb{R})$. The spin group $\mathrm{Spin}(1)$ is its subgroup $\mathrm{O}(1)$, and the covering is the same.

By Table 1.2, $\mathrm{Cl}_{1,2} = \mathbb{R}(2) \oplus \mathbb{R}(2)$. By Table 1.3, $\mathrm{Cl}_{1,2}^+ = \mathbb{R}(2)$. The check involution maps the matrix

$$\alpha 1_{\mathrm{Cl}_{1,2}} + \beta\sigma_1\sigma_2 + \gamma\sigma_1\sigma_3 + \delta\sigma_2\sigma_3 \in \mathrm{Cl}_{1,2}^+$$

to the matrix $\alpha 1_{\mathrm{Cl}_{1,2}} - \beta\sigma_1\sigma_2 + \gamma\sigma_1\sigma_3 + \delta\sigma_2\sigma_3 \in \mathrm{Cl}_{1,2}^+$, where the matrices γ_i are given by Equation (1.18). Easy calculations show that the group $\mathrm{Spin}(1, 2)$ is the set of all 2×2 matrices with determinant ± 1. Equation (1.28) gives $\mathrm{Spin}^0(1, 2) = \mathrm{SL}(2, \mathbb{R})$. We choose the basis in $\sigma(\mathbb{R}^{1,2})$ give by the matrices σ_1, σ_2 of Equation (1.11) and $\sigma_1\sigma_2$. The non-trivial covering $\pi \colon \mathrm{SL}(2, \mathbb{R}) \to \mathrm{SO}_0(1, 2)$ is

$$\pi\begin{pmatrix} \alpha & \beta \\ \gamma & \delta \end{pmatrix} = \frac{1}{2}\begin{pmatrix} \alpha^2+\beta^2+\gamma^2+\delta^2 & -2(\alpha\beta+\gamma\delta) & -\alpha^2+\beta^2-\gamma^2+\delta^2 \\ -2(\alpha\gamma+\beta\delta) & 2(\alpha\delta+\beta\gamma) & 2(\alpha\gamma-\beta\delta) \\ -\alpha^2-\beta^2+\gamma^2+\delta^2 & 2(\alpha\beta-\gamma\delta) & \alpha^2-\beta^2-\gamma^2+\delta^2 \end{pmatrix}. \tag{1.34}$$

The same result can be obtained starting from $\mathrm{Cl}_{2,1} = \mathbb{C}(2)$ and $\mathrm{Cl}_{2,1}^+ = \mathbb{R}(2)$. For calculations, we use the basis (1.19) in $\sigma(\mathbb{R}^{2,1})$.

Example 25. By Table 1.2, $\mathrm{Cl}_{0,3} = \mathbb{C}(2)$. The linear space Λ^0 of the elements of degree 0 is generated by the identity matrix, while the space Λ^2 is generated by the products

$$\sigma_1\sigma_2 = \begin{pmatrix} i & 0 \\ 0 & -i \end{pmatrix}, \quad \sigma_1\sigma_3 = \begin{pmatrix} 0 & -1 \\ 1 & 0 \end{pmatrix}, \quad \sigma_2\sigma_3 = \begin{pmatrix} 0 & i \\ i & 0 \end{pmatrix}$$

of the classical Pauli matrices (1.14). A quaternion $q = \alpha + \beta i + \gamma i + \delta k \in \mathbb{H} = \mathrm{Cl}_{0,3}^+$ is embedded into $\mathrm{Cl}_{0,3}$ as the matrix

$$\alpha 1_{\mathrm{Cl}_{0,3}} + \beta\sigma_1\sigma_2 + \gamma\sigma_1\sigma_3 + \delta\sigma_2\sigma_3 = \begin{pmatrix} \alpha+\beta i & -\gamma+\delta i \\ \gamma+\delta i & \alpha-\beta i \end{pmatrix}.$$

The hat involution maps q to $\bar{q}$. By Equation (1.28), the spin group $\mathrm{Spin}(3)$ is $\mathrm{Sp}(1)$. The covering (1.27) becomes

$$\pi(q)\mathbf{a} = \begin{pmatrix} \alpha+\beta i & -\gamma+\delta i \\ \gamma+\delta i & \alpha+\beta i \end{pmatrix} \mathbf{a} \begin{pmatrix} \alpha-\beta i & \gamma-\delta i \\ -\gamma-\delta i & \alpha-\beta i \end{pmatrix}, \quad \mathbf{a} \in \sigma(\mathbb{C}^3).$$

It follows that the non-trivial covering $\pi \colon \mathrm{Spin}(3) \to \mathrm{SO}(3)$ is

$$\pi(q) = \begin{pmatrix} 2\alpha^2+2\beta^2-1 & 2(\beta\gamma-\alpha\delta) & 2(\alpha\gamma+\beta\delta) \\ 2(\alpha\delta+\beta\gamma) & 2\alpha^2+2\gamma^2-1 & 2(\gamma\delta-\alpha\beta) \\ 2(\beta\delta-\alpha\gamma) & 2(\alpha\beta+\gamma\delta) & 2\alpha^2+2\delta^2-1 \end{pmatrix}. \tag{1.35}$$

The same result can be obtained starting from $\mathrm{Cl}_{3,0} = \mathbb{H} \oplus \mathbb{H}$ and $\mathrm{Cl}_{3,0}^+ = \mathbb{H}$. For calculations, we use the basis (1.17) in $\sigma(\mathbb{R}^3)$.

Example 26. By Table 1.2, $\mathrm{Cl}_{1,1} = \mathbb{R}(2)$. The subalgebra $\mathrm{Cl}_{1,1}^+ = \mathbb{R} \oplus \mathbb{R}$ is the set of diagonal matrices. The invertible elements in $\mathrm{Cl}_{1,1}^+$ form the group $\mathrm{GL}(1,\mathbb{R}) \times \mathrm{GL}(1,\mathbb{R})$.

The hat involution maps the matrix $\alpha 1_{\mathrm{Cl}_{1,1}} + \beta \gamma_1 \gamma_2 = \begin{pmatrix} \alpha - \beta & 0 \\ 0 & \alpha + \beta \end{pmatrix} \in \mathrm{Cl}_{1,1}$ to the matrix $\begin{pmatrix} \alpha + \beta & 0 \\ 0 & \alpha - \beta \end{pmatrix}$, where the Dirac matrices γ_1 and γ_2 are given by Equation (1.10). By Equation (1.28), the spin group $\mathrm{Spin}(1,1)$ is the subgroup of the matrices $\mathbf{a} \in \mathrm{GL}(1,\mathbb{R}) \times \mathrm{GL}(1,\mathbb{R})$ with $\mathbf{a}\hat{\mathbf{a}} = \pm 1_{\mathrm{Cl}_{1,1}}$, which becomes the group of matrices of the form $\begin{pmatrix} \alpha & 0 \\ 0 & \pm \alpha^{-1} \end{pmatrix}$ with $\alpha \in \mathbb{R} \setminus \{0\}$. The reduced spin group $\mathrm{Spin}^0(1,1)$ becomes the group of matrices of the form $\begin{pmatrix} \alpha & 0 \\ 0 & \alpha^{-1} \end{pmatrix}$. The trivial covering (1.27) $\pi \colon \mathrm{Spin}^0(1,1) \to \mathrm{SO}_0(1,1)$ takes the form

$$\pi(\alpha)\mathbf{x} = \begin{pmatrix} \alpha & 0 \\ 0 & \alpha^{-1} \end{pmatrix} \mathbf{x} \begin{pmatrix} \alpha & 0 \\ 0 & \alpha^{-1} \end{pmatrix} \tag{1.36}$$

that is, $\pi(\alpha) = \alpha^2$.

By Table 1.3, $\mathrm{Cl}_{2,2}^+ = \mathbb{R}(2) \oplus \mathbb{R}(2) \subset \mathrm{Cl}_{2,2} = \mathbb{R}(4)$. The basis of $\mathrm{Cl}_{2,2}^+$ is given by the identity matrix, six products of the matrices (1.22), and the volume element. The hat involution multiplies the above six products by -1. Equation (1.28) gives $\mathrm{Spin}^0(2,2) = \mathrm{SL}(2,\mathbb{R}) \times \mathrm{SL}(2,\mathbb{R})$. This group acts on $\gamma(\mathbb{R}^{2,2})$ by $g \cdot \mathbf{a} = g a g^{-1}$. Simple calculations give

$$\pi\left(\begin{pmatrix} \alpha_1 & \beta_1 \\ \gamma_1 & \delta_1 \end{pmatrix}, \begin{pmatrix} \alpha_2 & \beta_2 \\ \gamma_2 & \delta_2 \end{pmatrix} \right) = \frac{1}{2} \begin{pmatrix} \alpha_1\delta_2 + \alpha_2\delta_1 - \beta_1\gamma_2 - \beta_2\gamma_1 & \alpha_1\delta_2 - \alpha_2\delta_1 + \beta_1\gamma_2 - \beta_2\gamma_1 \\ \alpha_1\delta_2 - \alpha_2\delta_1 - \beta_1\gamma_2 + \beta_2\gamma_1 & \alpha_1\delta_2 + \alpha_2\delta_1 + \beta_1\gamma_2 + \beta_2\gamma_1 \\ -\alpha_1\beta_2 + \alpha_2\beta_1 + \gamma_1\delta_2 - \gamma_2\delta_1 & -\alpha_1\beta_2 - \alpha_2\beta_1 + \gamma_1\delta_2 + \gamma_2\delta_1 \\ -\alpha_1\beta_2 + \alpha_2\beta_1 - \gamma_1\delta_2 + \gamma_2\delta_1 & -\alpha_1\beta_2 - \alpha_2\beta_1 - \gamma_1\delta_2 - \gamma_2\delta_1 \end{pmatrix}$$

$$\begin{pmatrix} -\alpha_1\gamma_2 + \alpha_2\gamma_1 - \beta_1\delta_2 + \beta_2\delta_1 & -\alpha_1\gamma_2 + \alpha_2\gamma_1 + \beta_1\delta_2 - \beta_2\delta_1 \\ -\alpha_1\gamma_2 - \alpha_2\gamma_1 + \beta_1\delta_2 + \beta_2\delta_1 & -\alpha_1\gamma_2 - \alpha_2\gamma_1 + \beta_1\delta_2 - \beta_2\delta_1 \\ \alpha_1\alpha_2 + \beta_1\beta_2 - \gamma_1\gamma_2 - \delta_1\delta_2 & \alpha_1\alpha_2 - \beta_1\beta_2 - \gamma_1\gamma_2 + \delta_1\delta_2 \\ \alpha_1\alpha_2 + \beta_1\beta_2 + \gamma_1\gamma_2 + \delta_1\delta_2 & \alpha_1\alpha_2 - \beta_1\beta_2 + \gamma_1\gamma_2 - \delta_1\delta_2 \end{pmatrix}. \tag{1.37}$$

Example 27. We have $\mathrm{Cl}_{0,4} = \mathrm{Cl}_{4,0} = \mathbb{H}(2)$, and $\mathrm{Cl}_{0,4}^+ = \mathrm{Cl}_{4,0}^+ = \mathbb{H} \oplus \mathbb{H}$. The basis of $\mathrm{Cl}_{0,4}^+$ is given by the identity matrix, six products of the matrices (1.15), and the volume element. The hat involution multiplies the above six products by -1. Equation (1.28) gives $\mathrm{Spin}(4) = \mathrm{Sp}(1) \times \mathrm{Sp}(1)$. This group acts on the matrices (1.15) by $(q_1, q_2) \cdot \gamma = \begin{pmatrix} q_1 & q_2 \\ q_2 & q_1 \end{pmatrix} \gamma \begin{pmatrix} \overline{q_1} & \overline{q_2} \\ \overline{q_2} & \overline{q_1} \end{pmatrix}$. Direct calculations prove the following: for all $q_1, q_2 \in \mathbb{H}$, and for all $q_0 \in \{i, j, k\}$ we have $\mathrm{Re}(q_1 q_0 \overline{q_1}) = 0$ and $\mathrm{Im}(q_1 q_0 \overline{q_2} - q_2 q_0 \overline{q_1}) = 0$. Then, we obtain

$$\pi(q_1, q_2) = \mathrm{Re} \begin{pmatrix} q_2 \overline{i q_2} i - q_1 \overline{i q_1} i & q_2 \overline{j q_2} i - q_1 \overline{j q_1} i & q_2 k \overline{q_2} i - q_1 k \overline{q_1} i & q_1 \overline{q_2} i - q_2 \overline{q_1} i \\ q_2 \overline{i q_2} j - q_1 \overline{i q_1} j & q_2 \overline{j q_2} j - q_1 \overline{j q_1} j & q_2 k \overline{q_2} j - q_1 k \overline{q_1} j & q_1 \overline{q_2} j - q_2 \overline{q_1} j \\ q_2 \overline{i q_2} k - q_1 \overline{i q_1} k & q_2 \overline{j q_2} k - q_1 \overline{j q_1} k & q_2 k \overline{q_2} k - q_1 k \overline{q_1} k & q_1 \overline{q_2} k - q_2 \overline{q_1} k \\ q_2 \overline{i q_1} - q_1 \overline{i q_2} & q_2 \overline{j q_1} - q_1 \overline{j q_2} & q_2 k \overline{q_1} - q_1 k \overline{q_2} & |q_1|^2 - |q_2|^2 \end{pmatrix}.$$

Example 28. By Table 1.2, $\mathrm{Cl}_{0,2} = \mathbb{R}(2)$. The subalgebra $\mathrm{Cl}_{0,2}^+ = \mathbb{C}$ is embedded into $\mathrm{Cl}_{0,2}$ by

$$\zeta \mapsto \mathbf{a} = \mathrm{Re}\,\zeta 1_{\mathrm{Cl}_{0,2}} + \mathrm{Im}\,\zeta \gamma_1 \gamma_2 = \begin{pmatrix} \mathrm{Re}\,\zeta & -\mathrm{Im}\,\zeta \\ \mathrm{Im}\,\zeta & \mathrm{Re}\,\zeta \end{pmatrix},$$

where γ_1 and γ_2 are the Dirac matrices of the basis (1.11). Its invertible elements form the group $\mathrm{GL}(1,\mathbb{C})$. The hat involution maps $\mathbf{a}$ to $\mathrm{Re}\,\zeta 1_{\mathrm{Cl}_{0,2}} -$

$\operatorname{Im} \zeta \gamma_1 \gamma_2$, and the spin group $\operatorname{Spin}(1, \mathbb{C})$ is the subgroup $U(1)$ that consists of matrices with $|\zeta| = 1$.

The Clifford involution in $\operatorname{Cl}_{0,2}$ acts on $\operatorname{Spin}(1)$ by the identical map. The nontrivial covering (1.27) $\pi \colon \operatorname{Spin}(2) \to SO(2)$ takes the form

$$\pi(\mathbf{a})\mathbf{x} = \begin{pmatrix} \operatorname{Re}\zeta & -\operatorname{Im}\zeta \\ \operatorname{Im}\zeta & \operatorname{Re}\zeta \end{pmatrix} \begin{pmatrix} -\alpha & \beta \\ \beta & \alpha \end{pmatrix} \begin{pmatrix} \operatorname{Re}\zeta & \operatorname{Im}\zeta \\ -\operatorname{Im}\zeta & \operatorname{Re}\zeta \end{pmatrix} = \begin{pmatrix} \operatorname{Re}(\zeta^2) & -\operatorname{Im}(\zeta^2) \\ \operatorname{Im}(\zeta^2) & \operatorname{Re}(\zeta^2) \end{pmatrix} \begin{pmatrix} -\alpha & \beta \\ \beta & \alpha \end{pmatrix},$$
(1.38)

that is, $\pi(\zeta) = \zeta^2$.

The same result can be obtained starting from $\operatorname{Cl}_{2,0} = \mathbb{H}$ and $\operatorname{Cl}_{2,0}^+ = \mathbb{C}$. The invertible elements in $\mathbb{C}$ form the group $GL(1, \mathbb{C})$. The hat involution maps ζ to $\bar{\zeta}$, and the spin group $\operatorname{Spin}(2)$ is the subgroup $U(1)$. The Clifford involution in $\operatorname{Cl}_{2,0}$ acts on $\operatorname{Spin}(2)$ by the identical map. The nontrivial covering (1.27) $\pi \colon \operatorname{Spin}(2) \to SO(2)$ takes the form

$$\pi(\zeta)(\alpha \mathrm{j} + \beta \mathrm{k}) = \zeta(\alpha \mathrm{j} + \beta \mathrm{k})\zeta,$$

that is, $\pi(\zeta) = \zeta^2$.

By Table 1.2, $\operatorname{Cl}_{1,3} = \mathbb{R}(4)$. By Table 1.3, $\operatorname{Cl}_{1,3}^+ = \mathbb{C}(2)$. The basis of $\operatorname{Cl}_{1,3}^+$ is given by the identity matrix, six products of the matrices (1.21), and the volume element. The hat involution multiplies the above six products by -1. Equation (1.28) gives $\operatorname{Spin}^0(1,3) = SL(2, \mathbb{C})$. This group acts on $\gamma(\mathbb{R}^{1,3})$ by $g \cdot \mathbf{a} = g a g^{-1}$. Simple calculations give

$$\pi\begin{pmatrix} \alpha & \beta \\ \gamma & \delta \end{pmatrix} = \frac{1}{2} \begin{pmatrix} 2\operatorname{Re}(\alpha\bar{\delta}-\beta\bar{\gamma}) & -2\operatorname{Im}(\bar{\alpha}\beta+\gamma\bar{\delta}) & 2\operatorname{Im}(\alpha\bar{\gamma}-\beta\bar{\delta}) & 2\operatorname{Im}(\alpha\bar{\gamma}+\beta\bar{\delta}) \\ 2\operatorname{Im}(\alpha\bar{\delta}+\beta\bar{\gamma}) & 2\operatorname{Re}(\alpha\bar{\delta}+\beta\bar{\gamma}) & 2\operatorname{Re}(-\alpha\bar{\gamma}+\beta\bar{\delta}) & -2\operatorname{Re}(\alpha\bar{\gamma}+\beta\bar{\delta}) \\ 2\operatorname{Im}(\bar{\alpha}\beta+\gamma\bar{\delta}) & 2\operatorname{Re}(-\alpha\bar{\beta}+\gamma\bar{\delta}) & |\alpha|^2-|\beta|^2-|\gamma|^2+|\delta|^2 & |\alpha|^2+|\beta|^2-|\gamma|^2-|\delta|^2 \\ 2\operatorname{Im}(\bar{\alpha}\beta-\gamma\bar{\delta}) & -2\operatorname{Re}(\alpha\bar{\beta}+\gamma\bar{\delta}) & |\alpha|^2-|\beta|^2+|\gamma|^2-|\delta|^2 & |\alpha|^2+|\beta|^2+|\gamma|^2+|\delta|^2 \end{pmatrix}.$$
(1.39)

The same result can be obtained starting from $\operatorname{Cl}_{3,1} = \mathbb{H}(2)$ and $\operatorname{Cl}_{3,1}^+ = \mathbb{C}(2)$. For calculations, we use the basis (1.20) in $\sigma(\mathbb{R}^{3,1})$.

Observe!

In our approach, the groups $U(1)$ and $SO(2)$ are *not* isomorphic. Instead, the former group covers the latter one twice according to Equation (1.38).

It turns out that the *spin* groups $\operatorname{Spin}(n, \mathbb{C})$ with $n \leq 6$ and the groups $\operatorname{Spin}^0(p, q)$ with $p + q \leq 6$ are isomorphic to certain non-spin groups. Sometimes these isomorphisms are called *sporadic*. This is not true any more if $n \geq 7$ or $p + q \geq 7$.

1.7 Vector and Tensor Fields

1.7.1 Bundles: A Group Action Approach

Let G be a Lie group, and let M be such a manifold, that the manifolds G and M are similar.

Definition 48. A real (resp., complex) Lie group G acts from the left on a real (resp., complex) manifold M, if there is a real-analytic (resp., holomorphic) map $G \times M \to M$, $(g, p) \mapsto g \cdot p$, that the identity element e of the group G acts trivially, $e \cdot p = p$ for all $p \in M$, and $g \cdot (h \cdot p) = (gh) \cdot p$ for all g, $h \in G$ and $p \in M$.

Definition 49. A real (resp., complex) Lie group G acts from the right on a real-analytic (resp., holomorphic) manifold M, if there is a real-analytic (resp., holomorphic) map $M \times G \to M$, $(p, g) \mapsto p \cdot g$, that the identity element e of the group G acts trivially, $p \cdot e = p$ for all $p \in M$, and $(p \cdot g) \cdot h = p \cdot (gh)$ for all g, $h \in G$ and $p \in M$.

Remark 5. It is not difficult to extend these too restrictive definitions. For example, the manifold of a real group G can be considered as smooth, and one may define an action of a real Lie group on a smooth manifold, and so on. In what follows, we left such extensions to the reader.

Example 29. For a manifold M, a Lie group G acts from the right on the Cartesian product $M \times G$ by $(p, h)g = (p, hg)$, $p \in M$, g, $h \in G$.

Example 30. The classical groups, except the groups $\mathrm{Spin}(n, \mathbb{C})$ with $n \geq 7$ and the groups $\mathrm{Spin}^0(p, q)$ with $p + q \geq 7$ act from the left on a finite-dimensional linear space L of suitable dimension by matrix-vector multiplication.

Example 31. Let H be a closed subgroup of a real (resp., complex) Lie group G. The group H acts on G from the right by group multiplication.

Definition 50. A left action is called *transitive* if for any points p and q in M there is $g \in G$ with $q = g \cdot p$.

Definition 51. A left action is called *free* if for any $p \in M$ we have $g \cdot p = p$ if and only if $g = e$.

Definition 52. A left action is called *faithful* if for any $g \in G$, $g \neq e$ there is an $p \in M$ such that $g \cdot p \neq p$.

Definition 53. An *orbit* of a left action is an equivalence class with respect to the relation $p \sim q$ if and only if there exists $g \in G$ with $q = g \cdot p$.

The formulation of similar definitions for a right action may be left to the reader. We denote by $G \cdot p$ (resp., $p \cdot G$) the orbit of a point $p \in X$ under a left (resp., right) action of G.

Example 32. Fix a finite-dimensional real linear space U. Let P be a set and $A \in P$. Assume that the group U acts on P, and denote by $A + \mathbf{x}$ the result of the action of an element $\mathbf{x} \in U$ on a point A. The triple $(P, U, +)$ is called an *affine space* modelled upon U if the above action is transitive and faithful.

The map $A \mapsto A + \mathbf{x}$ is called the *translation* by the vector $\mathbf{x}$. The vector in U that translates the element $A \in P$ to the element $B \in B$ is denoted by $B - A$.

Non-formally, according to (Berger 2009),

> An affine space is nothing more than a vector space whose origin we try to forget about, by adding translations to the linear maps.

Before giving the next definition, consider a motivating example.

Example 33. Define the holomorphic map $\pi \colon \mathbb{C}^2 \setminus \{\mathbf{0}\} \to \mathbb{C}P^1$ by $\pi(\zeta_0, \zeta_1) = [\zeta_0 : \zeta_1]$. The group $G = \mathrm{GL}(1, \mathbb{C})$ acts on $P = \mathbb{C}^2 \setminus \{\mathbf{0}\}$ from the right by $(\zeta_0, \zeta_1)\lambda = (\zeta_0\lambda, \zeta_1\lambda)$. This action is holomorphic. As a set, each orbit of G is a copy of G (but lacks the group structure). Moreover, G acts freely and transitively on each orbit.

For the chart domains $\mathcal{U}_0$ and $\mathcal{U}_1$ of Example 18, we have:

$$\pi^{-1}(\mathcal{U}_0) = \{ (\zeta_0, \zeta_1) \in P \colon \zeta_0 \neq 0 \}, \qquad \pi^{-1}(\mathcal{U}_1) = \{ (\zeta_0, \zeta_1) \in P \colon \zeta_1 \neq 0 \}.$$

The one-to-one holomorphic map

$$\varphi_0 \colon \pi^{-1}(\mathcal{U}_0) \to \mathcal{U}_0 \times G, \qquad (\zeta_0, \zeta_1) \mapsto ([\zeta_0 : \zeta_1], \zeta_0)$$

with holomorphic inverse 'trivialises' the projection π in the following sense: $\pi = \Pi \circ \varphi_0$, where $\Pi \colon \mathcal{U}_0 \times G \to \mathcal{U}_0$ is the projection map of the Cartesian product to the first coordinate. Similarly, the one-to-one holomorphic map $\varphi_1 \colon \pi^{-1}(\mathcal{U}_1) \to \mathcal{U}_0 \times G$, $(\zeta_0, \zeta_1) \mapsto ([\zeta_0 : \zeta_1], \zeta_1)$ with holomorphic inverse trivialises the projection π. However, *globally P is not a Cartesian product* $M \times G$.

This example leads to the following definition.

Definition 54. A quadruple $\xi = (P, \pi, M, G)$ is called a smooth (resp., real-analytic, resp., holomorphic) *principal bundle* with *total space P*, *base M*, *bundle projection π*, and *fibres $\pi^{-1}(p)$*, $p \in M$, if

- P and M are smooth (resp., real-analytic, resp., holomorphic) manifolds.

- $\pi \colon P \to M$ is a smooth (resp., real-analytic, resp., holomorphic) map with $\pi(P) = M$.

- G is a Lie group with a smooth (resp., real-analytic, resp., holomorphic) action on P from the right, that preserves the fibres $\pi^{-1}(p)$, $p \in M$, and acts freely and transitively on each fibre.

- for each $p \in M$ there is a *trivialising chart* of M with domain $\mathcal{U}$ containing p, and a one-to-one smooth (resp., real-analytic, resp., holomorphic) map $\varphi \colon \pi^{-1}(\mathcal{U}) \to \mathcal{U} \times G$ with smooth (resp., real-analytic, resp., holomorphic) inverse that trivialises π: over $\pi^{-1}(\mathcal{U})$ the map π can be decomposed into the composition $\Pi_1 \circ \varphi$, where Π_1 is the projection of the Cartesian product $\mathcal{U} \times G$ to $\mathcal{U}$.

We see that the principal bundle of Example 33 is holomorphic.

Example 34. The action described in Example 29 determines a principal fibre bundle $(B \times G, \pi, B, G)$ by defining $\pi(p, g) = p$. Such a bundle is called *trivial*.

Example 35. The coverings of Examples 22–28 are principal bundles, and their fibres are copies of the group $G = O(1)$.

Definition 55. Two principal bundles $\xi_1 = (P_1, \pi_1, M, G)$ and $\xi_2 = (P_2, \pi_2, M, G)$ with the same base and fibre are called equivalent if there is a smooth one-to-one map $\Pi \colon P_1 \to P_2$ with smooth inverse that commutes with bundle projections: $\pi_1 = \pi_2 \circ \Pi$.

The following example can be considered as a 'compact' version of Example 33.

Example 36. The set $\{ (\zeta_0, \zeta_1)^\top \in \mathbb{C}^2 \colon |\zeta_0|^2 + |\zeta_1|^2 = 1 \}$ is a real-analytic manifold, a copy of the sphere S^3. The Lie group $G = U(1)$ acts on S^3 on the right by $(\zeta_0, \zeta_1)^\top \lambda = (\zeta_0 \lambda, \zeta_1 \lambda)^\top$, $\lambda \in \mathbb{C}^1$, $|\lambda| = 1$. Define the map $\pi \colon S^3 \to \mathbb{C}P^1$ by $\pi(\zeta_0, \zeta_1) = [\zeta_0 : \zeta_1]$. The reader may verify that the map π realises a one-to-one correspondence between the orbits of G and the points of $\mathbb{C}P^1$. We have

$$\pi^{-1}(\mathcal{U}_0) = \{ (\zeta_0, \zeta_1)^\top \in S^3 \colon \zeta_0 \neq 0 \},$$
$$\pi^{-1}(\mathcal{U}_1) = \{ (\zeta_0, \zeta_1)^\top \in S^3 \colon \zeta_1 \neq 0 \}.$$

The map

$$\varphi_+ \colon \pi^{-1}(\mathcal{U}_0) \to \mathcal{U}_0 \times G, \qquad (\zeta_0, \zeta_1)^\top \mapsto (\mathbf{x}_+^{-1}(\zeta_1/\zeta_0), \zeta_0/|\zeta_0|)$$

trivialises the map π over $\pi^{-1}(\mathcal{U}_+)$. Similarly, the map

$$\varphi_- \colon \pi^{-1}(\mathcal{U}_1) \to \mathcal{U}_1 \times G, \qquad (\zeta_0, \zeta_1)^\top \mapsto (\mathbf{x}_-^{-1}(\zeta_0/\zeta_1), \zeta_1/|\zeta_1|)$$

trivialises the map π over $\pi^{-1}(\mathcal{U}_1)$. The real-analytic coordinate change $\varphi_-^{-1} \circ \varphi_+$ maps a point $(\zeta, \lambda) \in (\mathbb{C}^1 \setminus \{0\}) \times G$ to the point $(\zeta^{-1}, \lambda\zeta/|\zeta|)$. The constructed real-analytic principal bundle $(S^3, \pi, \mathbb{C}P^1, U(1))$ is called the *Hopf bundle*. In what follows, we weaken this structure and consider the Hopf bundle as a *smooth* principal bundle.

The circles in Fig. 1.3 are the inverse images of singletons under the map π.[3]

---

3. Source: https://tex.stackexchange.com/questions/521253/how-we-can-draw-the-hopf-fibration.

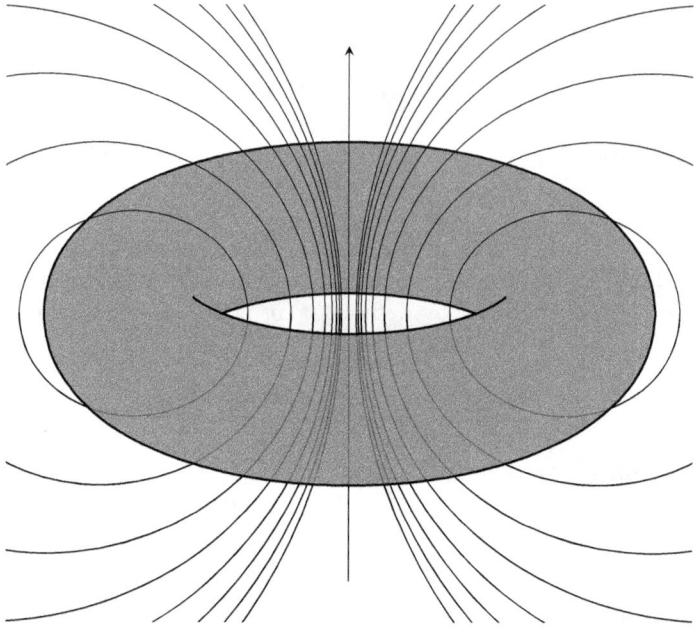

FIGURE 1.3
The Hopf bundle.

Example 37. The action described in Example 31 determines the principal fibre bundle $(G, \pi, G/H, H)$, where G/H is the set of left cosets

$$gH = \{ gh \colon h \in H \} \subset G,$$

and where $\pi(g) = gH$. The case of $G = \mathrm{SU}(2)$, $H = \mathrm{U}(1)$ reproduces Example 36. To see this, observe that a point $(\zeta_0, \zeta_1)^\top \in \mathbb{C}^2$ may be identified with the quaternion $q = \zeta_0 + \zeta_1 \mathrm{j}$. The map $\mathrm{Sp}(1) \to \mathrm{SU}(2)$,

$$\alpha + \beta \mathrm{i} + \gamma \mathrm{j} + \delta \mathrm{k} \mapsto \alpha \sigma_0 - \mathrm{i}(\beta \sigma_1 + \gamma \sigma_2 + \delta \sigma_3) = \begin{pmatrix} \alpha - \delta \mathrm{i} & -\gamma - \beta \mathrm{i} \\ \gamma - \beta \mathrm{i} & \alpha + \delta \mathrm{i} \end{pmatrix}, \qquad (1.40)$$

where σ_0 is the identity matrix and σ_1, $1 \leq i \leq 3$, are the Pauli matrices (1.14), is the real-analytic group isomorphism. The map π takes the form

$$\pi \begin{pmatrix} \alpha - \delta \mathrm{i} & -\gamma - \beta \mathrm{i} \\ -\gamma - \beta \mathrm{i} & \alpha + \delta \mathrm{i} \end{pmatrix} = \begin{pmatrix} 2(\alpha\gamma + \beta\delta) \\ 2(\gamma\delta - \alpha\beta) \\ 2\alpha^2 + 2\delta^2 - 1 \end{pmatrix}.$$

The composition of the inverse map and the map (1.35) gives the nontrivial covering $\mathrm{SU}(2) \to \mathrm{SO}(3)$ as

$$\begin{pmatrix} \alpha - \delta \mathrm{i} & -\gamma - \beta \mathrm{i} \\ -\gamma - \beta \mathrm{i} & \alpha + \delta \mathrm{i} \end{pmatrix} \mapsto \begin{pmatrix} 2\alpha^2 + 2\beta^2 - 1 & 2(\beta\gamma - \alpha\delta) & 2(\alpha\gamma + \beta\delta) \\ 2(\alpha\delta + \beta\gamma) & 2\alpha^2 + 2\gamma^2 - 1 & 2(\gamma\delta - \alpha\beta) \\ 2(\beta\delta - \alpha\gamma) & 2(\alpha\beta + \gamma\delta) & 2\alpha^2 + 2\delta^2 - 1 \end{pmatrix}.$$

The following principal bundles play a special role in this book: the principal bundle $(\mathbb{C}^2 \setminus \{\mathbf{0}\}, \pi, \mathbb{C}P^1, \mathrm{GL}(1, \mathbb{C}))$ of Example 33, the Hopf bundle $(S^3, \pi, \mathbb{C}P^1, \mathrm{U}(1))$ of Example 36, and two particular cases of Example 37. The first one is $(\mathrm{SU}(2), \pi, S^2, \mathrm{U}(1))$, just another realisation of the Hopf bundle, the second one is described in what follows.

Example 38. Let $G = \mathrm{SL}(2, \mathbb{C}) \boxtimes \mathbb{R}^{1,3}$ be the Cartesian product $\mathrm{SL}(2, \mathbb{C}) \times \mathbb{R}^{1,3}$ with the group operation $(g_1, \mathbf{x}_1)(g_2, \mathbf{x}_2) = (g_1 g_2, \mathbf{x}_1 + \pi(g_1)\mathbf{x}_2)$, where $\pi \colon \mathrm{SL}(2, \mathbb{C}) \to \mathrm{SO}_0(1, 3)$ is the covering (1.39). Denote by $\mathbf{p}$ a point in $\hat{\mathbb{R}}^{1,3}$, a copy of $\mathbb{R}^{1,3}$. Introduce the following notation: $\chi_{\mathbf{p}}(\mathbf{x}) = \exp(-iB(\mathbf{p}, \mathbf{x}))$, where B is the bilinear form (1.1).

The group G acts on $\hat{\mathbb{R}}^{1,3}$ by

$$((g, \mathbf{x}) \cdot \chi)_{\mathbf{p}}(\mathbf{y}) = \chi_{\mathbf{p}}(\pi(g^{-1})\mathbf{y}), \qquad g \in G, \quad \mathbf{x} \in \mathbb{R}^{1,3}, \quad \mathbf{y} \in \hat{\mathbb{R}}^{1,3}.$$

The stationary subgroup of the point $\mathbf{p} = (1, 0, 0, 1)^\top \in \hat{\mathbb{R}}^{1,3}$ is the group $H = E(2) \boxtimes \mathbb{R}^{1,3}$, where $E(2)$ is the subgroup of the group $\mathrm{SL}(2, \mathbb{C})$ consisting of the matrices of the form $\begin{pmatrix} \exp(i\theta/2) & \exp(-i\theta/2)\zeta \\ 0 & \exp(-i\theta/2) \end{pmatrix}$ with $0 \le \theta < 4\pi$ and $\zeta \in \mathbb{C}$. The homogeneous space $\hat{\mathcal{C}}_+ = G/H$ has the form

$$\hat{\mathcal{C}}_+ = \{\, \mathbf{p} \in \hat{\mathbb{R}}^{1,3} \colon B(\mathbf{p}, \mathbf{p}) = 0, p_0 > 0 \,\}$$

and is called the *mantle of the forward light cone*, see its two-dimensional version at Fig. 1.4.[4] The principal bundle is

$$(\mathrm{SL}(2, \mathbb{C}) \boxtimes \mathbb{R}^{1,3}, \pi, \hat{\mathcal{C}}_+, E(2) \boxtimes \mathbb{R}^{1,3}).$$

Definition 56. A *cross-section* or a *global gauge* of a principal bundle (P, π, M, G) is a map $s \colon M \to P$ such that the composition $\pi \circ s$ is the identity map on M.

Let $(\mathcal{U}, \mathbf{h})$ be a trivialising chart on M.

Definition 57. A *local cross-section* or a *local gauge* of a principal bundle (P, π, M, G) is a map $s \colon \mathcal{U} \to P$ such that the composition $\pi \circ s$ is the identity map on $\mathcal{U}$.

A local gauge s determines the one-to-one map $\mathcal{U} \times G \to \pi^{-1}(\mathcal{U})$, $(p, g) \mapsto s(p) \cdot g$. The inverse map trivialises π. If the gauge s is global, then the principal bundle (P, π, M, G) is trivial. A choice of a local gauge is a choice of local coordinates.

Definition 58. A *global gauge transformation* of a principal bundle $\xi = (P, \pi, M, G)$ is a smooth one-to-one map $\sigma \colon P \to P$ with smooth inverse that preserves the fibres of ξ: $\pi \circ \sigma = \pi$, and commutes with the action of G: $\sigma(x \cdot g) = \sigma(x) \cdot g$ for all $x \in P$ and $g \in G$.

---

4. Source: https://tex.stackexchange.com/questions/28802/cut-off-cone-in-tikz.

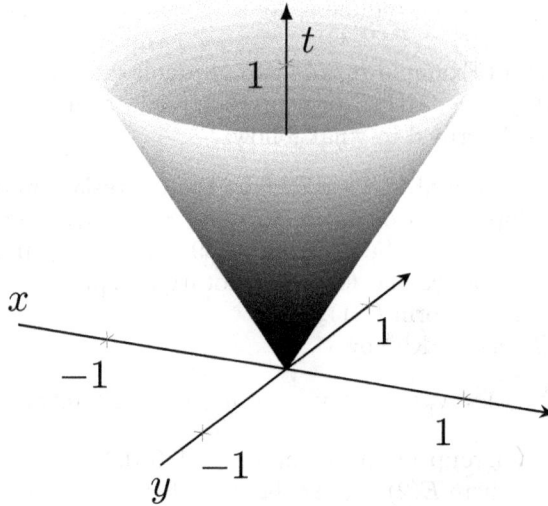

FIGURE 1.4
A part of the mantle of the forward light cone, the two-dimensional version.

Definition 59. A *local gauge transformation* of a principal bundle ξ with a trivialising chart $(\mathcal{U}, \mathbf{h})$ is a global gauge transformation of the principal bundle $(\pi^{-1}(\mathcal{U}), \pi, \mathcal{U}, G)$.

Definition 60. A *physical gauge transformation* of a principal bundle (P, π, M, G) over a trivialising chart $(\mathcal{U}, \mathbf{h})$ is a smooth map $\tau \colon \mathcal{U} \to G$.

After a choice of a local gauge $s \colon \mathcal{U} \to P$, we may identify local gauge transformations with physical ones. Indeed, let $\sigma \colon \pi^{-1}(\mathcal{U}) \to \pi^{-1}(\mathcal{U})$ be a local gauge transformation. Then the map $\sigma \circ s$ is a physical gauge transformation. Conversely, if $\tau \colon \mathcal{U} \to G$ is a physical gauge transformation, then the map $\sigma_\tau \colon \pi^{-1}(\mathcal{U}) \to \pi^{-1}(\mathcal{U})$ given by

$$\sigma_\tau(s(p) \cdot g) = g^{-1} \tau(p) g, \qquad p \in \mathcal{U}, \quad g \in G$$

is a local gauge transformation.

Let $\xi = (P, \pi, M, G)$ be a smooth (resp., real-analytic, resp., holomorphic) principal bundle. Assume that G acts from the left on a smooth (resp., real-analytic, resp., holomorphic) manifold F and the action is smooth (resp., real-analytic, resp., holomorphic). Define a right action of G on the Cartesian product $P \times F$ by

$$(p, q) \cdot g = (p \cdot g, g^{-1} \cdot q), \qquad (p, q) \in P \times F, \quad g \in G.$$

Denote by E the set of G-orbits in $P \times F$.

Definition 61. A *fibre bundle* over M with fibre F and associated principal G-bundle $\xi = (P, \pi, M, G)$ is the triple $\xi[F] = (E, \pi_F, M)$, where the bundle projection π_F acts by $\pi_F((p, q)G) = \pi(p)$, $p \in M$, $q \in F$.

Example 39. Let the group $O(1)$ acts on the circle

$$S^1 = \{(x, y)^\top \in \mathbb{R}^2 : x^2 + y^2 = 1\}$$

from the right by multiplication: $(x, y)^\top \cdot \pm 1 = \pm(x, y)^\top$, and on the interval $[-1, 1]$ from the left, also by multiplication. The total space $E = S^1 \times [-1, 1]/O(1)$ of the fibre bundle $\xi[-1, 1])$ over $S^1/O(1)$ with fibre $[-1, 1]$ and associated principal $O(1)$-bundle $\xi = (S^1, \pi, S^1/O(1))$ is the Möbius strip!

1.7.2 Bundles: A Classical Approach

Assume that $\mathbb{F} = \mathbb{R}$ and let $f \colon M \to \mathbb{R}^1$ be a smooth map. We define the partial derivatives of f at a point $p \in M$ in a chart $(\mathcal{U}, \mathbf{x})$ by

$$\frac{\partial f}{\partial x^\mu}(p) = \frac{\partial (f \circ \mathbf{x}^{-1})}{\partial x_\mu}(\mathbf{x}(p)),$$

where the right hand side is the usual partial derivative. If another chart, $(\mathcal{V}, \mathbf{y})$, satisfies the condition $\mathcal{U} \cap \mathcal{V} \neq \varnothing$, then for each $p \in \mathcal{U} \cap \mathcal{V}$ the Chain Rule gives

$$\frac{\partial f}{\partial y^\mu}(p) = \frac{\partial f}{\partial x^\nu}(p) \frac{\partial x^\nu}{\partial y^\mu}(\mathbf{x}(p)).$$

Let $\mathbf{v}, \mathbf{w} \in \mathbb{R}^n$. We introduce an equivalence relation: $(\mathbf{x}, \mathbf{v}) \sim_p (\mathbf{y}, \mathbf{w})$ if and only if

$$w_\mu = \frac{\partial x^\mu}{\partial y^\nu}(\mathbf{x}(p)) v_\nu.$$

Definition 62. A *tangent vector* at a point $p \in M$ is an equivalence class $[\mathbf{x}, \mathbf{v}]_p$ of the above equivalence relation.

Let $T_p M$ be the set of all tangent vectors at a point $p \in M$. It is a real linear space with respect to the addition $[\mathbf{x}, \mathbf{v}]_p + [\mathbf{x}, \mathbf{w}]_p = [\mathbf{x}, \mathbf{v} + \mathbf{w}]_p$ and the scalar-vector multiplication $\alpha[\mathbf{x}, \mathbf{v}]_p = [\mathbf{x}, \alpha \mathbf{v}]_p$, the *tangent space* of M at p. The correspondence $[\mathbf{x}, \mathbf{v}]_p \mapsto \mathbf{v}$ defines an isomorphism between the linear spaces $T_p M$ and $\mathbb{R}^n$. Denote by $\frac{\partial}{\partial x^\mu}\big|_p$ the inverse image of the vector $\mathbf{e}^\mu \in \mathbb{R}^n$ under the above map.

Denote by TM the set of all tangent vectors at all points $p \in M$. Define the map $\pi \colon TM \to M$ by $\pi[\mathbf{x}, \mathbf{v}]_p = p$. Observe that the map $t \colon \pi^{-1}(\mathcal{U}) \to \mathcal{U} \times \mathbb{R}^n$ defined by $t([\mathbf{x}, \mathbf{v}]_p) = (p, \mathbf{v})$ is one-to-one. We define a topology on TM by the requirement that the sets $t^{-1}(A)$, where A runs over all open subsets of all sets $\mathcal{U} \times \mathbb{R}^n$, constitute the base of the above topology. Moreover, the system of charts t can be extended to a maximal atlas and determines a structure of a smooth manifold on TM.

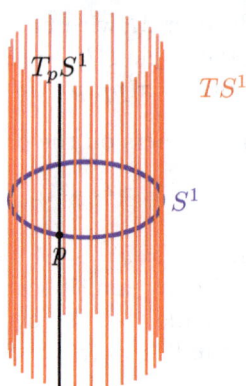

FIGURE 1.5
The tangent bundle of the circle.

Definition 63. The *tangent bundle* of M is the triple (TM, π, M).

We see that the tangent bundle is *locally trivial*: for any chart $(\mathcal{U}, \mathbf{x})$ the map t is a homeomorphism between $\pi^{-1}(\mathcal{U})$ and $\mathcal{U} \times \mathbb{R}^n$.

You may see the tangent bundle to the circle S^1 on Fig. 1.5.[5]

Definition 64. A *vector field* on M is a function $X \colon M \to TM$ satisfying $\pi(X(p)) = p$ for all $p \in M$.

In other words, a vector $X(p)$ lies in T_pM. In a chart $(\mathcal{U}, \mathbf{x})$, we have

$$X(p) = X^\mu(p) \left. \frac{\partial}{\partial x^\mu} \right|_p.$$

In what follows, the symbol $X^\mu(p)$ denotes the coordinates of a vector $X(p) \in T_pM$ in a particular chart.

A vector field X acts on a smooth function $f \colon M \to \mathbb{R}^1$ by

$$X(f)(p) = X^\mu(p) \frac{\partial f}{\partial x^\mu}(p).$$

The *bracket* of two vector fields X and Y is given by $[X, Y] = XY - YX$.

The tangent bundle is just a typical example of a vector bundle.

Definition 65. A triple (E, π, M) is called a *vector bundle* if

- E and M are similar manifolds;

- π is a smooth map, and for each $p \in M$, the *fibre* $\pi^{-1}(p)$ is a copy of a linear space L;

5. Source: https://tex.stackexchange.com/questions/286745/ how-to-draw-the-tangent-bundle-of-a-circle.

- there is a *trivialising atlas* A of the manifold M such that for any chart $(\mathcal{U}, \mathbf{x}) \in$ A there exists a homeomorphism $t\colon \pi^{-1}(\mathcal{U}) \to \mathcal{U} \times L$.

Definition 66. A *cross-section* of a vector bundle (E, π, M) is a map $\mathbf{s}\colon M \to E$ such that the composition $\pi \circ \mathbf{s}$ is the identity map on M.

A typical method for constructing new vector bundles from old ones is as follows. Consider any of the operations over linear spaces described in Subsection 1.2.2. As an example, let this operation be $U \to cU$.

Assume that $\mathbb{F} = \mathbb{C}$. Let $(\mathcal{U}, \zeta)$ be a chart. Let $\zeta^1(p)$, ..., $\zeta^n(p)$ be the coordinates of a point $p \in \mathcal{U}$ in a chart $(\mathcal{U}, \zeta)$. The real-valued local coordinates $x^\mu(p) = \operatorname{Re} \zeta^\mu(p)$ and $y^\mu(p) = \operatorname{Im} \zeta^\mu(p)$, $1 \le \mu \le n$, determine the structure of a real-analytic manifold on the set M as $(\mathcal{U}, \zeta)$ runs over $\mathsf{A}_{\max}$. Let $M_\mathbb{R}$ be the set M together with the above structure, and let $(E, \pi, M_\mathbb{R})$ be the tangent bundle of $M_\mathbb{R}$.

Define cE as the union of the linear spaces $c(\pi^{-1}(p))$ over all $p \in M_\mathbb{R}$, and let $c\pi\colon cE \to M_\mathbb{R}$ maps all elements of the linear space $cU = c(\pi^{-1}(p))$ to p. Let $t\colon \pi^{-1}(\mathcal{U}) \to \mathcal{U} \times U$ be a trivialisation. The restriction t_p of the map t to a fibre $(c\pi)^{-1}(p)$ is an isomorphism between the fibre and the linear space $\{p\} \times U$. The extension of this isomorphism to the linear space $c(\pi^{-1}(p))$ given by $ct((\mathbf{x}, \mathbf{y}) \otimes \zeta) = t(\mathbf{x}, \mathbf{y}) \otimes \zeta$, $\zeta \in \mathbb{C}^1$, gives the trivialisation ct of a new vector bundle $(cE, c\pi, M_\mathbb{R})$ over the same base $M_\mathbb{R}$.

In particular, the vectors $\left.\frac{\partial}{\partial \zeta^\mu}\right|_p = \frac{1}{2}\left(\left.\frac{\partial}{\partial x^\mu}\right|_p - \mathrm{i}\left.\frac{\partial}{\partial y^\mu}\right|_p\right)$ generate a subspace $T_p^{1,0}M$ of the complex linear space $cT_pM_\mathbb{R}$. The elements of this subspace are called the *holomorphic tangent vectors*. Similarly, the vectors $\left.\frac{\partial}{\partial \bar{\zeta}^\mu}\right|_p = \frac{1}{2}\left(\left.\frac{\partial}{\partial x^\mu}\right|_p + \mathrm{i}\left.\frac{\partial}{\partial y^\mu}\right|_p\right)$ generate a subspace $T_p^{0,1}M$ of the complex linear space $cT_pM_\mathbb{R}$. The elements of this subspace are called the *antiholomorphic tangent vectors*.

Apply the above construction to the operation $U \to U^*$. A tangent bundle (TM, π, M) with the fibre U becomes the *cotangent bundle* (T^*M, π^*, M). For a chart $(\mathcal{U}, \mathbf{x})$ and the basis $\left.\frac{\partial}{\partial x^\mu}\right|_p$, $1 \le \mu \le n$, define the vectors of the dual basis $\mathrm{d}x^\mu(p) \in U^*$ by $\mathrm{d}x^\mu(p)\left(\left.\frac{\partial}{\partial x^\nu}\right|_p\right) = \delta^\mu{}_\nu$. A cross-section of this bundle is a *covector field* of the form $X(p) = X_\mu(p)\left.\mathrm{d}x^\mu\right|_p$.

Similarly, we may construct tensor bundles and their cross-sections: tensor fields. Specifically, we construct the tensor bundle $(T_l^k M, \pi_l^k, M)$ with fibres $T_l^k(U)$. In a chart $(\mathcal{U}, \mathbf{x})$, a cross-section of this bundle has the form

$$T(p) = X^{\nu_1 \cdots \nu_l}{}_{\mu_1 \cdots \mu_k}(p)\mathrm{d}x^{\mu_1} \otimes \cdots \otimes \mathrm{d}x^{\mu_k} \otimes \frac{\partial}{\partial x^{\nu_1}} \otimes \cdots \otimes \frac{\partial}{\partial x^{\nu_l}}, \qquad p \in \mathcal{U},$$

while in a chart $(\mathcal{V}, \mathbf{y})$ with $\mathcal{U} \cap \mathcal{V} \ne \varnothing$ it takes the form

$$T(p) = Y^{\nu_1 \cdots \nu_l}{}_{\mu_1 \cdots \mu_k}(p)\mathrm{d}y^{\mu_1} \otimes \cdots \otimes \mathrm{d}y^{\mu_k} \otimes \frac{\partial}{\partial y^{\nu_1}} \otimes \cdots \otimes \frac{\partial}{\partial y^{\nu_l}}, \qquad p \in \mathcal{U} \cap \mathcal{V},$$

where

$$Y^{\beta_1\cdots\beta_l}{}_{\alpha_1\cdots\alpha_k}(p) = X^{\nu_1\cdots\nu_l}{}_{\mu_1\cdots\mu_k}(p)\frac{\partial x^{\mu_1}}{\partial y^{\alpha_1}}\cdots\frac{\partial x^{\mu_k}}{\partial y^{\alpha_k}}\frac{\partial y^{\beta_1}}{\partial x^{\nu_1}}\cdots\frac{\partial y^{\beta_l}}{\partial x^{\nu_l}}.$$

1.7.3   Bundles: A Cocycle Approach

Let (E, π, M) be a vector bundle with fibre L. Let $(\mathcal{U}_\alpha, \mathbf{x}_\alpha)$ and $(\mathcal{U}_\beta, \mathbf{x}_\beta)$ be two charts of a trivialising atlas with $\mathcal{U}_\alpha \cap \mathcal{U}_\beta \neq \varnothing$. Let t_α and t_β be the trivialisations that correspond to the above two charts.

Let $\mathrm{GL}(L)$ be the group of all invertible linear operators in L. For a point $p \in \mathcal{U}_\alpha \cap \mathcal{U}_\beta$, and for a vector $\mathbf{l} \in L$, the overlap $t_\alpha \circ t_\beta^{-1}$ maps the point $(p, \mathbf{l}) \in \mathcal{U}_\beta \times L$ to the point $(p, f_{\alpha\beta}(p)\mathbf{l}) \in \mathcal{U}_\alpha \times L$ for a suitable function $f_{\alpha\beta} : \mathcal{U}_\alpha \cap \mathcal{U}_\beta \to \mathrm{GL}(L)$, that is, the L-factors in the Cartesian products are identified with the help of an invertible linear operator.

Definition 67. A *cocycle* on a manifold M is a collection $\{\mathcal{U}_\alpha, f_{\alpha\beta}\}$, where $\mathcal{U}_\alpha$ runs over the domains of the charts of an atlas of M, and where $f_{\alpha\beta} : \mathcal{U}_\alpha \cap \mathcal{U}_\beta \to \mathrm{GL}(L)$ are maps satisfying the *cocycle condition*: for all $p \in \mathcal{U}_\alpha \cap \mathcal{U}_\beta \cap \mathcal{U}_\gamma \neq \varnothing$ we have $f_{\alpha\beta}(p)f_{\beta\gamma}(p)f_{\gamma\alpha}(p) = I \in \mathrm{GL}(L)$.

It is clear that the maps $g_{\alpha\beta}$ constructed from the overlaps $t_\alpha \circ t_\beta^{-1}$, satisfy the cocycle condition. Conversely, if $\{\mathcal{U}_\alpha, f_{\alpha\beta}\}$ is a cocycle, then consider the union of the Cartesian products $\mathcal{U}_\alpha \times L$ over all values of α, and let E be the quotient set of the above union with respect to the following equivalence relation: the points $(p, \mathbf{l}) \in \mathcal{U}_\alpha \times L$ and $(p, \mathbf{m}) \in \mathcal{U}_\beta \times L$ are equivalent if and only if $\mathbf{l} = f_{\alpha\beta}\mathbf{m}$.

Example 40. Let $(\mathcal{U}_\alpha, \mathbf{x}_\alpha)$ and $(\mathcal{U}_\beta, \mathbf{x}_\beta)$ be two charts of an atlas of a smooth manifold M with $\mathcal{U}_\alpha \cap \mathcal{U}_\beta \neq \varnothing$. Define the map $f_{\alpha\beta} : \mathcal{U}_\alpha \cap \mathcal{U}_\beta \to \mathrm{CL}(n, \mathbb{R})$ by

$$(f_{\alpha\beta}(p))_{\mu\nu} = \frac{\partial(\mathbf{x}_\beta^{-1} \circ \mathbf{x}_\alpha)_\mu}{\partial u_\nu}(p).$$

This cocycle determines the tangent bundle (TM, π, M).

Example 41. Let $\{\mathcal{U}_\alpha, f_{\alpha\beta}\}$ be a cocycle on a manifold M, and let $\mathbf{s}_\alpha : \mathcal{U}_\alpha \to L$ be a collection of local maps satisfying $\mathbf{s}_\alpha(p) = f_{\alpha\beta}(p)\mathbf{s}_\beta(p)$ for all $p \in \mathcal{U}_\alpha \cap \mathcal{U}_\beta \neq \varnothing$. These local maps define a section $\mathbf{s}$ of the vector bundle described by the above cocycle.

Let G be a subgroup of the group $\mathrm{GL}(L)$.

Definition 68. A vector bundle (E, π, M) defined by a cocycle $\{\mathcal{U}_\alpha, f_{\alpha\beta}\}$ is called *reducible* to a subgroup G if there are maps $f'_{\alpha\beta} : \mathcal{U}_\alpha \cap \mathcal{U}_\beta \to G$ such that $\{\mathcal{U}_\alpha, f'_{\alpha\beta}\}$ is a cocycle on M that determines the same vector bundle as $\{\mathcal{U}_\alpha, f_{\alpha\beta}\}$ does.

Example 42. It is possible to prove that a vector bundle (E, π, M) is reducible to the group $\{I\} \subset \mathrm{GL}(L)$ if and only if $E = M \times L$ and $\pi(p, \mathbf{l}) = p$. Such a bundle is called *trivial*.

Under very mild conditions, any vector bundle over a smooth manifold M with m-dimensional real (resp., complex, resp., quaternionic) fibres L is reducible to the subgroup $\mathrm{O}(m)$ (resp., $\mathrm{U}(m)$, resp., $\mathrm{Sp}(m)$), see Husemoller (1994).

1.7.4 Riemannian Manifolds and Geodesics

Let (E, π, M) be a vector bundle over a smooth manifold M with real fibres U. Construct the bundle $(E^* \otimes E^*, \pi^* \otimes \pi^*, M)$.

Definition 69. A *Riemannian metric* on (E, π, M) is a smooth section $s\colon M \to E^* \otimes E^*$ such that for any $p \in M$ the bilinear form $s(p)$ is an inner product on $U = \pi^{-1}(p)$, that is, $s(p)$ is symmetric and positive-definite.

In what follows, we restrict our attention to the case of the tangent bundle (TM, π, M) and say that a Riemannian metric on (TM, π, M) is just a metric on M. In a local chart $(\mathcal{U}, \mathbf{x})$ on a smooth manifold M of dimension n, we have

$$s(p) = g_{\mu\nu}(p)\mathrm{d}x^{\mu} \otimes \mathrm{d}x^{\nu},$$

where the matrix $g_{\mu\nu}(p)$ is symmetric and positive-definite, and the real-valued functions $g_{\mu\nu}(p)\colon \mathcal{U} \to \mathbb{R}^1$ are smooth.

Example 43. Let $M = \mathcal{U} = \mathbb{R}^n$, and let $\mathbf{h}(\mathbf{x}) = \mathbf{x}$ for all $\mathbf{x} \in M$. Define

$$g_{\mu\nu}(p) = \begin{cases} 1, & \text{if } \mu = \nu, \\ 0, & \text{otherwise.} \end{cases}$$

Example 44. Let $(\mathcal{U}, \mathbf{h})$ be the angular spherical coordinates on S^2. In this chart, the matrix $g(\theta, \varphi) = \left(\begin{smallmatrix} 1 & 0 \\ 0 & \sin^2\theta \end{smallmatrix}\right)$ determines a Riemannian metric. Indeed, at a point $p = (\sin\theta\cos\varphi, \sin\theta\sin\varphi, \cos\theta)^{\top} \in S^2 \subset \mathbb{R}^3$ the basis vectors of the tangent plane T_pS^2 are as follows:

$$\frac{\partial p}{\partial \theta} = (\cos\theta\cos\varphi, \cos\theta\sin\varphi, -\sin\theta)^{\top},$$

$$\frac{\partial p}{\partial \varphi} = (-\sin\theta\sin\varphi, \sin\theta\cos\varphi, 0)^{\top}.$$

The inner products of these vectors form the matrix $g(\theta, \varphi)$.

Let M and N be two smooth manifolds, and let $f\colon M \to N$ be a smooth map. By definition, the map $f_*\colon TM \to TN$ maps a point $[\mathbf{x}, \mathbf{v}]_p \in TM$ to the point $\left[\mathbf{y}, \frac{\partial(\mathbf{y}\circ f\circ \mathbf{x}^{-1})_{\alpha}}{\partial x_{\beta}}(\mathbf{x}(p))\mathbf{v}\right]_{f(p)} \in TN$, where $(\mathcal{U}, \mathbf{x})$ (resp., $(\mathcal{V}, \mathbf{y})$) is a chart in M (resp., N) with $p \in \mathcal{U}$ (resp., $f(p) \in \mathcal{V}$).

Consider the smooth manifold $\mathbb{R}^1$ with the chart $(\mathcal{U}, x)$ given by $\mathcal{U} = \mathbb{R}^1$ and $x(t) = t$. The vector $\frac{\partial}{\partial t}\big|_{t_0}$ is traditionally denoted by $\frac{d}{dt}\big|_{t_0}$ and constitutes a basis for the tangent space $T_{t_0}\mathbb{R}^1$.

Definition 70. A *smooth curve* is a smooth map $\gamma\colon \mathbb{R}^1 \to M$. The *tangent vector* to γ at t_0 is the element of the tangent space $T_{\gamma(t_0)}M$ given by $\frac{d\gamma}{dt}\big|_{t_0} = \gamma_*\left(\frac{d}{dt}\big|_{t_0}\right)$.

Denote the tangent vector $\frac{d\gamma}{dt}$ by $\gamma'(t)$. Its length is

$$\|\gamma'(t)\| = \sqrt{s(\gamma(t))(\gamma'(t), \gamma'(t))}.$$

The length of the curve γ from a to b is

$$L_a^b(\gamma) = \int_a^b \|\gamma'(t)\| \, dt.$$

Assume that M is connected, and introduce the distance $d(p, q)$ between the points p and $q \in M$ as the infimum of the lengths of all smooth curves connecting p and q. Is it possible to find a curve that realises the above infimum?

To investigate this question, we use classical variational calculus. For a smooth curve $\gamma\colon \mathbb{R}^1 \to M$ with $\gamma(a) = p$ and $\gamma(b) = q$, consider a *variation* of γ with fixed endpoints: a function $\alpha\colon (-\varepsilon, \varepsilon) \times [a, b] \to M$ which is smooth on $(-\varepsilon, \varepsilon) \times (a, b)$ and satisfies $\alpha(0, t) = \gamma(t)$, and $\alpha(u, a) = p$, $\alpha(u, b) = q$ for all $u \in (-\varepsilon, \varepsilon)$. Instead of minimising the value of $L_a^b(\gamma)$, we try to find critical points of a general functional of the form

$$J(\gamma) = \int_a^b F(t, \gamma(t), \gamma'(t)) \, dt,$$

Definition 71. The *first variation* of the functional J is given by

$$\delta J(\gamma) = \frac{d}{du}\bigg|_{u=0} J(\overline{\alpha}(u)),$$

where $\overline{\alpha}(u)(t) = \alpha(u, t)$. A *critical point* of the functional J is a function γ with $\delta J(\gamma) = 0$.

For the variation of a special form $\alpha(u, t) = \gamma(t) + u\boldsymbol{\eta}(t)$, where $\boldsymbol{\eta}\colon [a, b] \to \mathbb{R}^n$ with $\boldsymbol{\eta}(a) = \boldsymbol{\eta}(b) = \mathbf{0}$, the first variation has the form

$$\delta J_\mu = \int_a^b \eta^\mu(t) \left[\frac{\partial F}{\partial x^\mu} - \frac{d}{dt}\frac{\partial F}{\partial y^\mu} \right] dt.$$

Theorem 5 (The Euler–Lagrange equations). *A function γ is a critical point of J if and only if it satisfies the Euler–Lagrange equations*

$$\frac{\partial F}{\partial x^\mu}(t, \gamma(t), \gamma'(t)) - \frac{d}{dt}\frac{\partial F}{\partial y^\mu}(t, \gamma(t), \gamma'(t)) = 0, \qquad 1 \le \mu \le n.$$

Example 45. Instead of calculating the critical point of the functional $L_a^b(\gamma)$, which involves the square root, we first do that for another functional:

$$E(\gamma) = \frac{1}{2} \int_a^b \|\gamma'(t)\|^2 \, dt.$$

The factor $\frac{1}{2}$ refers to the physical equation for kinetic energy.

For simplicity, we assume that the curve γ lies in a domain $\mathcal{U}$ of one chart, say $(\mathcal{U}, \mathbf{x})$, otherwise we partition γ into pieces that fill into the domain of one chart each, work in each domain separately, and glue everything together. This may be left to the reader.

In the above chart, the functional $E(\gamma)$ takes the form

$$E(\gamma) = \frac{1}{2} \int_a^b g_{\alpha\beta}(\gamma(t)) \frac{d\gamma^\alpha(t)}{dt} \frac{d\gamma^\beta(t)}{dt} = \int_a^b F(\gamma(t), \gamma'(t)) \, dt,$$

where we used the Einstein summation convention, and where $F(\mathbf{x}, \mathbf{y}) = \frac{1}{2} g_{\alpha\beta}(\mathbf{x}) y^\alpha y^\beta$. Calculating the partial derivatives, we obtain

$$\frac{\partial F}{\partial x^\mu}(\gamma(t), \gamma'(t)) = \frac{1}{2} \frac{\partial g_{\alpha\beta}}{\partial x^\mu}(\gamma(t)) \frac{d\gamma^\alpha(t)}{dt} \frac{d\gamma^\beta(t)}{dt},$$

$$\frac{\partial F}{\partial y^\mu}(\gamma(t), \gamma'(t)) = g_{\mu\nu}(\gamma(t)) \frac{d\gamma^\nu(t)}{dt}.$$

Differentiating the second equation, we have

$$\frac{d}{dt} \frac{\partial F}{\partial y^\mu}(\gamma(t), \gamma'(t)) = g_{\mu\nu}(\gamma(t)) \frac{d^2\gamma^\nu(t)}{dt^2} + \frac{\partial g_{\gamma\nu}}{\partial x^\beta}(\gamma(t)) \frac{d\gamma^\beta(t)}{dt} \frac{d\gamma^\nu(t)}{dt}.$$

Observe that the second term can be written in another form:

$$\frac{\partial g_{\mu\nu}}{\partial x^\beta}(\gamma(t)) \frac{d\gamma^\beta(t)}{dt} \frac{d\gamma^\nu(t)}{dt} = \frac{1}{2} \frac{\partial g_{\alpha\mu}}{\partial x^\beta}(\gamma(t)) \frac{d\gamma^\alpha(t)}{dt} \frac{d\gamma^\beta(t)}{dt}$$
$$+ \frac{1}{2} \frac{\partial g_{\beta\mu}}{\partial x^\alpha}(\gamma(t)) \frac{d\gamma^\alpha(t)}{dt} \frac{d\gamma^\beta(t)}{dt}.$$

The Euler–Lagrange equations take the form

$$g_{\mu\nu}(\gamma(t)) \frac{d^2\gamma^\nu(t)}{dt^2} + \frac{1}{2} \left(\frac{\partial g_{\alpha\mu}}{\partial x^\beta} + \frac{\partial g_{\beta\mu}}{\partial x^\alpha} - \frac{\partial g_{\alpha\beta}}{\partial x^\mu} \right) (\gamma(t)) \frac{d\gamma^\alpha(t)}{dt} \frac{d\gamma^\beta(t)}{dt} = 0.$$

Introduce the Christoffel symbols:

$$\Gamma^\mu_{\alpha\beta} = \frac{1}{2} g^{\mu\nu} \left(\frac{\partial g_{\alpha\nu}}{\partial x^\beta} + \frac{\partial g_{\beta\nu}}{\partial x^\alpha} - \frac{\partial g_{\alpha\beta}}{\partial x^\nu} \right). \tag{1.41}$$

We obtain: a critical point of the functional $E(\gamma)$ satisfies the system

$$\frac{d^2\gamma^\mu(t)}{dt^2} + \Gamma^\mu_{\alpha\beta}(\gamma(t)) \frac{d\gamma^\alpha(t)}{dt} \frac{d\gamma^\beta(t)}{dt} = 0. \tag{1.42}$$

Such a critical point is called a *geodesic*. It turns out that the length functional $L_a^b(\gamma)$ has the same critical points, see, e.g., Spivak (1979).

The standard results about existence and uniqueness of solutions to differential equations show that any point $p \in M$ has a neighbourhood with the following property: an arbitrary point $q \neq p$ of that neighbourhood can be connected with p by a unique geodesic line. As an example: let $p = (0,0,1)^\top \in S^2 \subset \mathbb{R}^3$ be the North Pole. Any point $q \neq (0,0,-1)^\top$ can be connected with p by the unique meridian, the neighbourhood is big enough. Infinitely many meridians connect the two poles.

1.7.5   The Riemannian Measure

Let $(\mathcal{U}, \mathbf{x})$ be a local chart on a Riemannian manifold M, and let g be the determinant of the metric g_{ij}. For a closed and bounded subset K such that $\mathbf{x}(K)$ is measurable subset of $\mathbb{R}^m$, define

$$\mu(K) = \int \mathbf{x}(K)\sqrt{g \circ \mathbf{x}^{-1}}\, d\mathbf{x}.$$

It can be proved that this number does not depend on the choice of a chart and can be extended to a measure on the Borel σ-field of M called the *Riemannian measure*.

1.7.6   Differentiation of Vector and Tensor Fields

Let Y be a smooth vector field on a smooth manifold M. How the derivative of Y looks like? An attempt to define a derivative as the limit of an expression like $\frac{Y(q)-Y(p)}{d(p,q)}$ fails, because $Y(p) \in T_pM$, $Y(q) \in T_qM$, and their difference makes no sense. In fact, we need to 'connect' in some way the vectors lying in different tangent spaces.

The axiomatic approach does help. Let X and Y be two smooth vector fields.

Definition 72. A *Koszul connection* $\nabla_X Y$ is a smooth vector field that satisfies the following conditions:

$$\nabla_{X_1+X_2}Y = \nabla_{X_1}Y + \nabla_{X_2}Y, \quad \nabla_X(Y_1+Y_2) = \nabla_X Y_1 + \nabla_X Y_2,$$
$$\nabla_{fX}(Y) = f \cdot \nabla_X Y, \qquad\qquad \nabla_X(fY) = f \cdot \nabla_X Y + X(f) \cdot Y.$$

A standard example of a connection on the manifold $M = \mathbb{R}^n$ is the directional derivative.

Denote $X_p = X(p)$ and $\nabla_{X_p}Y = \nabla_X Y(p)$. The map $X_p \to \nabla_{X_p}Y$ is a linear operator in T_pM.

In a chart $(\mathcal{U}, \mathbf{x})$, the Koszul connection $\nabla_{\partial/\partial x^\alpha}\frac{\partial}{\partial x^\beta}$ takes the form

$$\nabla_{\partial/\partial x^\alpha}\frac{\partial}{\partial x^\beta}(p) = \Gamma_{\alpha\beta}^\mu(p)\frac{\partial}{\partial x^\mu}$$

for some coefficients $\Gamma^{\mu}_{\alpha\beta}(p)$. Conversely, let $\Gamma^{\mu}_{\alpha\beta}(p)$ be smooth functions. Do they define a Koszul connection? It turns out that they do define such a connection if and only if

$$\Gamma^{\mu'}_{\alpha'\beta'}(p) = \Gamma^{\mu}_{\alpha\beta}(p)\frac{\partial x^{\alpha}}{\partial y^{\alpha'}}\frac{\partial x^{\beta}}{\partial y^{\beta}}\frac{\partial y^{\mu'}}{\partial x^{\mu}} + \frac{\partial^2 x^{\nu}}{\partial y^{\alpha'}\partial y^{\beta'}}\frac{\partial y^{\mu'}}{\partial x^{\nu}}, \qquad p \in \mathcal{U} \cap \mathcal{V}.$$

A system of functions $\Gamma^{\mu}_{\alpha\beta}(p)$ that satisfy to the above equations in all local charts is called a *classical connection*. As a simple exercise, the reader can prove that the Christoffel symbols (1.41) on a Riemannian manifold constitute a classical connection. This special classical connection is called the *Levi-Civita connection*.

If in a chart $(\mathcal{U}, \mathbf{x})$ we have $X = X^{\alpha}$ and $Y = Y^{\beta}$, then

$$\nabla_X Y^{\beta} = X^{\alpha}\nabla_{\alpha}Y^{\beta},$$

where

$$\nabla_{\alpha}Y^{\beta} = \frac{\partial Y^{\beta}}{\partial x^{\alpha}} + \Gamma^{\beta}_{\mu\alpha}Y^{\mu}$$

is called the *covariant derivative* of Y.

The tensor $T^{\mu}_{\alpha\beta} = \Gamma^{\mu}_{\alpha\beta} - \Gamma^{\mu}_{\beta\alpha}$ is called the *torsion tensor*. A classical connection is called *symmetric* if its torsion tensor vanishes.

Let $\gamma \colon [a, b] \to M$ be a smooth curve, and let X be a smooth vector field. The function $\frac{dX}{dt} = \nabla_{d\gamma/dt}X$ has the property $\frac{dX}{dt} \in T_{\gamma(t)}M$. It is called the *covariant derivative* of X along γ.

Is is possible to define the covariant derivative for a vector field X defined *only* along a curve γ? It turns out that, for such a vector field X, there exists a unique operation $X \mapsto \frac{dX}{dt}$ that satisfies the following three conditions: $\frac{d(X+Y)}{dt} = \frac{dX}{dt} + \frac{dY}{dt}$, $\frac{d(fX)}{dt} = \frac{df}{dt}X + f\frac{dX}{dt}$, and $\frac{dX}{dt} = \nabla_{d\gamma/dt}Y$ for an arbitrary vector field Y with $Y(\gamma(s)) = X(s)$ for all s in some neighbourhood of t. The above operation is given by

$$\frac{dX}{dt} = \left(\frac{dx^{\mu}}{dt} + \Gamma^{\mu}_{\alpha\beta}(\gamma(t))\frac{d\gamma^{\alpha}}{dt}X^{\beta}(t)\right)\frac{\partial}{\partial x^{\mu}}\bigg|_{\gamma(t)}.$$

If $\frac{dX}{dt} = 0$ along γ, we say that X is *parallel* along γ with respect to the classical connection $\Gamma^{\mu}_{\alpha\beta}$.

Given a smooth curve $\gamma \colon [a, b] \to M$ and a vector $X(a) \in T_{\gamma(a)}M$, there exists a unique vector field $X(t) = X^{\alpha}(t)$ which is parallel along γ. Indeed, the coefficients $X^{\alpha}(t)$ constitute the unique solution to the system of linear differential equations $\frac{dX^{\mu}}{dt} + \Gamma^{\mu}_{\alpha\beta}(\gamma(t))\frac{d\gamma^{\alpha}}{dt}X^{\beta}(t) = 0$ with the given initial condition $X(a)$.

The vector $X(t)$ is said to be obtained from $X(a)$ by *parallel translation* along γ with respect to the classical connection $\Gamma^{\mu}_{\alpha\beta}$. Varying the initial condition over the entire tangent space $T_{\gamma(a)}M$, we obtain an isomorphism $\tau_t \colon T_{\gamma(a)}M \to T_{\gamma(t)}M$, or a 'connection' between any two tangent spaces at the points of the curve γ. Conversely, given a parallel translation τ_t and a curve γ with $\gamma(0) = p$ and $\gamma'(0) = X_p$, we can restore the Koszul connection by defining

$$\nabla_{X_p} Y = \lim_{h \to 0} h^{-1}(\tau_h^{-1} Y_{\gamma(h)} - Y_p)$$

and $\nabla_X Y(p) = \nabla_{X_p} Y$. Similarly, for a tensor field Y of type $\binom{k}{l}$, we define a family of parallel translations $T_l^k(\tau_h): T_l^k(T_p M) \to T_l^k(T_p M)$, and the Koszul connection

$$\nabla_{X_p} Y = \lim_{h \to 0} ((T_l^k(\tau_h))^{-1} Y_{\gamma(h)} - Y_p)$$

and $\nabla_X Y(p) = \nabla_{X_p} Y$.

In coordinates, the covariant derivative of a tensor field Y of type $\binom{k}{l}$ is given by

$$
\begin{aligned}
\nabla_\mu Y_{\beta_1 \cdots \beta_l}^{\alpha_1 \cdots \alpha_k} &= \frac{\partial Y_{\beta_1 \cdots \beta_l}^{\alpha_1 \cdots \alpha_k}}{\partial x^\mu} + \sum_{i=1}^{k} Y_{\beta_1 \cdots \beta_l}^{\alpha_1 \cdots \alpha_{i-1} \nu \alpha_{i+1} \cdots \alpha_k} \Gamma_{\mu\nu}^{\alpha_i} \\
&\quad - \sum_{j=1}^{l} Y_{\beta_1 \cdots \beta_{j-1} \nu \beta_{j+1} \cdots \beta_l}^{\alpha_1 \cdots \alpha_k} \Gamma_{\mu\beta_j}^{\nu}.
\end{aligned}
\tag{1.43}
$$

If M is a Riemannian manifold, we say that a connection ∇ is compatible with the Riemannian structure if and only if the isomorphisms τ_t are isometries. The fundamental theorem of Riemannian geometry says that such a connection is unique.

Theorem 6. *A symmetric connection ∇ is compatible with the Riemannian structure if and only if ∇ is the Levi-Civita connection.*

The covariant derivatives do not commute. The difference of covariant derivatives of the second order may serve as a measure of difference in curvature in the points $p \in M$ and $\mathbf{0} \in \mathbb{R}^n$.

Indeed, let X_α be a covector field on M, that is, a smooth section of the cotangent bundle (T^*M, π^*, M). Then, in a chart $(\mathcal{U}, \mathbf{x})$ we have

$$
\begin{aligned}
\nabla_\gamma(\nabla_\beta X_\alpha) &= \frac{\partial \nabla_\beta X_\alpha}{\partial x^\gamma} + \Gamma_{\delta\gamma}^{\beta} \nabla_\gamma X_\alpha - \Gamma_{\alpha\gamma}^{\delta} \nabla_\beta X_\delta \\
&= \frac{\partial^2 X_\alpha}{\partial x^\gamma \partial x^\beta} - \frac{\partial \Gamma_{\alpha\beta}^{\varepsilon}}{\partial x^\gamma} X_\varepsilon - \Gamma_{\alpha\beta}^{\varepsilon} \frac{\partial X_\varepsilon}{\partial x^\gamma} - \Gamma_{\alpha\gamma}^{\delta} \frac{\partial X_\delta}{\partial x^\beta} + \Gamma_{\alpha\gamma}^{\delta} \Gamma_{\delta\beta}^{\varepsilon} X_\varepsilon \\
&\quad - \Gamma_{\beta\gamma}^{\delta} \frac{\partial X_\alpha}{\partial x^\delta} + \Gamma_{\beta\gamma}^{\delta} \Gamma_{\alpha\delta}^{\varepsilon} X_\varepsilon.
\end{aligned}
$$

Interchanging β and γ, we obtain

$$
\begin{aligned}
\nabla_\beta(\nabla_\gamma X_\alpha) &= \frac{\partial^2 X_\alpha}{\partial x^\beta \partial x^\gamma} - \frac{\partial \Gamma_{\alpha\gamma}^{\varepsilon}}{\partial x^\beta} X_\varepsilon - \Gamma_{\alpha\gamma}^{\varepsilon} \frac{\partial X_\varepsilon}{\partial x^\beta} - \Gamma_{\alpha\beta}^{\delta} \frac{\partial X_\delta}{\partial x^\gamma} + \Gamma_{\alpha\beta}^{\delta} \Gamma_{\delta\gamma}^{\varepsilon} X_\varepsilon \\
&\quad - \Gamma_{\gamma\beta}^{\delta} \frac{\partial X_\alpha}{\partial x^\delta} + \Gamma_{\gamma\beta}^{\delta} \Gamma_{\alpha\delta}^{\varepsilon} X_\varepsilon.
\end{aligned}
$$

After subtracting, the first terms of the right hand sides of both equations cancel each other because partial derivatives commute. The sixth and the seventh terms cancel each other because the Christoffel symbols are symmetric in their lower indices. The third (resp., the fourth) term of the first equation cancels with the fourth (resp., the third) term of the second one because repeated indices are dummy. Finally, we obtain

$$\nabla_\gamma(\nabla_\beta X_\alpha) - \nabla_\beta(\nabla_\gamma X_\alpha) = R^\varepsilon{}_{\alpha\beta\gamma} X_\varepsilon$$

where

$$R^\varepsilon{}_{\alpha\beta\gamma} = \Gamma^\delta_{\alpha\gamma}\Gamma^\varepsilon_{\delta\beta} - \Gamma^\delta_{\alpha\beta}\Gamma^\varepsilon_{\delta\gamma} + \frac{\partial \Gamma^\varepsilon_{\alpha\gamma}}{\partial x^\beta} - \frac{\partial \Gamma^\varepsilon_{\alpha\beta}}{\partial x^\gamma}. \tag{1.44}$$

The tensor $R^\varepsilon{}_{\alpha\beta\gamma}$ is called the *Riemann curvature tensor*.

Observe!

Different authors choose one of the two different *sign conventions* for the Riemann curvature tensor. We express both of them in the form of the following equation:

$$R^\varepsilon{}_{\alpha\beta\gamma} = S_2 \left(\Gamma^\delta_{\alpha\gamma}\Gamma^\varepsilon_{\delta\beta} - \Gamma^\delta_{\alpha\beta}\Gamma^\varepsilon_{\delta\gamma} + \frac{\partial \Gamma^\varepsilon_{\alpha\gamma}}{\partial x^\beta} - \frac{\partial \Gamma^\varepsilon_{\alpha\beta}}{\partial x^\gamma} \right),$$

where $S_2 \in \{-1, 1\}$.

Consider the tensor $R_{\varepsilon\alpha\beta\gamma} = g_{\alpha\varepsilon} R^\varepsilon{}_{\alpha\beta\gamma}$. It has the skew symmetry

$$R_{[\varepsilon\alpha]\beta\gamma} = R_{\varepsilon\alpha[\beta\gamma]} = 0 \tag{1.45}$$

and the *interchange symmetry*

$$R_{\varepsilon\alpha\beta\gamma} = R_{\beta\gamma\varepsilon\alpha}. \tag{1.46}$$

The reader may check that in coordinate-free form, the torsion tensor is a linear function of two smooth vector fields X and Y given by $T(X,Y) = \nabla_X Y - \nabla_Y X - [X, Y]$. In what follows we consider only symmetric connections with $T(X, Y) = 0$.

The curvature tensor in a coordinate-free form is a linear function of three smooth vector fields X, Y, and Z, given by

$$R(X, Y, Z) = \nabla_X(\nabla_Y Z) - \nabla_Y(\nabla_X Z) - \nabla_{[X,Y]} Z.$$

Theorem 7. *The first* Bianchi identity *or the* cyclic identity *is*

$$R_{\varepsilon\alpha\beta\gamma} + R_{\varepsilon\gamma\alpha\beta} + R_{\varepsilon\beta\gamma\alpha} = 0.$$

The second Bianchi identity *is*

$$\nabla_\delta R_{\varepsilon\alpha\beta\gamma} + \nabla_\gamma R_{\varepsilon\alpha\delta\beta} + \nabla_\beta R_{\varepsilon\alpha\gamma\delta} = 0.$$

The symmetries (1.45) and (1.46) together with the first Bianchi identity imply that the dimension of the linear space of the Riemann curvature tensors on a manifold of dimension n is less than n^4, the dimension of the space of all rank 4 tensors. In fact, the former space has dimension $\frac{1}{12}n^2(n^2-1)$. In particular, on a 4-dimensional manifold it has dimension 20.

Definition 73. The *Ricci tensor* is given by the contraction of the first and third indices of the Riemann curvature tensor: $R_{\alpha\gamma} = R^\varepsilon{}_{\alpha\varepsilon\gamma}$. The *Ricci scalar* is $R = g^{\alpha\gamma}R_{\alpha\gamma}$.

Observe!

Some authors define a Ricci tensor by contraction on a different pair of indices. This amounts to a sign choice in the definition. We express various choices in the form of the following equation:

$$R_{\alpha\gamma} = S_2 S_3 R^\varepsilon{}_{\alpha\varepsilon\gamma},$$

where $S_3 \in \{-1, 1\}$.

Equation (1.44) gives

$$R_{\alpha\gamma} = R^\varepsilon{}_{\alpha\varepsilon\gamma} = \Gamma^\delta_{\alpha\gamma}\Gamma^\varepsilon_{\delta\varepsilon} - \Gamma^\delta_{\alpha\varepsilon}\Gamma^\varepsilon_{\delta\gamma} + \frac{\partial\Gamma^\varepsilon_{\alpha\gamma}}{\partial x^\varepsilon} - \frac{\partial\Gamma^\varepsilon_{\alpha\varepsilon}}{\partial x^\gamma}. \tag{1.47}$$

1.7.7 Isometries

Let M_1 and M_2 be two manifolds, let $\psi\colon M_1 \to M_2$ be a smooth map, and let T be a tensor field on M_2.

The tensor field T can be 'pulled back' by the map φ.

Definition 74. The *pull back* of T by ψ is the map $\varphi^*T\colon M_1 \to \mathbb{R}^1$ given by $(\varphi^*T)(p) = T(\psi(p))$, $p \in M_1$.

A vector field X on M can be 'pushed forward' by φ.

Definition 75. The *push forward* of a vector field X is the vector field ψ_*X on M_2 given by $(\psi_*X)(f) = X(\psi^*f)$.

Example 46. Let $M_1 = S^2$, $M_2 = \mathbb{R}^3$, let $(\mathcal{U}, x^\mu)$ be the spherical coordinates on S^2, and let $(\mathbb{R}^3, y^\nu)$ be the Cartesian coordinates. The map $\mathcal{U} \to \mathbb{R}^3$, $(\varphi, \theta) \to (\sin\theta\cos\varphi, \sin\theta\sin\varphi, \cos\theta)^\top$ may be extended by continuity to a smooth map $\psi\colon S^2 \to \mathbb{R}^3$. We pull the Riemannian metric $g = \mathrm{d}x^2 + \mathrm{d}y^2 + \mathrm{d}z^2$ back to S^2. In the introduced charts, we have

$$(\psi^*g)_{\mu\nu} = \frac{\partial y^\alpha}{\partial x^\mu}\frac{\partial y^\beta}{\partial x^\nu}g_{\alpha\beta},$$

The matrix $\frac{\partial y^\alpha}{\partial x^\mu}$ has the form $\begin{pmatrix} \cos\theta\cos\varphi & \cos\theta\sin\varphi & -\sin\theta \\ -\sin\theta\sin\varphi & \sin\theta\cos\varphi & 0 \end{pmatrix}$. Multiplying the matrices, we obtain $(\psi^* g)_{\mu\nu} = \begin{pmatrix} 1 & 0 \\ 0 & \sin^2\theta \end{pmatrix}$.

Example 46 gives rise to a definition. Let (M_1, g_1) and (M_2, g_2) be two Riemannian spaces.

Definition 76. A smooth one-to-one map $\psi\colon M_1 \to M_2$ with smooth inverse is called an *isometry* if $\psi^* g_2 = g_1$.

We see that the map ψ of Example 46 is an isometry between S^2 and $\psi(S^2) \subset \mathbb{R}^3$. The set $\psi(S^2)$ is an example of a *submanifold*: an image of an isometric embedding.

Remark 6. Riemannian geometry makes sense in the case when the bilinear form $g_{\alpha\beta}$ is still non-degenerate but not necessarily positive-definite. Such a metric is called *pseudo-Riemannian*. All equations shown above still hold true.

1.8 Group Representations

A group representation is just a special kind of group action. Let G be a Lie group, and let L be a right linear space over $\mathbb{K}$. In this Section, we suppose that L is finite-dimensional.

1.8.1 Basic Definitions

Definition 77. A left action $G \times L \to L$, $(g, \mathbf{x}) \to g \cdot \mathbf{x}$ is called a *representation* if $g \cdot \mathbf{x}$ is a $\mathbb{K}$-linear function of $\mathbf{x}$ and a continuous function of g and $\mathbf{x}$. The space L is called a *G-space* or a representation of G if the action is thought. The representation is called *real* (resp., *complex*, resp., *quaternionic*) if $\mathbb{K} = \mathbb{R}$ (resp., $\mathbb{K} = \mathbb{C}$, resp., $\mathbb{K} = \mathbb{H}$).

Alternatively, there is a continuous homomorphism that maps an element $g \in G$ to a linear operator $\theta_g \in \mathrm{Hom}_{\mathbb{K}}(L)$.

Example 47. Let G be a closed subgroup of the Lie group $\mathrm{GL}(n, \mathbb{K})$. The matrix-vector multiplication determines a representation of G in the G-space $L = \mathbb{K}^n$.

Example 48. The map (1.27) is a representation of the group $\mathrm{Pin}(L, B))$. Its restriction to the subgroup $\mathrm{Spin}(L, B))$ (resp., $\mathrm{Spin}^0(L, B))$ is a representation of the above subgroups called a *spinor representation*.

Let L_1 and L_2 be two G-spaces over the same (skew) field $\mathbb{K}$. Introduce the following notation:

$$\mathrm{Hom}_{\mathbb{K}G}(L_1, L_2) = \{\, f \in \mathrm{Hom}_{\mathbb{K}}(L_1, L_2)\colon f(g \cdot \mathbf{x}) = g \cdot (f(\mathbf{x})) \,\}$$

for all $\mathbf{x} \in L_1$ and $g \in G$, where $\mathrm{Hom}(L_1, L_2)$ is the set of all linear operators from L_1 to L_2. The above set is a linear space over $\mathbb{K}'$. Its elements are called the *intertwining operators* between the G-spaces L_1 and L_2.

Definition 78. The representations L_1 and L_2 are called *equivalent* if the linear space $\mathrm{Hom}_{\mathbb{K}G}(L_1, L_2)$ contains an invertible operator.

It is possible to regard a real or quaternionic representation as a complex representation with an additional structure. It turns out that real and quaternionic structures do the job if we assume in advance that they commute with the representation.

Definition 79. A real (resp., quaternionic) structure j on a complex representation V is called a *G-structure* if it commutes with the representation, that is, $j(g \cdot \boldsymbol{\zeta}) = g \cdot (j(\boldsymbol{\zeta}))$ for all $\boldsymbol{\zeta} \in V$ and $g \in G$.

Remark 7. The maps s_+ that maps a pair (V, j) to the $(+1)$ eigenspace of j, and c that maps a real finite-dimensional linear space U to the complex linear space $U \otimes_{\mathbb{R}} \mathbb{C}$, given in Example 98, give the equivalence of real representations with complex representations equipped by a real structure. Similarly, the maps s_- that maps a pair (V, j) to the right quaternionic linear space V^-, where $V^- = V$ as sets, and the product $\boldsymbol{\zeta}q$ of $\boldsymbol{\zeta} \in V$ and $q = \alpha + \beta\mathrm{i} + \gamma\mathrm{j} + \delta\mathrm{k}$ is defined by

$$\boldsymbol{\zeta}q = \alpha\boldsymbol{\zeta} + \beta\boldsymbol{\zeta}\mathrm{i} + \gamma j(\boldsymbol{\zeta}) + \delta j(\boldsymbol{\zeta}\mathrm{i}), \tag{1.48}$$

and c' that maps a quaternionic finite-dimensional linear space W to the complex linear space $c'W$, where $c'W = W$ as sets, but the skew field of scalars is restricted from $\mathbb{H}$ to $\mathbb{C}$, given in Example 99, give the equivalence of quaternionic representations with complex representations equipped by a quaternionic structure.

1.8.2 Operations over Representations

All operations over linear spaces described in Subsection 1.2.2, can be extended to representations. All we need is to specify the group action.

For the direct sum $L_1 \oplus L_2$ of two representation over the same (skew) field $\mathbb{K}$, the action is defined by $g \cdot (\mathbf{x}, \mathbf{y}) = (g \cdot \mathbf{x}, g \cdot \mathbf{y})$. It is also possible to consider two complex representations V_1 and V_2 together with two G-structures j_{V_1} and j_{V_2} satisfying $j_{V_1}^2 = j_{V_2}^2$ and to give a structure $j_{V_1} \oplus j_{V_2}$ to the representation $V_1 \oplus V_2$.

The action of the group G on the linear space $\mathrm{Hom}_{\mathbb{K}}(L_1, L_2)$ is given by

$$(g \cdot f)\mathbf{x} = g \cdot (f(g^{-1} \cdot \mathbf{x})), \qquad f \in \mathrm{Hom}_{\mathbb{K}}(L_1, L_2).$$

The restriction of this action to the subspace $\mathrm{Hom}_{\mathbb{K}G}(L_1, L_2) \subset \mathrm{Hom}_{\mathbb{K}}(L_1, L_2)$ acts trivially. If we consider two complex representations V_1 and V_2 together with two G-structures j_{V_1} with $j_{V_1}^2 = \varepsilon_1 \in \{-1, 1\}$ and j_{V_2} with $j_{V_2}^2 = \varepsilon_2$, then the map $j\colon \mathrm{Hom}_{\mathbb{K}}(L_1, L_2) \to \mathrm{Hom}_{\mathbb{K}}(L_1, L_2)$ acting by $jf = j_{V_2} f j_{V_1}^{-1}$ is a G-structure with $j^2 = \varepsilon_1\varepsilon_2$. Three different cases are possible there.

- $\varepsilon_1 = \varepsilon_2 = 1$. If U_1 (resp., U_2) denotes the $+1$-eigenspace of j_{V_1} (resp., j_{V_2}), then $cU_1 = V_1$, $cU_2 = V_2$, and the $+1$-eigenspace of j is the space $\operatorname{Hom}_{\mathbb{R}}(U_1, U_2)$, and finally we obtain $\operatorname{Hom}_{\mathbb{C}}(cU_1, cU_2) = c \operatorname{Hom}_{\mathbb{R}}(U_1, U_2)$.

- $\varepsilon_1 \neq \varepsilon_2$. Consider one of two similar cases: $\varepsilon_1 = 1$, $\varepsilon_2 = -1$. Let U be the $+1$-eigenspace of j_{V_1}, then $cU = V_1$. Let $W = V_2$ as sets, and the right action of $\mathbb{H}$ is defined by Equation (1.48). The i-eigenspace of j is the space $\operatorname{Hom}_{\mathbb{R}}(U, W)$, and finally we obtain $\operatorname{Hom}_{\mathbb{C}}(cU, c'W) = c \operatorname{Hom}_{\mathbb{R}}(U, W)$.

- $\varepsilon_1 = \varepsilon_2 = -1$. The $+1$-eigenspace of j is the space $\operatorname{Hom}_{\mathbb{H}}(W_1, W_2)$, where $W_1 = V_1$ (resp., $W_2 = V_2$) as sets, and the right action of $\mathbb{H}$ is defined by Equation (1.48). Finally, $\operatorname{Hom}_{\mathbb{C}}(c'W_1, c'W_2) = c \operatorname{Hom}_{\mathbb{H}}(W_1, W_2)$.

In particular, the representation $V^* = \operatorname{Hom}_{\mathbb{C}}(V, \mathbb{C}^1)$, where G acts on $\mathbb{C}^1$ trivially, $g \cdot \zeta = \zeta$, is called the representation *dual* to V.

Observe that the $\mathbb{C}$-linear map $\alpha \colon V_1^* \otimes V_2 \to \operatorname{Hom}_{\mathbb{C}}(V_1, V_2)$ defined by

$$(\alpha(\mathbf{x}_1^* \otimes \mathbf{x}_2))\mathbf{y}_1 = \mathbf{x}_1^*(\mathbf{y}_1)\mathbf{x}_2. \tag{1.49}$$

is an invertible intertwining operator. Using the above map, we may transfer the properties of a representation $\operatorname{Hom}_{\mathbb{K}}(L_1, L_2)$ to the corresponding properties of the representation $L_1^* \otimes_{\mathbb{K}} L_2$. In particular:

- the tensor product of two real representations is real;

- the tensor product of a real and quaternionic representation is quaternionic;

- the tensor product of two quaternionic representations is real.

Finally, the Adams operations of Subsection 1.2.2 act on representations as well. All we need is to specify the group action. The group G acts in cU by $g \cdot (\zeta \otimes \mathbf{x}) = \zeta \otimes (g \cdot \mathbf{x})$, in qV by $g \cdot (q \otimes \zeta) = q \otimes (g \cdot \zeta)$, in $c'W$ in the same way as in W, in rV in the same way as in V, and in tV in the same way as in V. We have

$$c(U_1 \otimes U_2) = cU_1 \otimes cU_2,$$

the skew field $\mathbb{H}$ acts on the tensor product $U \otimes_{\mathbb{R}} W$ by $(\mathbf{x} \otimes \mathbf{q})q = \mathbf{x} \otimes (\mathbf{q}q)$, and the -1-eigenspace of the G-map $j_{W_1} \otimes j_{W_2}$ on $V_1 \otimes_{\mathbb{C}} V_2$ is equal to $W_1 \otimes_{\mathbb{H}} W_2$.

Moreover, we have

$$\operatorname{Hom}_{\mathbb{C}}(cU_1, cU_2) = c \operatorname{Hom}_{\mathbb{R}}(U_1, U_2),$$
$$\operatorname{Hom}_{\mathbb{C}}(c'W_1, c'W_2) = c \operatorname{Hom}_{\mathbb{H}}(W_1, W_2).$$

Remark 8. Example 95 shows how to extend the maps $U \to cU$ and $V \to qV$ to covariant functors. The maps $W \to c'W$, $V \to rV$, and $V \to tV$ can be extended similarly.

Example 97 shows that the functors $V \to V \oplus tV$ and $V \to crV$ are naturally isomorphic. We use short notation $cr = 1 + t$. Similarly, $rc = 2$, $qc' = 2$, $c'q = 1 + t$, $tc = c$, $rt = r$, $tc' = c'$, $qt = q$, and $t^2 = 1$.

1.8.3 Integration and Characters

In this subsection, we suppose that G is a *compact* topological group. There exists a unique probabilistic measure $\mathrm{d}g$ on the Borel σ-field $\mathfrak{G}$ of the group G, called the *Haar measure* such that

$$\int_G f(hg)\,\mathrm{d}g = \int_G f(gh)\,\mathrm{d}g = \int_G f(g)\,\mathrm{d}g$$

for all $h \in G$.

It turns out that there is a positive-definite G-invariant Hermitian bilinear form $(\cdot, \cdot)$ on an arbitrary G-space V of dimension n. Moreover, if V is equipped by a G-structure j, one may achieve the equality $(j\mathbf{x}, j\mathbf{y}) = (\mathbf{y}, \mathbf{x})$. In this case, the representation automorphism θ maps G to $\mathrm{O}(n)$ (resp., to $\mathrm{U}(n)$, resp., to $\mathrm{Sp}(n)$), and the representation is called *orthogonal* (resp., *unitary*, resp., *symplectic*).

Definition 80. A nonzero G-space L is called *reducible* if some proper nonzero subspace of V is again a G-space. Otherwise, if such a space does not exist, then the representation V is called *irreducible*.

Every representation of a compact group G is equivalent to the direct sum of irreducible representations. Let L_0 be an irreducible representation of G, and let L be its finite-dimensional representation.

Definition 81. The L_0-*isotypical component* of L is the subspace of L generated by all subspaces where a representation equivalent to L_0 acts.

It turns out that a finite-dimensional representation of G uniquely decomposes into a direct sum of isotypical components.

Lemma 1 (Schur). *If the representations L_1 and L_2 of a group G over a (skew) field $\mathbb{K}$ are irreducible and equivalent (resp., non-equivalent), then the $\mathbb{K}'$-linear space $\mathrm{Hom}_{\mathbb{K}G}(L_1, L_2)$ is a division algebra over $\mathbb{K}'$ with respect to the composition of linear operators (resp., is equal to $\{\mathbf{0}\}$).*

Let $\varepsilon \colon V^* \otimes V \to \mathbb{C}$ be the *evaluation map* defined by $\varepsilon(\mathbf{x}^* \otimes \mathbf{y}) = \mathbf{x}^*(\mathbf{y})$. Let α be the isomorphism (1.49).

Definition 82. For $f \in \mathrm{Hom}_{\mathbb{C}}(V)$, the number $\mathrm{tr}\, f = \varepsilon\alpha^{-1}f$ is called the *trace* of f.

The reader may check that this definition is equivalent to the usual definition of the trace of a matrix.

Definition 83. The *character* of a complex representation V of a group G is given by $\chi_V(g) = \mathrm{tr}\,\theta g$, $g \in G$. The character of a real representation U (resp., quaternionic representation W) is equal to that of cU (resp., $c'W$).

Theorem 8. *We have*

$$\chi_L(hgh^{-1}) = \chi_L(g), \qquad \chi_{L_1 \oplus L_2}(g) = \chi_{L_1}(g) + \chi_{L_2}(g),$$
$$\chi_{L_1 \otimes L_2}(g) = \chi_{L_1}(g)\chi_{L_2}(g), \qquad \chi_{V^*}(g) = \chi_V(g^{-1}),$$
$$\chi_{tV}(g) = \overline{\chi_V(g)}, \qquad \chi_V(e) = \dim_{\mathbb{C}} V.$$

Moreover, the function $\chi_L \colon G \to \mathbb{C}$ is continuous, the character of a complex representation in a space V with G-structure is real-valued. If, in addition, the group G is compact, then $\chi_V(g^{-1}) = \chi_{V^*}(g) = \chi_{tV}(g) = \overline{\chi_V(g)}$.

1.8.4 The Adams Notation

Let V be a complex irreducible representation of a compact group G.

Definition 84. The representation V is called the representation of *complex type* if V and tV are not equivalent.

The representation V is called the representation of *real type* if there exists a real G-structure j on V that commutes with the representation.

The representation V is called the representation of *quaternionic type* if there exists a quaternionic G-structure j on V that commutes with the representation.

For a complex irreducible representation V of real type, there exists a real irreducible representation U such that $V = cU$. The representation U is called a representation of *real type*. The quaternionic representation $qcU = qV$ is irreducible, it is also called a representation of *real type*.

For a complex irreducible representation V of complex type, the representations rV and rtV are real, irreducible, and equivalent. They are called representation of *complex type*. The representations qV and qtV are quaternionic, irreducible, and equivalent. They are also called the representations of *complex type*.

For a complex irreducible representation V of quaternionic type, there exists a quaternionic irreducible representation W such that $V = c'W$. The representation W is called a representation of *quaternionic type*. The real representation $rc'W = rV$ is irreducible, it is also called a representation of *quaternionic type*.

Different characterisations of irreducible representations of various types are collected in the following results.

Theorem 9. *The irreducible real representations of a compact group G fall into three non-intersecting classes as follows. A representation U is of real type if and only if any of the following equivalent conditions are satisfied.*

- *The representation cU is irreducible.*

- *The division algebra $\mathrm{Hom}_{\mathbb{R}G}(U)$ is isomorphic to $\mathbb{R}$.*

A representation U is of complex type if and only if any of the following equivalent conditions are satisfied.

- *The representation cU is equivalent to the direct sum of two non-equivalent irreducible representations.*

- *The division algebra* $\mathrm{Hom}_{\mathbb{R}G}(U)$ *is isomorphic to* $\mathbb{C}$.

A representation U is of quaternionic type if and only if any of the following equivalent conditions are satisfied.

- *The representation cU is equivalent to the direct sum of two equivalent irreducible representations.*

- *The division algebra* $\mathrm{Hom}_{\mathbb{R}G}(U)$ *is isomorphic to* $\mathbb{H}$.

Theorem 10. *The irreducible complex representations of a compact group G fall into three non-intersecting classes as follows. A representation V is of real type if and only if any of the following equivalent conditions are satisfied.*

- *The representation rV is equivalent to the direct sum of two equivalent irreducible representations, while the representation qV is irreducible.*

- $\int_G \chi_V(g^2)\,\mathrm{d}g = 1$.

- *There exists a non-degenerate symmetric G-invariant bilinear form on V.*

A representation V is of complex type if and only if any of the following equivalent conditions are satisfied.

- *The representations rV and qV are irreducible.*

- $\int_G \chi_V(g^2)\,\mathrm{d}g = 0$.

- *There exists neither a non-degenerate symmetric nor a non-degenerate skew-symmetric G-invariant bilinear forms on V.*

A representation V is of quaternionic type if and only if any of the following equivalent conditions are satisfied.

- *The representation rV is irreducible, while the representation qV is equivalent to the direct sum of two equivalent irreducible representations.*

- $\int_G \chi_V(g^2)\,\mathrm{d}g = -1$.

- *There exists a non-degenerate skew-symmetric G-invariant bilinear form on V.*

The integral $\int_G \chi_V(g^2)\,\mathrm{d}g$ is called the *Frobenius–Schur indicator.*

Theorem 11. *The irreducible quaternionic representations of a compact group G fall into three non-intersecting classes as follows. A representation W is of real type if and only if any of the following equivalent conditions are satisfied.*

TABLE 1.5

The Adams Notation

Representation of	Representation over		
	$\mathbb{R}$	$\mathbb{C}$	$\mathbb{H}$
real type	U	cU	qcU
complex type	rV	V	qV
quaternionic type	$rc'W$	$c'W$	W

- *The representation $c'W$ is equivalent to the direct sum of two equivalent irreducible representations.*

- *The division algebra $\mathrm{Hom}_{\mathbb{R}G}(W)$ is isomorphic to $\mathbb{H}$.*

A representation W is of complex type if and only if any of the following equivalent conditions are satisfied.

- *The representation $c'W$ is equivalent to the direct sum of two non-equivalent irreducible representations.*

- *The division algebra $\mathrm{Hom}_{\mathbb{R}G}(W)$ is isomorphic to $\mathbb{C}$.*

A representation W is of quaternionic type if and only if any of the following equivalent conditions are satisfied.

- *The representation $c'W$ is irreducible.*

- *The division algebra $\mathrm{Hom}_{\mathbb{R}G}(W)$ is isomorphic to $\mathbb{R}$.*

In what follows, we denote the irreducible representation according to Table 1.5 and call this the *Adams notation* after Adams (1969).

Specifically, the capital letter in this notation determines the type of the representation (real for U, complex for V, quaternionic for W). The first small letter determines the skew field, over which the representation acts (no letter for the same name for the field and the type, r for representation over $\mathbb{R}$, c or c' for representation over $\mathbb{C}$, q for representation over $\mathbb{H}$).

1.8.5 Classical Spinors

Irrespectively of a group G there are its *trivial irreducible representations* U_0, cU_0, and qcU_0 with $U_0 = \mathbb{R}^1$ and $g \cdot x = x$, $x \in \mathbb{R}^1$.

Observe that if H is a normal subgroup of a group G, then there is a one-to-one correspondence between the class of irreducible representations of G that become trivial upon restriction to H and the class of irreducible representations of the quotient group G/H. In particular, exactly one half of the collection of irreducible representations of the group $\mathrm{Spin}^0(n, \mathbb{C})$ (resp., $\mathrm{Spin}^0(p, q)$) descends to the covered group $\mathrm{SO}(n, \mathbb{C})$ (resp., $\mathrm{SO}_0(p, q)$).

It turns out that the restriction of any irreducible representation of the Clifford algebra Cl_n^+ (resp., $\mathrm{Cl}_{p,q}^+$) to the group $\mathrm{Spin}(n, \mathbb{C})$ (resp., $\mathrm{Spin}^0(p, q)$) is an irreducible representation of this group.

TABLE 1.6

Classical Spinors of Small Dimension

G	S	Name
$\mathrm{Spin}^0(1,\mathbb{C}) = \mathrm{O}(1,\mathbb{C})$	$S = \mathbb{C}^1$	Dirac
$\mathrm{Spin}(2,\mathbb{C}) = \mathrm{GL}(1,\mathbb{C})$	$S_+ = \mathbb{C}^1$	left-handed Weyl
$\mathrm{Spin}(2,\mathbb{C}) = \mathrm{GL}(1,\mathbb{C})$	$S_- = \mathbb{C}^1$	right-handed Weyl
$\mathrm{Spin}(3,\mathbb{C}) = \mathrm{SL}(2,\mathbb{C})$	$S = \mathbb{C}^2$	Dirac
$\mathrm{Spin}(4,\mathbb{C}) = \mathrm{SL}(2,\mathbb{C}) \times \mathrm{SL}(2,\mathbb{C})$	$S_+ = \mathbb{C}^2$	left-handed Weyl
$\mathrm{Spin}(4,\mathbb{C}) = \mathrm{SL}(2,\mathbb{C}) \times \mathrm{SL}(2,\mathbb{C})$	$S_- = \mathbb{C}^2$	right-handed Weyl
$\mathrm{Spin}(1) = \mathrm{O}(1)$	$S = \mathbb{R}^1$	Majorana
$\mathrm{Spin}(2) = \mathrm{U}(1)$	$S = \mathbb{R}^2$	Majorana
$\mathrm{Spin}(2) = \mathrm{U}(1)$	$S = \mathbb{C}^1$	left-handed Weyl
$\mathrm{Spin}(2) = \mathrm{U}(1)$	$tS = t\mathbb{C}^1$	right-handed Weyl
$\mathrm{Spin}(3) = \mathrm{Sp}(1)$	$S = \mathbb{H}^1$	symplectic Majorana
$\mathrm{Spin}(4) = \mathrm{Sp}(1) \times \mathrm{Sp}(1)$	$S_+ = \mathbb{H}^1$	left-handed symplectic Majorana–Weyl
$\mathrm{Spin}(4) = \mathrm{Sp}(1) \times \mathrm{Sp}(1)$	$S_- = \mathbb{H}^1$	right-handed symplectic Majorana–Weyl
$\mathrm{Spin}^0(1,1) = \mathrm{GL}(1,\mathbb{R})$	$S_+ = \mathbb{R}^1$	left-handed Majorana–Weyl
$\mathrm{Spin}^0(1,1) = \mathrm{GL}(1,\mathbb{R})$	$S_- = \mathbb{R}^1$	right-handed Majorana–Weyl
$\mathrm{Spin}^0(1,2) = \mathrm{SL}(2,\mathbb{R})$	$S = \mathbb{R}^2$	Majorana
$\mathrm{Spin}^0(1,3) = \mathrm{SL}(2,\mathbb{C})$	$S = \mathbb{R}^4$	Majorana
$\mathrm{Spin}^0(1,3) = \mathrm{SL}(2,\mathbb{C})$	$S = \mathbb{C}^2$	left-handed Weyl
$\mathrm{Spin}^0(1,3) = \mathrm{SL}(2,\mathbb{C})$	$tS = t\mathbb{C}^2$	right-handed Weyl
$\mathrm{Spin}^0(2,2) = \mathrm{SL}(2,\mathbb{R}) \times \mathrm{SL}(2,\mathbb{R})$	$S_+ = \mathbb{R}^2$	left-handed Majorana–Weyl
$\mathrm{Spin}^0(2,2) = \mathrm{SL}(2,\mathbb{R}) \times \mathrm{SL}(2,\mathbb{R})$	$S_- = \mathbb{R}^2$	right-handed Majorana–Weyl

Definition 85. The elements of the linear space S, where the above group representation acts, is called a *classical spinor*.

Table 1.4 produces Table 1.6 of classical spinors.

Example 49. According to Table 1.4, the complex Clifford algebra $\mathrm{Cl}_1^+ = \mathbb{C}$ has the irreducible representation $\zeta \to \zeta$ in the complex linear space $S = \mathbb{C}^1$ of Dirac spinors. The restriction of this representation to the group $\mathrm{Spin}^0(1,\mathbb{C}) = \mathrm{O}(1,\mathbb{C})$ is the Dirac spinor representation $g \mapsto g$ of the above group. This complex representation is of real type, that is, $S = cU$ with $U = \mathbb{R}^1$. The corresponding quaternionic representation is $qS = \mathbb{H}^1$. The remaining irreducible representation of the group $\mathrm{Spin}^0(1,\mathbb{C})$ is the covering (1.29) in the space $U_0 = \mathbb{R}^1$. This representation as well as its companions, cU_0 and qcU_0, is trivial.

Among the above described two representations, only U_0 and its companions can be descended to the group $\mathrm{SO}(1,\mathbb{C})$.

According to Table 1.4, the complex Clifford algebra $\mathrm{Cl}_3^+ = \mathbb{C}(2)$ has the irreducible representation in the complex linear space $S = \mathbb{C}^2$ of Dirac spinors. The restriction of this representation to the group $\mathrm{Spin}(3,\mathbb{C}) = \mathrm{SL}(2,\mathbb{C})$ has the form of the Dirac spinor representation $g \mapsto g$.

The same representation can also be considered as the left-handed Weyl representation of the group $\mathrm{Spin}^0(1,3)$. The complex conjugate right-handed Weyl representation acts in the linear space tS. Denote $V_{1/2,0} = S$, $V_{0,1/2} = tS$.

If the irreducible representation $V_{\ell,0}$ is already constructed, then the tensor product $V_{\ell,0} \otimes V_{1/2,0}$ is equivalent to the direct sum of two irreducible components $V_{\ell-1/2,0} \oplus V_{\ell+1/2,0}$. We have $V_{0,\ell} = tV_{\ell,0}$. The complex irreducible representations of complex type are $V_{\ell_1,\ell_2} = V_{\ell_1,0} \otimes tV_{\ell_2,0}$ for $\ell_1 \neq \ell_2$, of real type $cU_{\ell,\ell} = V_{\ell,0} \otimes tV_{\ell,0}$. The real irreducible representations are $U_{\ell,\ell}$ and rV_{ℓ_1,ℓ_2} with $\ell_1 < \ell_2$, the quaternionic are $qcU_{\ell,\ell}$ and qV_{ℓ_1,ℓ_2} with $\ell_1 < \ell_2$. The representation $U_{1/2,1/2}$ is the Majorana representation of $\mathrm{Spin}^0(1,3)$.

The covering (1.30) is the representation $V_{1,0}$. Only the representations with integer $\ell_1 + \ell_2$ have value 1 at the element $-I \in \mathrm{SL}(2,\mathbb{C})$ and can be descended to an irreducible representation of the group $\mathrm{SO}(3,\mathbb{C})$.

Example 50. According to Table 1.4, the Clifford algebra $\mathrm{Cl}_2^+ = \mathbb{C} \oplus \mathbb{C}$ has two irreducible representations $(\alpha, \beta) \mapsto \alpha$ and $(\alpha, \beta) \mapsto \beta$ in the linear spaces $S_{1,1} = S_{-1,-1} = \mathbb{C}^1$ of left-handed and right-handed Weyl spinors. The restrictions of these representations to the group $\mathrm{Spin}(2,\mathbb{C}) = \mathrm{GL}(1,\mathbb{C})$ have the form $\left(\begin{smallmatrix} \alpha & 0 \\ 0 & \alpha^{-1} \end{smallmatrix}\right) \mapsto \alpha$ and $\left(\begin{smallmatrix} \alpha & 0 \\ 0 & \alpha^{-1} \end{smallmatrix}\right) \mapsto \alpha^{-1}$. These representations are complex of complex type. The first (resp., the second) one is the positive (resp., negative), or left-handed (resp., right-handed) Weyl spinor representation.

The irreducible representations of the form $\left(\begin{smallmatrix} re^{i\varphi} & 0 \\ 0 & r^{-1}e^{-i\varphi} \end{smallmatrix}\right) \mapsto r^\lambda e^{im\varphi} \in S_{\lambda,m}$, $\lambda \in \mathbb{C}$, $m \in \mathbb{Z}$ are real of real type if $\lambda \in \mathbb{R}$ and $m = 0$, and complex of complex type otherwise. In the former case, we denote them by $U_{\lambda,0}$ and in the latter by $V_{\lambda,m}$.

If $\lambda \in \mathbb{R}$ and $m = 0$, then the remaining complex (resp., quaternionic) irreducible representations are $cU_{\lambda,0}$ (resp., $qcU_{\lambda,0}$). Otherwise, when $\lambda \in \mathbb{C} \setminus \mathbb{R}$ or $m \neq 0$, then the remaining real (resp., quaternionic) irreducible representations are $rV_{\lambda,m}$ (resp., $qV_\lambda^\pm$) with $\mathrm{Im}\,\lambda > 0$ and $m \neq 0$.

The two-fold covering $\pi\colon \mathrm{Spin}(2,\mathbb{C}) \to \mathrm{SO}(2,\mathbb{C})$ has the form (1.31). Only the representations with even m can be descended to $\mathrm{SO}(2,\mathbb{C})$.

According to Table 1.4, the Clifford algebra $\mathrm{Cl}_4^+ = \mathbb{C}(2) \oplus \mathbb{C}(2)$ has two irreducible representations $(A, B) \mapsto A$ and $(A, B) \mapsto B$ in the linear spaces $S_{1/2,0,0,0} = S_{0,0,1/2,0} = \mathbb{C}^2$ of left-handed and right-handed Weyl spinors. The restrictions of these representations to the group $\mathrm{Spin}(4,\mathbb{C}) = \mathrm{SL}(2,\mathbb{C}) \times \mathrm{SL}(2,\mathbb{C})$ have the form $\left(\begin{smallmatrix} A & 0 \\ 0 & B \end{smallmatrix}\right) \mapsto A$ and $\left(\begin{smallmatrix} A & 0 \\ 0 & B \end{smallmatrix}\right) \mapsto B$. These representations are complex of complex type. The first (resp., the second) one is the positive (resp., negative), or left-handed (resp., right-handed) Weyl spinor representation.

The remaining irreducible representations are tensor products $V_{\ell_1,\ell_2} \otimes V_{\ell_3,\ell_4}$, where the terms of the tensor product are irreducible representations of the group $SL(2,\mathbb{C})$. If $\ell_1 \neq \ell_2$ or $\ell_3 \neq \ell_4$, then the representation $V_{\ell_1,\ell_2} \otimes V_{\ell_3,\ell_4}$ is of complex type. The representations $V_{\ell_1,\ell_1} \otimes V_{\ell_2,\ell_2} = cU_{\ell_1,\ell_1} \otimes U_{\ell_2,\ell_2}$ are of real type. The real irreducible representations are $U_{\ell_1,\ell_1} \otimes U_{\ell_2,\ell_2}$ and $rV_{\ell_1,\ell_2} \otimes V_{\ell_3,\ell_4}$, the quaternionic are $qcU_{\ell_1,\ell_1} \otimes U_{\ell_2,\ell_2}$ and $qV_{\ell_1,\ell_2} \otimes V_{\ell_3,\ell_4}$.

The two-fold covering (1.32) is the representation $V_{1,0} \otimes V_{1,0}$. Only the representations with integer value of the sum $\ell_1 + \ell_2 + \ell_3 + \ell_4$ are the representations of the group $SO(4,\mathbb{C})$.

Example 51. According to Table 1.4, the Clifford algebra $Cl_{0,1}^+ = Cl_{1,0}^+ = \mathbb{R}$ has the unique obvious irreducible representation in the real linear space $S = \mathbb{R}^1$ of Majorana spinors. Its restriction to the group $Spin(1) = O(1)$ is the non-trivial irreducible Majorana spinor representation. The remaining nontrivial irreducible representations of $O(1)$ are cS and qcS. The two-fold covering (1.33) is the real trivial representation of $Spin(1)$.

The quotient group $SO(1) = O(1)/O(1)$ has only trivial irreducible representations.

We have $Cl_{1,2}^+ = Cl_{2,1}^+ = \mathbb{R}(2)$. This Clifford algebra has the unique obvious irreducible representation in the linear space $S = \mathbb{R}^2$ of Majorana spinors. The restriction of this representation to the group $Spin^+(1,2) = SL(2,\mathbb{R})$ has the form $A \mapsto A$, a real Majorana representation of real type. Denote it by $U_{1/2}$. The remaining irreducible real representations are constructed by induction: if U_ℓ is already constructed, then the tensor product $U_\ell \otimes U_{1/2}$ is equivalent to the direct sum of two irreducible components $U_{\ell-1/2} \oplus U_{\ell+1/2}$. The complex irreducible representations are cU_ℓ, and the quaternionic are qcU_ℓ.

Only the representations with integer ℓ have value 1 at the element $-I \in Spin^+(1,2)$ and can be descended to an irreducible representation of the group $SO_0(1,2)$. In particular, the two-fold covering (1.34) is U_1.

Example 52. According to Table 1.4, the Clifford algebra $Cl_{0,3}^+ = Cl_{3,0}^+ = \mathbb{H}$ has the unique obvious irreducible representation in the linear space $S = \mathbb{H}^1$ of symplectic Majorana spinors. The restriction of this representation to the group $Spin(3) = Sp(1)$ has the form $q \mapsto q$, a quaternionic symplectic Majorana representation of quaternionic type. We have also the complex irreducible representations $c'S$ acting by

$$q = \alpha + \beta i + \gamma j + \delta k \mapsto \begin{pmatrix} \alpha + \beta i & -\gamma + \delta i \\ \gamma + \delta i & \alpha - \beta i \end{pmatrix}, \tag{1.50}$$

and the real irreducible representation $rc'S$ acting by

$$q \mapsto \begin{pmatrix} \alpha & -\delta & -\gamma & -\beta \\ \delta & \alpha & \beta & -\gamma \\ \gamma & -\beta & \alpha & \delta \\ \beta & \gamma & -\delta & \alpha \end{pmatrix}.$$

Denote $W_{1/2} = S$. It is possible to prove that the real representation $W_{1/2} \otimes W_{1/2}$ is equivalent to the direct sum $U_0 \oplus U_1$, where U_1 is a real irreducible representation of real type, and the quaternionic representation $U_1 \otimes W_{1/2}$ is equivalent to the direct sum of two irreducible representations $W_{1/2} \oplus W_{3/2}$.

By induction, we construct the irreducible real representations U_ℓ of dimension $2\ell + 1$ with $\ell = 0, 1, \ldots$, and the irreducible quaternionic representations W_ℓ of dimension $\ell + 1/2$ with $\ell = 1/2, 3/2, \ldots$. The real (resp., complex, resp., quaternionic) irreducible representations of Spin(3) are U_ℓ (resp., cU_ℓ, resp., qcU_ℓ) with integer ℓ and $rc'W_\ell$ (resp., $c'W_\ell$, resp. W_ℓ) with half-integer ℓ.

Only the representations with integer ℓ have value 1 at the element $-I \in$ Spin(3) and can be descended to an irreducible representation of the group SO(3). In particular, the two-fold covering (1.35) is U_1.

Example 53. According to Table 1.4, the Clifford algebra $\mathrm{Cl}_{1,1}^+ = \mathbb{R} \oplus \mathbb{R}$ has two irreducible representations $(\alpha, \beta) \mapsto \alpha$ and $(\alpha, \beta) \mapsto \beta$ in the linear spaces $S_+ = \mathbb{R}^1$ of left-handed Majorana–Weyl spinors and $S_- = \mathbb{R}^1$ of right-handed Majorana–Weyl spinors. The restrictions of these representations to the group $\mathrm{Spin}_+(1,1) = \mathrm{GL}(1,\mathbb{R})$ have the form $\left(\begin{smallmatrix} \alpha & 0 \\ 0 & \alpha^{-1} \end{smallmatrix} \right) \mapsto \alpha$ and $\left(\begin{smallmatrix} \alpha & 0 \\ 0 & \alpha^{-1} \end{smallmatrix} \right) \mapsto \alpha^{-1}$. These Majorana–Weyl representations are real of real type.

The irreducible representations of the form $\left(\begin{smallmatrix} \pm|\alpha| & 0 \\ 0 & \pm|\alpha|^{-1} \end{smallmatrix} \right) \mapsto |\alpha|^\lambda \in S_\lambda^+$ and $\left(\begin{smallmatrix} \pm|\alpha| & 0 \\ 0 & \pm|\alpha|^{-1} \end{smallmatrix} \right) \mapsto \pm|\alpha|^\lambda \in S_\lambda^-$ with $\alpha > 0$ are real of real type if $\lambda \in \mathbb{R}$ and complex of complex type if $\lambda \in \mathbb{C} \setminus \mathbb{R}$. In the former case, we denote them by $U_\lambda^\pm$, and in the latter case by $V_\lambda^\pm$. We have $tV_\lambda^\pm = V_{-\lambda}^\pm$.

If $\lambda \in \mathbb{R}$, then the remaining complex (resp., quaternionic) irreducible representations are $cU_\lambda^\pm$ (resp., $qcU_\lambda^\pm$). Otherwise, when $\lambda \in \mathbb{C} \setminus \mathbb{R}$, then the remaining real (resp., quaternionic) irreducible representations are $rV_\lambda^\pm$ with $\mathrm{Im}\,\lambda > 0$ (resp., $qV_\lambda^\pm$ with $\mathrm{Im}\,\lambda > 0$).

Write down the two-fold covering $\pi\colon \mathrm{Spin}_+(1,1) \to \mathrm{SO}_0(1,1)$, Equation (1.36), in the form

$$\pi \begin{pmatrix} \alpha & 0 \\ 0 & \alpha^{-1} \end{pmatrix} = \frac{1}{2} \begin{pmatrix} \alpha^2 + \alpha^{-2} & \alpha^{-2} - \alpha^2 \\ \alpha^{-2} - \alpha^2 & \alpha^2 + \alpha^{-2} \end{pmatrix}. \tag{1.51}$$

Only the representations without upper index $-$ can be descended to the group $\mathrm{SO}_0(1,1)$.

We have $\mathrm{Cl}_{2,2}^+ = \mathbb{R}(2) \oplus \mathbb{R}(2)$. This Clifford algebra has two irreducible representations $(A_1, A_2) \mapsto A_1$ and $(A_1, A_2) \mapsto A_2$ in the linear spaces $S_+ = S_- = \mathbb{R}^2$ of left-handed and right-handed Majorana–Weyl spinors, where A_1, $A_2 \in \mathbb{R}(2)$. The restrictions of these representations to the group $\mathrm{Spin}^+(2,2) = \mathrm{SL}(2,\mathbb{R}) \times \mathrm{SL}(2,\mathbb{R})$ have the same form. These Majorana–Weyl representations are real of real type.

Denote $U_{1/2,0} = S_+$. It is possible to prove that the real representation $U_{1/2,0} \otimes U_{1/2,0}$ is equivalent to the direct sum $U_{0,0} \oplus U_{1,0}$, where $U_{1,0}$ is a real irreducible representation of real type. Here we introduced notation $U_{0,0} = U_0$. By induction, we construct the real irreducible representations $U_{\ell_1,0}$ and U_{0,ℓ_2}.

Tensor products produce real irreducible representations $U_{\ell_1,\ell_2} = U_{\ell_1,0} \otimes U_{0,\ell_2}$. The complex irreducible representations of $\text{Spin}^+(2,2)$ are cU_{ℓ_1,ℓ_2}, the quaternionic are qcU_{ℓ_1,ℓ_2}.

The two-fold covering (1.37) is $U_{1,0}$. Only the representations with integer $\ell_1 + \ell_2$ have value 1 at the element $(-I,-I) \in \text{Spin}^+(2,2)$ and can be descended to an irreducible representation of the group $\text{SO}_0(2,2)$.

Example 54. According to Table 1.4, the Clifford algebra $\text{Cl}_{0,4}^+ = \text{Cl}_{4,0}^+ = \mathbb{H} \oplus \mathbb{H}$ has two irreducible representations $(q_1, q_2) \mapsto q_1$ and $(q_1, q_2) \mapsto q_2$ in the linear spaces $S_+ = S_- = \mathbb{H}^1$ of left-handed and right-handed symplectic Majorana–Weyl spinors. The restrictions of these representations to the group $\text{Spin}(4) = \text{Sp}(1) \times \text{Sp}(1)$ have the same form. These representations are quaternionic of quaternionic type. The first (resp., the second) one is the left-handed (resp., right-handed) symplectic Majorana–Weyl representation.

Denote $W_{1/2,0} = S_+$. It is possible to prove that the real representation $W_{1/2,0} \otimes W_{1/2,0}$ is equivalent to the direct sum $U_{0,0} \oplus U_{1,0}$, where $U_{1,0}$ is a real irreducible representation of real type, and the quaternionic representation $U_{1,0} \otimes W_{1/2,0}$ is equivalent to the direct sum of two irreducible representations $W_{1/2,0} \oplus W_{3/2,0}$. Here we introduced notation $U_{0,0} = U_0$. By induction, we construct the real irreducible representations $U_{\ell_1,0}$ and U_{0,ℓ_2} with nonnegative integers ℓ_1 and ℓ_2 and quaternionic irreducible representations $W_{\ell_1,0}$ and W_{0,ℓ_2} with positive half-integers ℓ_1 and ℓ_2.

Tensor products produce real irreducible representations $U_{\ell_1,\ell_2} = U_{\ell_1,0} \otimes U_{0,\ell_2}$ with integer ℓ_1 and ℓ_2, and $U_{\ell_1,\ell_2} = W_{\ell_1,0} \otimes W_{0,\ell_2}$ with half-integer ℓ_1 and ℓ_2, as well as quaternionic irreducible representations $W_{\ell_1,\ell_2} = U_{\ell_1,0} \otimes W_{0,\ell_2}$ with integer ℓ_1 and half-integer ℓ_2, and $W_{\ell_1,\ell_2} = W_{\ell_1,0} \otimes U_{0,\ell_2}$ with half-integer ℓ_1 and integer ℓ_2.

The real irreducible representations of $\text{Spin}(4)$ are U_{ℓ_1,ℓ_2}, where both indices are either integers or half-integers, and $rc'W_{\ell_1,\ell_2}$, where one of the indices is integer, while another one is half-integer. The complex irreducible representations are cU_{ℓ_1,ℓ_2} and $c'W_{\ell_1,\ell_2}$, the quaternionic are qcU_{ℓ_1,ℓ_2} and W_{ℓ_1,ℓ_2}.

Only the representations with integer $\ell_1 + \ell_2$ have value 1 at the element $(-I,-I) \in \text{Spin}(4)$ and can be descended to an irreducible representation of the group $\text{SO}(4)$. In particular, the two-fold covering (1.37) is $U_{1,1}$.

Example 55. According to Table 1.4, we have $\text{Cl}_{0,2}^+ = \text{Cl}_{2,0}^+ = \mathbb{C}$. Its irreducible representations were described in Example 12. Their restrictions to the group $\text{Spin}(2) = \text{U}(1)$ define the complex irreducible representations $\zeta \mapsto \zeta$ and $\zeta \mapsto \overline{\zeta} = \zeta^{-1}$ of complex type of the above group in complex linear spaces $S = \mathbb{C}^1$ of positive, or left-handed Weyl spinors and tS of negative, or right-handed Weyl spinors.

The remaining complex irreducible representations are the representations $\zeta \mapsto \zeta^m$ and $\zeta \mapsto \overline{\zeta}^m = \zeta^{-m}$ with $m \geq 2$ in the linear spaces $V_{m/2} = S^{\otimes m}$ and $tV_{m/2}$. The real representations $rV_{m/2}$ and the quaternionic representations $qV_{m/2}$ are irreducible. The number $m \in \mathbb{Z}$ in the representation $\zeta \mapsto \zeta^m$ is called the *winding number*. The representation $rV_{1/2}$ is the Majorana spinor representation.

Only the representations with even winding number have value 1 at the element $-1 \in \mathrm{Spin}(2)$ and can be descended to an irreducible representation of the group $\mathrm{SO}(2)$. In particular, the two-fold covering (1.38) is the representation V_1.

1.8.6 Spin-Tensors

We denote the classical spinors by lowercase bold Greek letters $\boldsymbol{\xi}$, $\boldsymbol{\eta}$, $\dots$. In the abstract index notation, we use the *uppercase* Latin indices: ξ^A. If a basis of S is chosen, we index the components of a spinor $\boldsymbol{\xi}$ by uppercase Greek letters: ξ^{Γ}.

How to rise and lower spinor indices? Recall that we rise and lower tensor indices with the help of a non-degenerate real symmetric bilinear form g_{ij}. For spinor indices, we need to use all eight types of non-degenerate bilinear forms described in Subsection 1.2.2. We will consider such a form in details only for the case of left- and right-handed Weyl spinors. But first, we need a definition.

Let U be a real linear space. Consider two real algebras: the algebra $\mathbb{C}$ of Example 3 and the algebra $\mathrm{Hom}_{\mathbb{R}}(U)$ of Example 4.

Definition 86. A *complex structure* on U is a $\mathbb{R}$-linear map $j\colon \mathbb{C} \to \mathrm{Hom}_{\mathbb{R}}(U)$ that respects multiplication and maps $1 \in \mathbb{C}$ to the identity operator $I \in \mathrm{Hom}_{\mathbb{R}}(U)$.

Example 56. Let U be a real linear space of dimension 4. Fix a complex structure $j\colon \mathbb{C} \to \mathrm{Hom}_{\mathbb{R}}(U)$. Introduce the scalar-vector multiplication in U by $\zeta \mathbf{x} = j(\zeta)\mathbf{x}$, $\zeta \in \mathbb{C}$, $\mathbf{x} \in U$. Denote by S the set U with the above multiplication. S becomes a complex linear space of dimension 2. Define a map $i\colon U \to S$ by $i\mathbf{x} = \mathbf{x}$.

By definition, the *conjugate complex structure* tj acts by $(tj)(\zeta) = j(\bar{\zeta})$, $\zeta \in \mathbb{C}$. Consider the corresponding complex linear space tS and the map $ti\colon U \to tS$, $(ti)\mathbf{x} = \mathbf{x}$. There exist two unique conjugate-linear maps $c\colon S \to tS$ and $tc\colon tS \to S$ such that $ti = c \circ i$ and $i = tc \circ ti$.

Observe that the group $\mathrm{Spin}^0(1,3) = \mathrm{SL}(2,\mathbb{C})$ acts in S (resp., tS) by the irreducible complex representation $V_{1/2,0}$ (resp., $V_{0,1/2}$) of Example 49. By this reason, we call the elements of S and tS left-handed and right-handed Weyl spinors. For a spinor $\boldsymbol{\pi} \in S$ (resp., $\dot{\boldsymbol{\pi}} \in tS$), we denote $\overline{\boldsymbol{\pi}} = c(\boldsymbol{\pi})$ (resp., $\overline{\dot{\boldsymbol{\pi}}} = c(\dot{\boldsymbol{\pi}})$).

Let ε be a nonzero skew-symmetric bilinear form on S. The form $\overline{\varepsilon}$ on tS is given by

$$\overline{\varepsilon}(\dot{\boldsymbol{\pi}}_1, \dot{\boldsymbol{\pi}}_2) = \overline{\varepsilon(\dot{\boldsymbol{\pi}}_1, \dot{\boldsymbol{\pi}}_2)}.$$

The tensor product $j \otimes (tj)$ of the complex structure j and conjugate complex structure tj is a real structure in the complex linear space $S \otimes (tS)$. Denote by

T the real eigenspace of $j \otimes (tj)$ that corresponds to the eigenvalue 1. It has dimension 4, and the restriction of the bilinear form $\varepsilon \otimes \bar{\varepsilon}$ to T is a symmetric real-valued bilinear form with one positive and three negative eigenvalues.

Now, we introduce abstract indices: undotted (resp., dotted) uppercase Latin for the elements of S (resp., tS). In this notation, a left-handed (resp., right-handed) Weyl spinor $\boldsymbol{\pi} \in S$ (resp., $\dot{\boldsymbol{\pi}} \in tS$) becomes an undotted (resp., dotted) spinor π_A (resp., $\pi_{\dot{A}}$). The nonzero skew-symmetric bilinear form ε becomes a *skew spinor* ε^{AB}, while $\bar{\varepsilon}$ becomes $\varepsilon^{\dot{A}\dot{B}}$. Their inverses become ε_{AB} and $\varepsilon_{\dot{A}\dot{B}}$. We normalise them so that one has

$$\varepsilon^{AB}\varepsilon_{AB} = \varepsilon^{\dot{A}\dot{B}}\varepsilon_{\dot{A}\dot{B}} = 2.$$

The spinor indices are raising and lowering according to the rules

$$\pi^B = \pi_A \varepsilon^{BA}, \qquad \pi^{\dot{B}} = \pi_{\dot{A}} \varepsilon^{\dot{B}\dot{A}}, \qquad \pi_B = \pi^A \varepsilon_{AB}, \qquad \pi_{\dot{B}} = \pi^{\dot{A}} \varepsilon_{\dot{A}\dot{B}}.$$

1.8.7 The Clebsch–Gordan Decomposition

In Example 49, we classified the finite-dimensional irreducible complex (resp., real) representations $cU_{\ell,\ell}$ and V_{ℓ_1,ℓ_2} (resp., $U_{\ell,\ell}$ and rV_{ℓ_1,ℓ_2}) of the group $\mathrm{SL}(2,\mathbb{C})$. The indices ℓ, ℓ_1, and ℓ_2 are nonnegative integer or half-integer numbers. In the real case, $\ell_1 < \ell_2$. We are interested in the structure of the tensor products of the above representations.

From now on, we introduce the standard physical notation:

$$D^{(\ell_1,\ell_2)} = \begin{cases} cU_{\ell_1,\ell_2}, & \text{if } \ell_1 = \ell_2, \\ V_{\ell_1,\ell_2}, & \text{otherwise} \end{cases}$$

Theorem 12. *For the finite-dimensional irreducible complex representations, we have the following* Clebsch–Gordan *decomposition:*

$$D^{(\ell_1,\ell_2)} \otimes D^{(\ell_1',\ell_2')} = \sum_{L=|\ell_1-\ell_1'|}^{\ell_1+\ell_1'} \sum_{L'=|\ell_2-\ell_2'|}^{\ell_2+\ell_2'} \oplus D^{(L,L')}.$$

By abuse of notation, denote

$$D_{\mathbb{R}}^{(\ell_1,\ell_2)} = \begin{cases} U_{\ell_1,\ell_2}, & \text{if } \ell_1 = \ell_2, \\ rV_{\ell_1,\ell_2}, & \text{if } \ell_1 < \ell_2. \end{cases}$$

We do not consider the case of the tensor product of two arbitrary irreducible real representations, but the most important example instead.

Example 57. Let $D_{\mathbb{R}}^{(1/2,1/2)}$ be the Majorana spinor representation of the group $\mathrm{SL}(2,\mathbb{C})$. To decompose its tensor square, we first decompose the tensor square of its real counterpart, $D^{(1/2,1/2)}$. Theorem 12 gives

$$D^{(1/2,1/2)} \otimes D^{(1/2,1/2)} = D^{(0,0)} \oplus (D^{(0,1)} \oplus D^{(1,0)}) \oplus D^{(1,1)}.$$

We observe that all three components of this tensor square are of real type, which gives

$$D_{\mathbb{R}}^{(1/2,1/2)} \otimes D_{\mathbb{R}}^{(1/2,1/2)} = D_{\mathbb{R}}^{(0,0)} \oplus D_{\mathbb{R}}^{(0,1)} \oplus D_{\mathbb{R}}^{(1,1)}.$$

In this decomposition, we have

$$\dim D_{\mathbb{R}}^{(0,0)} = 1,$$
$$\dim D_{\mathbb{R}}^{(1,1)} = (2 \cdot 1 + 1)(2 \cdot 1 + 1) = 9,$$
$$\dim D_{\mathbb{R}}^{(0,1)} = 2(2 \cdot 1 + 1) = 6.$$

Moreover, the symmetric part of the above tensor square is $D_{\mathbb{R}}^{(0,0)} \oplus D_{\mathbb{R}}^{(1,1)}$ of dimension 10, while the skew-symmetric part $D_{\mathbb{R}}^{(0,1)}$ has dimension 6. In coordinates, the representation $D_{\mathbb{R}}^{(0,0)}$ (resp., $D_{\mathbb{R}}^{(0,1)}$, resp., $D_{\mathbb{R}}^{(1,1)}$) acts in the linear space generated by the identity matrix (resp., of skew-symmetric matrices, resp., of symmetric traceless matrices).

1.9 Spin and Spin-Tensor Fields

We will not consider the subject of spin fields in its full generality. Instead, we develop several important examples.

Example 58. Let g be a pseudo-Riemannian metric on a smooth manifold M of dimension n. Then, the tangent space $T_p M$ is isomorphic to $\mathbb{R}^{p,q}$ with $p + q = n$. Define P as the set of pairs $(p, \mathbf{v}_\mu)$, where p runs over M, and $\mathbf{v}_\mu$ runs over the set of all orthonormal bases in $T_p M$.

Let $(\mathcal{U}, \mathbf{h})$ be a chart of M, and let $\frac{\partial}{\partial x^\mu}$ be the corresponding basis of the space $T_p M$ for a point $p \in \mathcal{U}$. Then, we have $\mathbf{v}_\mu = a_\mu{}^\nu \frac{\partial}{\partial x^\nu}$.

Let $\mathcal{V}$ be the subset of P which includes all pairs $(p, \mathbf{v}_\mu)$ with $p \in \mathcal{U}$. Define the map $\mathbf{k} \colon \mathcal{V} \to \mathbb{R}^{n+n^2}$ by $\mathbf{k}(p, \mathbf{v}_\mu) = (\mathbf{h}(p), a_\mu{}^\nu)$. Repeat this procedure for all charts of some atlas of M. We obtain a smooth structure on P.

The group $O(p, q)$ acts freely on P on the right by

$$(p, \mathbf{v}_1, \ldots, \mathbf{v}_n) \cdot g = (p, g^\top \mathbf{v}_1, \ldots, g^\top \mathbf{v}_n).$$

The tuple $(P, \pi, M, O(p, q))$ with $\pi(p, \mathbf{v}_\mu) = p$ is a principal bundle called the *bundle of orthonormal bases*.

If M is orientable, then there exists a sub-bundle $(P_0, \pi, M, SO_0(p, q))$ of *oriented orthonormal bases*.

1.9.1   Spin Bundles: A Group Action Approach

Definition 87. An orientable manifold M is called *spin* if there is a principal bundle $\tilde{\xi} = (\tilde{P}, \tilde{\pi}, M, \mathrm{Spin}^0(p, q))$ and a fibre-preserving two-fold covering $\Pi \colon \tilde{P} \to P_0$ that commutes with bundle projections: $\tilde{\pi} = \Pi \cdot \pi$.

In other words, the principal bundle ξ of oriented orthonormal bases can be lifted to the principal bundle $\tilde{\xi}$. The principal bundle $\tilde{\xi}$ is called the *spin structure*.

Example 59. Let $M = S^1$, the unit circle. The corresponding bundle of oriented orthonormal bases is $(P_0, \pi, S^1, \mathrm{SO}(1)) = (S^1, \pi, S^1, \mathrm{SO}(1))$, where π is the identity map. The trivial principal bundle $(S^1 \times \mathrm{O}(1), \pi, S^1, \mathrm{O}(1))$ with $\pi(p, g) = \Pi(p, g) = (p, e)$ is a spin structure on S^1.

Realise S^1 as the set of complex numbers w with $|w| = 1$. The non-trivial principal bundle $(S^1, \pi, S^1, \mathrm{O}(1))$ with $\pi(w) = \Pi(w) = w^2$ is another spin structure on S^1.

Example 60. Realise the unit sphere S^2 as the set of vectors $\mathbf{p} \in \mathbb{R}^3$ with $\|\mathbf{p}\| = 1$. Let $(P_0, \pi, S^2, \mathrm{SO}(2))$ be the corresponding bundle of oriented orthonormal bases. The map $P_0 \to \mathrm{SO}(3)$, $(\mathbf{p}, \mathbf{e}_1, \mathbf{e}_2) \mapsto (\mathbf{p} \ \ \mathbf{e}_1 \ \ \mathbf{e}_2)$ is a smooth one-to-one with smooth inverse. The Hopf bundle $(\mathrm{SU}(2), \pi, S^2, \mathrm{U}(1))$ is a spin structure on S^2, where $\Pi \colon \mathrm{SU}(2) \to \mathrm{SO}(3)$ is the covering (1.35).

Consider a principal bundle $\tilde{\xi} = (\tilde{P}, \tilde{\pi}, M, \mathrm{Spin}^0(p, q))$. A section of its restriction to a trivialising chart, say $(\mathcal{U}, \mathbf{h})$, is a *local spin gauge*. Let S be one of the spaces mentioned in Subsection 1.8.5, and let the group $G = \mathrm{Spin}^0(p, q)$ acts in S by its spinor representation. Let $\tilde{\xi}[S]$ be the fibre bundle (E, π_S, M) with fibre S, associated to $\tilde{\xi}$. It is called a *spinor bundle*. Its sections are called *spinor fields*.

Example 61. We continue to study the Hopf bundle. Map a point $(\varphi, \theta, \psi) \in (0, 2\pi) \times (0, \pi) \times (0, 4\pi) \subset \mathbb{R}^3$ to the point

$$\begin{pmatrix} \cos(\theta/2)\exp(-\mathrm{i}(\varphi + \psi)/2) & \sin(\theta/2)\exp(\mathrm{i}(\varphi - \psi)/2) \\ -\sin(\theta/2)\exp(-\mathrm{i}(\varphi - \psi)/2) & \cos(\theta/2)\exp(\mathrm{i}(\varphi + \psi)/2) \end{pmatrix} \in \mathrm{SU}(2). \quad (1.52)$$

This map is one-to-one, and its image, say $\mathcal{U}$, is a dense open subset of $\mathrm{SU}(2)$. The inverse map, call it $\mathbf{h}$, is a chart on the manifold $\mathrm{SU}(2)$ called the *Euler angles*. In this chart, the bundle projection π of the Hopf bundle becomes

$$\pi(\varphi, \theta, \psi) = (\cos\varphi\sin\theta, \sin\varphi\sin\theta, \cos\theta)^\top. \quad (1.53)$$

We recognise the right hand side as the chart on S^2 known as *angular spherical coordinates*. A function $\psi = \psi(\theta, \varphi)$ that maps this chart to the point $(\varphi, \theta, \psi) \in \mathcal{U}$, is a local spin gauge. We choose the following standard gauge:

$$\psi(\theta, \varphi) = (\varphi, \theta, 0) \in \mathcal{U}.$$

The complex irreducible representations $\zeta \mapsto z$ and $\zeta \mapsto \overline{\zeta}$ of complex type of the group $U(1)$ in complex linear spaces $S_{1/2} = \mathbb{C}^1$ and $tS_{1/2}$ of left-handed and right-handed Weyl spinors described in Example 55 define two complex spinor bundles $\tilde{\xi}[S_{1/2}]$ and $\tilde{\xi}[tS_{1/2}]$. Their sections are spinor fields on S^2.

1.9.2 Spin Bundles: A Classical Approach

Let (TM, π, M) be the tangent bundle of a smooth manifold M of dimension n. In Subsection 1.7.2, we constructed new bundles by applying the operations over linear spaces described in Subsection 1.2.2 to the tangent spaces. This time, we assume that M is pseudo-Riemannian, and its tangent spaces are copies of the space $\mathbb{R}^{p,q}$ with $p+q = n$. Apply the Clifford map $f: \mathbb{R}^{p,q} \to \mathrm{Cl}_{p,q}$. We obtain the Clifford bundle of algebras over M, denote it by $(\mathrm{Cl}, \pi_{\mathrm{Cl}}, M)$. Consider its sub-bundle $(\mathrm{Cl}^+, \pi_{\mathrm{Cl}^+}, M)$ with fibre $\mathrm{Cl}_{p,q}^+$

Similarly, for a vector bundle (E, π, M) with fibre L, apply the map $L \mapsto \mathrm{Hom}_{\mathbb{K}}(L)$, and denote the resulting bundle of algebras by $(\mathrm{Hom}, \pi_{\mathrm{Hom}}, M)$.

Definition 88. A vector bundle (E, π, M) is called *spin* if there is a fibre-preserving map $\tau: \mathrm{Cl}^+ \to \mathrm{Hom}$ that commutes with projections: $\pi_{\mathrm{Cl}^+} = \pi_{\mathrm{Hom}}\tau$ and such that, for any $p \in M$, the restriction of τ to $\pi_{\mathrm{Cl}^+}^{-1}(p)$ is an irreducible representation of the Clifford algebra $\mathrm{Cl}_{p,q}^+$.

Example 62. Consider the tangent bundle (TS^2, π, S^2). The Riemannian metric of Example 44 induces the structure of the space $\mathbb{R}^{2,0}$ on all tangent planes. The fibres of the Clifford bundle $(\mathrm{Cl}^+, \pi_{\mathrm{Cl}^+}, S^2)$ are the spaces $\mathbb{C}^1$.

Denote by $(\mathrm{Hom}_+, \pi_+, S^2)$ (resp., $(\mathrm{Hom}_-, \pi_-, S^2)$) the fibre bundle constructed from the line bundle $\tilde{\xi}[S_{1/2}]$ (resp., $\tilde{\xi}[tS_{1/2}]$) of Example 61 by applying the map $S_{1/2} \to \mathrm{Hom}_{\mathbb{C}}(S_{1/2})$ (resp., $tS_{1/2} \to \mathrm{Hom}_{\mathbb{C}}(tS_{1/2})$).

The map $\tau: \mathrm{Cl}^+ \to \mathrm{Hom}_+$, that maps a point $\zeta \in \pi_{\mathrm{Cl}^+}^{-1}(p)$ to the point $\zeta \in \pi_+^{-1}(p)$, introduces the structure of a left-handed Weyl spin vector bundle on $\tilde{\xi}[S_{1/2}]$. Similarly, the map $\tau: \mathrm{Cl}^+ \to \mathrm{Hom}_-$, that maps a point $\zeta \in \pi_{\mathrm{Cl}^+}^{-1}(p)$ to the point $\overline{\zeta} \in \pi_-^{-1}(p)$, introduces the structure of a right-handed Weyl spin vector bundle on $\tilde{\xi}[tS_{1/2}]$.

1.9.3 Spin Bundles: A Cocycle Approach

Again, assume that a manifold M is pseudo-Riemannian, and its tangent spaces are copies of the space $\mathbb{R}^{p,q}$ with $p + q = n$. For simplicity, we exclude the values of p and q for which the group $\mathrm{Spin}^0(p,q)$ is not connected.

Consider the cocycle of Example 40 that determines the tangent bundle of M. Assume there is a cocycle $\{\mathcal{U}_{ij}, f_{ij}\}$ that determines the same bundle and takes values in $\mathrm{SO}_0(p,q)$. Let $\pi: \mathrm{Spin}^0(p,q) \to \mathrm{SO}_0(p,q)$ be the map (1.27).

Definition 89. A manifold M is *spin* if there is a cocycle $\{\mathcal{U}_{ij}, \tilde{f}_{ij}\}$ that takes values in $\mathrm{Spin}^0(p,q)$ and satisfies $\pi\tilde{f}_{ij} = f_{ij}$.

In other words, it is possible to lift the cocycle $\{\mathcal{U}_{ij}, f_{ij}\}$ against the projection π. Since π is a two-fold cover, it may happen that on a triple intersection $\mathcal{U}_{ijk} = \mathcal{U} \cap \mathcal{U}_j \cap \mathcal{U}_k \neq \varnothing$ we have $\hat{f}_{ij}(p)\hat{f}_{jk}(p)\hat{f}_{ki}(p) = -I$.

1.9.4 Differentiation of Spin Fields

Example 63. In this example, we differentiate smooth spinor fields on a smooth pseudo-Riemannian manifold M of dimension 4 such that the metric tensor g_{ij} has one positive and three negative eigenvalues. We assume that there exists a spin bundle (E, π_E, M) with fibres S_p, the copies of a two-dimensional complex linear space S carrying the spinor representation of the group $\mathrm{SL}(2, \mathbb{C})$ described in Example 55.

Define the *Infeld–van der Waerden symbols* as a cross-section $\sigma^{A\dot{A}}{}_i$ of the bundle $E \otimes (tE) \otimes T^*M$ satisfying the condition

$$g_{ij} = \sigma_i{}^{A\dot{A}}\sigma_j{}^{B\dot{B}}\varepsilon_{AB}\varepsilon_{\dot{A}\dot{B}}.$$

The notation for the Infeld–van der Waerden symbols is explained as follows. In the particular case of $M = \mathbb{R}^{1,3}$ we may choose $\sigma^i{}_{A\dot{A}} = \frac{1}{\sqrt{2}}\sigma^i$, where σ^0 is the 2×2 identity matrix, and σ^i, $1 \leq i \leq 3$, are the classical Pauli matrices (1.14).

With the help of the introduced symbols, we can identify a spinor field $\xi^{A\dot{A}}(p)$ with the vector field

$$X^i(p) = \sigma^i{}_{A\dot{A}}(p)\xi^{A\dot{A}}(p). \tag{1.54}$$

By this reason, we denote the *spinor covariant derivative* in the direction of the vector field X by $\nabla_{A\dot{A}}$. Our task is to construct it.

The vectors

$$l_i = \sigma_i{}^{0\dot{0}}, \qquad n_i = \sigma_i{}^{1\dot{1}}$$

belong to T_pM, while the vectors

$$m_i = \sigma_i{}^{0\dot{1}}, \qquad \overline{m}_i = \sigma_i{}^{1\dot{0}}$$

belong to cT_pM. It follows easy from the rules of raising the spinor index that

$$l^i = n_i, \qquad n^i = l_i, \qquad m^i = -\overline{m}_i, \qquad \overline{m}^i = -m_i.$$

The constructed *null tetrad* has the following properties.

$$l^i n_i = 1, \qquad m^i \overline{m}_i = -1,$$

while the remaining products of vectors are equal to 0.

At every point $p \in M$, we choose a basis $\{o^A(p), \iota^A(p)\}$, satisfying the condition $o_A(p)\iota^A(p) = 1$. Such a pair of spinors is called a *spin-frame*.

We define the coordinate form of the spinor covariant derivative by

$$\nabla_{0\dot{0}} = o^A o^{\dot{A}} \nabla_{A\dot{A}} = l^a \nabla_a = \sigma^a{}_{1\dot{1}} \nabla_a, \qquad \nabla_{0\dot{1}} = o^A \iota^{\dot{A}} \nabla_{A\dot{A}} = m^a \nabla_a = \sigma^a{}_{1\dot{0}} \nabla_a,$$
$$\nabla_{1\dot{0}} = \iota^A o^{\dot{A}} \nabla_{A\dot{A}} = \overline{m}^a \nabla_a = \sigma^a{}_{0\dot{1}} \nabla_a, \qquad \nabla_{1\dot{1}} = \iota^A \iota^{\dot{A}} \nabla_{A\dot{A}} = n^a \nabla_a = \sigma^a{}_{0\dot{0}} \nabla_a.$$
$$(1.55)$$

The introduced spinor covariant derivative has the following properties.

- Linearity: $\nabla_{A\dot{A}}(\boldsymbol{\xi} + \boldsymbol{\eta}) = \nabla_{A\dot{A}}\boldsymbol{\xi} + \nabla_{A\dot{A}}\boldsymbol{\eta}$.

- The Leibnitz rule: $\nabla_{A\dot{A}}(\boldsymbol{\xi} \otimes \boldsymbol{\eta}) = (\nabla_{A\dot{A}}\boldsymbol{\xi}) \otimes \boldsymbol{\eta} + \boldsymbol{\xi} \otimes (\nabla_{A\dot{A}}\boldsymbol{\eta})$.

- $\nabla_{A\dot{A}}(t\boldsymbol{\xi}) = t\nabla_{A\dot{A}}\boldsymbol{\xi}$.

- $\nabla_{A\dot{A}}\varepsilon_{BC} = \nabla_{A\dot{A}}\varepsilon^{BC} = 0$.

- $\nabla_{A\dot{A}}$ commutes with any index substitution not involving A and $\dot{A}$.

- For all spinor fields $\boldsymbol{\xi}$ of type $\left(\begin{smallmatrix} 0 & 0 \\ 0 & 0 \end{smallmatrix}\right)$, we have $\nabla_{A\dot{A}}\nabla_{B\dot{B}}\boldsymbol{\xi} = \nabla_{B\dot{B}}\nabla_{A\dot{A}}\boldsymbol{\xi}$.

In fact, Penrose and Rindler (1987) formulate the above properties as *axioms*, then prove that such a spinor covariant derivative exists and is unique, and finally *prove* that the spinor components are differentiating according to the above formulae. Their method is called the *Newman–Penrose formalism*. We will use a part of it later on.

We continue with the following observation. Let $\xi = (P, \pi, M, G)$ be a principal bundle associated with a fibre bundle $\xi[F] = (E, \pi_\xi, M)$ by a smooth left action of G on a manifold F. Let $s(x)$ be a cross-section of the fibre bundle $\xi[F]$. For each point $p \in P$, there exists a unique point $\varphi_s(p) \in F$ such that $s(pG) = (p, \varphi_s(p))G$. For the function $\varphi_s \colon P \to F$, we have, by the definition of an orbit,

$$(p, \varphi_s(p))G = (p \cdot g, g^{-1} \cdot \varphi_s(p))G = (p \cdot g, \varphi_s(p \cdot g))G, \qquad g \in G.$$

Then the function φ_s satisfies the relation $\varphi_s(p \cdot g) = g^{-1} \cdot \varphi_s(p)$.

Conversely, if $\varphi \colon P \to F$ satisfies the above relation, then the function $s_\varphi(pG) = (p, \varphi(p))G$ is a cross-section of the fibre bundle $\xi[F]$. Indeed, we have $\pi_\xi(s_\varphi(pG)) = \pi_\xi((p, \varphi(p))G) = \pi(p) = pG$.

Example 64. In this example, we differentiate smooth spinor fields on the sphere S^2.

On the one hand, consider the holomorphic principal bundle $\xi = (\mathbb{C}^2 \setminus \{\mathbf{0}\}, \pi, \mathbb{C}P^1, \mathrm{GL}(1, \mathbb{C}))$ of Example 33. Put $F = \mathbb{C}^1$, and let the group $\mathrm{GL}(1, \mathbb{C})$ acts in F by $\lambda \cdot \zeta = \lambda^s(\overline{\lambda})^{-s}\zeta$, $\lambda \in \mathrm{GL}(1, \mathbb{C})$, $\zeta \in \mathbb{C}^1$, where s is an integer or a half-integer number. Let $\xi_s[\mathbb{C}^1]$ be the holomorphic line bundle with associated principal bundle ξ.

On the other hand, consider the smooth Hopf principal bundle $\eta = (S^3, \pi', \mathbb{C}P^1, \mathrm{U}(1))$. Let the group $\mathrm{U}(1)$ acts in $\mathbb{C}^1$ by $\lambda \cdot \zeta = \lambda^{2s}\zeta$, $\lambda \in \mathrm{U}(1)$,

$\zeta \in \mathbb{C}^1$. Let $\eta_s[\mathbb{C}^1]$ be the smooth line bundle with associated principal bundle η.

The line bundles $\xi_s[\mathbb{C}^1]$ and $\eta_s[\mathbb{C}^1]$ are holomorphic and smooth versions of the same line bundle. Indeed, let $(\mathbf{v}, \zeta)G$ be the G-orbit of a point $(\mathbf{v}, \zeta) \in (\mathbb{C}^2 \setminus \{\mathbf{0}\}) \times \mathbb{C}^1$ under the right action $(\mathbf{v}, \zeta) \cdot \lambda = (\mathbf{v}\lambda, \lambda^{-s}(\overline{\lambda})^s \zeta)$ of the group $GL(1, \mathbb{C})$. Define the map f from the total space of $\xi_s[\mathbb{C}^1]$ to that of $\eta_s[\mathbb{C}^1]$ by

$$f((\mathbf{v}, \zeta)G) = (\|\mathbf{v}\|^{-1}\mathbf{v}, \zeta)G,$$

where the right hand side is the U(1)-orbit of the point $(\|\mathbf{v}\|^{-1}\mathbf{v}, \zeta) \in S^3 \times \mathbb{C}^1$ under the right action $(\|\mathbf{v}\|^{-1}\mathbf{v}, \zeta) \cdot \lambda = (\mathbf{v}\lambda, \lambda^{-2s}\zeta)$. It is not difficult to check that f is a one-to-one smooth map with smooth inverse that respects the fibres. The condition of commuting with projections has the form $\pi_{\xi_s} = \pi_{\eta_s} \circ f$. Let $\mathbf{v} = (\zeta_0, \zeta_1) \in \mathbb{C}^2 \setminus \{\mathbf{0}\}$. By definition, $\pi_{\xi_s}((\mathbf{v}, \zeta)G) = \pi(\zeta_0, \zeta_1) = [\zeta_0 : \zeta_1]$. On the other hand,

$$(\pi_{\eta_s} \circ f)((\mathbf{v}, \zeta)G) = \pi_{\eta_s}(f((\mathbf{v}, \zeta)G)) = \pi_{\eta_s}(\|\mathbf{v}\|^{-1}\mathbf{v}, \zeta)G$$
$$= \pi'(\|\mathbf{v}\|^{-1}\mathbf{v}) = [\zeta_0 : \zeta_1],$$

what was to be shown.

As we will see soon, the line bundle $\xi_s[\mathbb{C}^1]$ is more convenient when one applies the methods of complex analysis, while the bundle $\eta_s[\mathbb{C}^1]$ is better when the methods of representation theory are applied.

Let $(\mathcal{U}_1, h_1)$ be the chart on $\mathbb{C}P^1$ constructed in Example 18. Let $\zeta = \frac{\zeta_0}{\zeta_1}$ be the coordinate of this chart. Following the standard practice of complex analysis, we denote by $f(\zeta)$ a *holomorphic* cross-section of the holomorphic line bundle $\xi_s[\mathbb{C}^1]$ in the above chart, and by $f(\zeta, \overline{\zeta})$ a *smooth* cross-section of the above line bundle.

On the one hand, denote by $\varphi_f(\zeta, \overline{\zeta})$ the function $\varphi_f \colon \mathbb{C}^2 \setminus \{\mathbf{0}\} \to \mathbb{C}^1$ that corresponds to a smooth cross-section $f(\zeta, \overline{\zeta})$. It satisfies the condition

$$\varphi_f(\zeta\lambda, \overline{\zeta\lambda}) = \lambda^{-s}(\overline{\lambda})^s \varphi_f(\zeta, \overline{\zeta}). \tag{1.56}$$

By definition, see (Eastwood and Tod 1982), a function that satisfies Equation (1.56) is called a *function of spin weight s*. Following the above paper, we use the symbol $\mathcal{B}(s)$ to denote the *smooth* line bundle obtained from the holomorphic bundle $\xi_s[\mathbb{C}^1]$ by weakening its structure.

On the other hand, let $\mathcal{O}(-2s)$ be the holomorphic line bundle over $\mathbb{C}P^1$ obtained by patching the Cartesian products $\mathcal{U}_+ \times \mathbb{C}^1$ and $\mathcal{U}_- \times \mathbb{C}^1$ with the help of the map $(\mathcal{U}_+ \setminus \{[1:0]\}) \times \mathbb{C}^1 \to \mathcal{U}_- \times \mathbb{C}^1$, $(\zeta, w) \mapsto (\zeta^{-1}, \zeta^{2s}w)$, and let $\mathcal{S}(-2s)$ is obtained from $\mathcal{O}(-2s)$ by weakening its structure.

The smooth line bundles $\mathcal{B}(s)$ and $\mathcal{S}(-2s)$ are isomorphic. Indeed, if $f(\zeta, \overline{\zeta})$ is a smooth cross-section of $\mathcal{B}(s)$, then φ_f satisfies Equation (1.56), the function $\psi_g(\zeta, \overline{\zeta}) = \varphi_f(\zeta, \overline{\zeta})|\zeta|^{-2s}$ satisfies the condition

$$\psi_g(\zeta\lambda, \overline{\zeta\lambda}) = \lambda^{-2s}\psi_g(\zeta, \overline{\zeta}), \tag{1.57}$$

which means that the corresponding cross-section g is indeed a cross-section of $\mathcal{S}(-2s)$. Conversely, if g is a smooth cross-section of $\mathcal{S}(-2s)$, then ψ_g satisfies Equation (1.57), the function $\varphi_f(\zeta, \overline{\zeta}) = \psi_g(\zeta, \overline{\zeta})|\zeta|^{2s}$ satisfies Equation (1.56), and f is a smooth cross-section of $\mathcal{B}(s)$.

Let $f(\zeta, \overline{\zeta})$ be a smooth cross-section of the bundle $\mathcal{B}(s)$, and let $g(\zeta, \overline{\zeta})$ be the smooth cross-section of $\mathcal{S}(-2s)$ constructed in the preceding paragraph. The operator ∂ is defined by $\partial g(\zeta, \overline{\zeta}) = \frac{\partial g(\zeta, \overline{\zeta})}{\partial \zeta} \, d\zeta$, see (Wells 2008). The right hand side is a smooth cross-section, say h, of the bundle $\mathcal{S}(-2s - 2)$, and the function $\varphi_h(\zeta, \overline{\zeta})|\zeta|^{2s-2}$ corresponds to some cross-section of the bundle $\mathcal{B}(s+1)$. We denote it by $\eth_s f$. Following (Eastwood and Tod 1982), we constructed the celebrated *spin-raising operator*, a differential operator $\eth_s$ that acts on smooth spinor fields of spin weight s (the cross-sections of $\mathcal{B}(s)$) and maps them to smooth spinor fields of spin weight $s+1$ (the cross-sections of $\mathcal{B}(s+1)$).

Similarly, the operator $\overline{\partial}$ is defined by $\partial g(\zeta, \overline{\zeta}) = \frac{\partial g(\zeta, \overline{\zeta})}{\partial \overline{\zeta}} \, d\overline{\zeta}$. The right hand side is a smooth cross-section, say h, of the bundle $\mathcal{S}(-2s + 2)$, and the function $\varphi_h(\zeta, \overline{\zeta})|\zeta|^{2s+2}$ corresponds to some cross-section $\overline{\eth}_s f$ of the bundle $\mathcal{B}(s - 1)$. The *spin-lowering operator*, $\overline{\eth}_s$ acts on smooth spinor fields of spin weight s (the cross-sections of $\mathcal{B}(s)$) and maps them to smooth spinor fields of spin weight $s - 1$ (the cross-sections of $\mathcal{B}(s - 1)$).

The reader may check that the operator $\eth_s$ satisfies the properties described in Example 63.

We use the result of Example 57 to decompose the Riemann curvature tensor.

Example 65. Let (M, g) be the pseudo-Riemannian manifold of dimension 4 such that its tangent bundle is spin. In particular, each tangent space $T_p M$ carries the Majorana representation $D_{\mathbb{R}}^{(1/2, 1/2)}$ of the Lorentz group $\mathrm{SL}(2, \mathbb{C})$. The tensor product of four copies of the tangent space has a linear subspace of the Riemann curvature tensors of dimension 20. Which representation of $\mathrm{SL}(2, \mathbb{C})$ does that space carry?

By Example 57, the tensor square of the tangent space $T_p M$ carries the representation

$$D_{\mathbb{R}}^{(1/2, 1/2)} \otimes D_{\mathbb{R}}^{(1/2, 1/2)} = D_{\mathbb{R}}^{(0,0)} \oplus D_{\mathbb{R}}^{(0,1)} \oplus D_{\mathbb{R}}^{(1,1)}.$$

The symmetries (1.45) and (1.46) of the Riemann curvature tensors show that it belongs to the symmetric tensor square $S^2(D_{\mathbb{R}}^{(0,1)})$ of the representation $D_{\mathbb{R}}^{(0,1)}$. Following the pattern of Example 57, we consider its complex counterpart, $D^{(0,1)} \oplus D^{(1,0)}$, and calculate its tensor square, using Theorem 12:

$$(D^{(0,1)} \oplus D^{(1,0)}) \otimes (D^{(0,1)} \oplus D^{(1,0)}) = D^{(0,1)} \otimes D^{(0,1)} \oplus D^{(0,1)} \otimes D^{(1,0)}$$
$$\oplus D^{(1,0)} \otimes D^{(0,1)} \oplus D^{(1,0)} \otimes D^{(1,0)}.$$

The first term is $D^{(0,1)} \otimes D^{(0,1)} = D^{(0,0)} \oplus D^{(0,1)} \oplus D^{(0,2)}$. Of these, the components $D^{(0,0)}$ and $D^{(0,2)}$ belong to the symmetric square $S^2(D^{(0,1)} \oplus$

$D^{(1,0)}$). Similarly, the components $D^{(0,0)}$ and $D^{(2,0)}$ of the fourth term belong to the above symmetric square. The second and the third terms give

$$D^{(0,1)} \otimes D^{(1,0)} \oplus D^{(1,0)} \otimes D^{(0,1)} = 2D^{(0,0)} \oplus 2D^{(0,1)} \oplus 2D^{(1,0)} \oplus 2D^{(1,1)}.$$

Of these, one component $D^{(1,1)}$ goes to the symmetric square. We obtain

$$S^2(D^{(0,1)} \oplus D^{(1,0)}) = 2D^{(0,0)} \oplus (D^{(0,2)} \oplus D^{(2,0)}) \oplus D^{(1,1)}.$$

The real Clebsch–Gordan expansion reads

$$S^2(D_{\mathbb{R}}^{(0,1)}) = 2D_{\mathbb{R}}^{(0,0)} \oplus D_{\mathbb{R}}^{(0,2)} \oplus D_{\mathbb{R}}^{(1,1)}.$$

The dimension of each hand side is 21, as it should be.

To subtract one more one-dimensional space, we use the first Bianchi identity, Theorem 7. Define the *Bianchi map* $b \colon S^2(D_{\mathbb{R}}^{(0,1)}) \to S^2(D_{\mathbb{R}}^{(0,1)})$ by

$$b(R_{\alpha\beta\gamma\delta}) = \frac{1}{3}(R_{\alpha\beta\gamma\delta} + R_{\beta\gamma\alpha\delta} + R_{\gamma\alpha\beta\delta}).$$

Using Theorem 7, it is easy to check that both the kernel and the image of the Bianchi map are $\mathrm{SL}(2, \mathbb{C})$-invariant, the image has dimension 1 and carries a copy of the representation $D_{\mathbb{R}}^{(0,0)}$. We obtain: the linear space of curvature tensors carries the direct sum $D_{\mathbb{R}}^{(0,0)} \oplus D_{\mathbb{R}}^{(0,2)} \oplus D_{\mathbb{R}}^{(1,1)}$. The elements of the linear space where the component $D_{\mathbb{R}}^{(0,2)}$ (resp., $D_{\mathbb{R}}^{(1,1)}$, resp., $D_{\mathbb{R}}^{(0,0)}$) acts, are denoted by $C_{\alpha\beta\gamma\delta}$ (resp., $E_{\alpha\beta\gamma\delta}$, resp., $G_{\alpha\beta\gamma\delta}$). The elements $C_{\alpha\beta\gamma\delta}$ are called the *Weyl tensors*. The three kinds of elements can be calculated by the following equations:

$$C_{\alpha\beta\gamma\delta} = R_{\alpha\beta\gamma\delta} + g_{\alpha[\delta}R_{\gamma]\beta} + \frac{R}{3}g_{\alpha[\gamma}g_{\delta]\beta}, \tag{1.58a}$$

$$E_{\alpha\beta\gamma\delta} = \frac{1}{2}(g_{\alpha\gamma}S_{\beta\delta} + g_{\beta\delta}S_{\alpha\gamma} - g_{\alpha\delta}S_{\beta\gamma} - g_{\beta\gamma}S_{\alpha\delta}), \tag{1.58b}$$

$$G_{\alpha\beta\gamma\delta} = \frac{R}{12}(g_{\alpha\gamma}g_{\beta\delta} - g_{\alpha\delta}g_{\beta\gamma}), \tag{1.58c}$$

where $R_{\alpha\delta}$ is the Ricci tensor, R the Ricci scalar, and

$$S_{\beta\delta} = R_{\beta\delta} - \frac{1}{4}g_{\beta\delta}R.$$

1.10   Induced Representations

Consider a principal fibre bundle $\xi = (G, \pi, G/H, H)$ of Example 37. Denote $M = G/H$. Assume H acts on a right finite-dimensional linear space L with an inner product by a representation, and consider the fibre bundle $\xi[L] = (E, \pi_L, M)$ over M with fibre L and associated principal bundle (G, π, M, H).

Define a left action of G on $G \times L$ by $g \cdot (h, 1) = (gh, 1)$. This action induces a left action of G on E.

Definition 90. A fibre bundle $\xi[L] = (E, \pi_L, M)$ is called *homogeneous* if G acts on E from the left, $g \cdot \pi_L^{-1}(p) = \pi_L^{-1}(g \cdot p)$, and the restriction of this action to any fibre $\pi_L^{-1}(p)$ is linear.

We observe that the action of G defined above turns $\xi[L]$ into a homogeneous fibre bundle. Conversely, according to (Wallach 1973, Lemma 5.2.3), any homogeneous fibre bundle can be obtained by the above described construction.

Denote by ΓE the Banach space of all continuous cross-sections of the homogeneous fibre bundle (E, π_L, M).

Definition 91. The representation $(g \cdot s)(x) = g \cdot s(g^{-1}x)$, $x \in M$, $s \in \Gamma E$, $g \in G$ is called *induced* by the representation L.

Observe that the Banach space ΓE is infinite-dimensional unless M is finite. For a representation in a Banach space L, we always assume that it is continuous, that is, for each $\mathbf{x} \in L$, the map $G \to L$, $g \mapsto \theta g \mathbf{x}$ is continuous. Here θg is the operator of the representation.

In many sources, an equivalent definition of an induced representation is used. Denote by $C(G, L)$ the linear space of all continuous functions $\mathbf{f} \colon G \to L$ satisfying the condition $\mathbf{f}(gh) = h^{-1} \cdot \mathbf{f}(g)$, $g \in G$, $h \in H$. Define a representation with the operator $\tilde{\theta} g$ acting by

$$(\tilde{\theta} g_0 \mathbf{f}) g = \mathbf{f}(g_0^{-1} g). \tag{1.59}$$

A one-to-one intertwining operator between the representations θ and $\tilde{\theta}$ maps a function $\mathbf{s} \in \Gamma E$ to the function $\mathbf{f} \in C(G, L)$ given by $\mathbf{f}(g) = g^{-1} \cdot \mathbf{s}(gH)$. In what follows we use the representation θ.

Example 66. Consider the linear space $\mathbb{R}^{1,3}$. Following physical tradition, call it the *Minkowski spacetime* or the *spacetime domain*. Denote by $\mathbf{x}$ a point in $\mathbb{R}^{1,3}$ with coordinates $x^0 = ct$, x^1, x^2, and x^3, where c is the speed of light in vacuum, and t is time. The bilinear form (1.1) becomes $B(\mathbf{x}, \mathbf{y}) = c^2 t_1 t_2 - x^1 y^1 - x^2 y^2 - x^3 y^3$.

Consider a second copy of $\mathbb{R}^{1,3}$, denote it by $\hat{\mathbb{R}}^{1,3}$, call it the *wavevector domain*. Denote by $\mathbf{k}$ a point in $\hat{\mathbb{R}}^{1,3}$ with coordinates $k_0 = c^{-1}\omega$, k_1, k_2, and k_3, where ω is the angular frequency. Denote $(\mathbf{k}, \mathbf{x}) = \omega t - x^i k_i$.

The functions $\mathbb{R}^{1,3} \to \mathbb{C}$, $\mathbf{x} \to \mathbf{k}(\mathbf{x}) = \exp(-i(\mathbf{k}, \mathbf{x}))$ are the characters of irreducible unitary representations of the group $\mathbb{R}^{1,3}$, or just characters.

Define $G = \mathrm{SL}(2, \mathbb{C}) \times \mathbb{R}^{1,3}$ with the following group multiplication:

$$(\Lambda_1, \mathbf{x}_1)(\Lambda_2, \mathbf{x}_2) = (\Lambda_1 \Lambda_2, \mathbf{x}_1 + \pi \Lambda_1 \mathbf{x}_2), \tag{1.60}$$

where π is the covering (1.39). This group is called the universal covering of the proper orthochronous Poincaré group. It acts on the wavevector domain by $((\Lambda, \mathbf{x}) \cdot \mathbf{k})(\mathbf{y}) = \mathbf{k}((\pi(\Lambda))^{-1}\mathbf{y})$. Let H be the stationary subgroup of the point $\mathbf{k}_0 = (1, 0, 0, 1)^\top \in \hat{\mathbb{R}}^{1,3}$. The set H is the Cartesian product $H = \mathcal{E} \times \mathbb{R}^{1,3}$, where $\mathcal{E}$ is the subgroup of the group $\mathrm{SL}(2, \mathbb{C})$ which consists of matrices of the form $\begin{pmatrix} \exp(\mathrm{i}\psi/2) & \exp(-\mathrm{i}\psi/2)z \\ 0 & \exp(-\mathrm{i}\psi/2) \end{pmatrix}$ with $0 \le \psi < 4\pi$ and $z \in \mathbb{C}$. The group multiplication is given by Equation (1.60), where $\Lambda_i \in \mathcal{E}$, $i = 1, 2$. The manifold $\mathcal{C}_0 = G/H$ is called the *mantle of the forward light cone* in the wavevector domain and has the form

$$\mathcal{C}_0 = \{\, \mathbf{k} \in \hat{\mathbb{R}}^{1,3} \colon g(\mathbf{k}, \mathbf{k}) = 0, \omega > 0 \,\}.$$

Put $V = \mathbb{C}^1$ and consider the representation of the group H in V with operator

$$\theta(L, \mathbf{x}) = \mathbf{k}_0(\mathbf{x}) \exp(\mathrm{i}\ell\psi/2),$$

where ℓ is a positive integer. In this book, we are especially interested in three particular cases, when $\ell \in \{1, 2, 4\}$.

The two representations of G induced by the representation of H act in the linear space of continuous cross-sections of the line bundle $\xi[V] = (E, \pi_V, \mathcal{C}_0)$. For an element $g = (\Lambda, \mathbf{x}) \in G$, their operators, $\theta_{\pm}g$ acts on a continuous cross-section $\mathbf{s}$ by

$$(\theta_{\pm}g\mathbf{s})(\mathbf{k}) = \mathbf{k}(\mathbf{x}) \exp(\pm \mathrm{i}\ell\psi(\Lambda, \mathbf{k})/2)\mathbf{s}((\pi\Lambda)^{-1}\mathbf{k}).$$

We continue to study this representation in Section 2.6.

Let $\mathrm{d}g$ be the probabilistic Haar measure on G, then there is a unique probabilistic measure $\mathrm{d}x$ on M such that for any continuous function f on M we have

$$\int_G f(gH) \, \mathrm{d}g = \int_M f(x) \, \mathrm{d}x.$$

The representation ΓE can be extended to the Hilbert space of all square-integrable cross-sections of the homogeneous fibre bundle $\xi[L]$. We denote it by $L \uparrow G$ and are interested in the structure of the representation $L \uparrow G$. More precisely, let R be an irreducible representation of the group G over the same (skew) field $\mathbb{K}$ as L is. We are interested in the structure of the $\mathbb{K}'$-linear space $\mathrm{Hom}_G(R, L \uparrow G)$ of all linear operators acting from the *linear space* R to the *Hilbert space* $L \uparrow G$ that intertwine the *representations* R and $L \uparrow G$.

The idea is as follows. We construct another $\mathbb{K}'$-linear space that consists of intertwining operators and has a simple structure. Then we establish an isomorphism between the two.

To construct the above mentioned linear space, we need to construct two representations. The second one is L, the representation of the group H. Denote the first one by $R \downarrow H$, the restriction of the representation R of the group G to the subgroup H. The structure of the $\mathbb{K}'$-linear space $\mathrm{Hom}_H(R \downarrow H, L)$

of the linear operators that intertwine the representations $R \downarrow H$ and L, is easy to find.

Indeed, let $\chi_R(g)$, $g \in G$, be the character of the representation R. The function $\chi_R(h)$, $h \in H$, can be uniquely written as a linear combination of finitely many irreducible characters of the group H with positive integer coefficients. By Theorem 8, these coefficients are equal to the multiplicities of the corresponding irreducible representations of H in the representation $R \downarrow H$.

Apply the same procedure to the representation L of the group H. Write down the list of irreducible representations L' of H that have nonzero multiplicities in both $R \downarrow H$ and L. For each pair of copies of L', one inside $R \downarrow H$, another inside L, the structure of the linear space of intertwining operators between the two is given by Schur's Lemma and Theorems 9–11. The linear space $\text{Hom}_H(R \downarrow H, L)$ is the direct sum of the above spaces over all above described pairs of copies and over all terms of the list.

It remains to construct an isomorphism. Let $\mathbf{r} \in R$, and define a function $\eta_R(\mathbf{r}) \colon G \to R$ by

$$\eta_R(\mathbf{r})(x) = x^{-1} \cdot \mathbf{r}, \qquad x \in G.$$

It is obvious, that the function $\eta_R(\mathbf{r})$ is continuous. Moreover, it is a continuous section of the fibre bundle $(E, \pi_R, G/H)$ with fibre R. Indeed, for any $h \in H$, we have

$$\eta_R(\mathbf{r})(x \cdot h) = (xh)^{-1} \cdot \mathbf{r} = h^{-1}x^{-1} \cdot \mathbf{r} = h^{-1} \cdot \eta_R(\mathbf{r}),$$

as it should be. That is, the constructed map η_R maps R to $(R \downarrow H) \uparrow G$.

Moreover, the map η_R intertwines the representation R of G with the representation $(R \downarrow H) \uparrow G$. Indeed, for any $g \in G$, we have

$$\eta_R(g \cdot \mathbf{r})(x) = x^{-1} \cdot (g^{-1} \cdot \mathbf{r}) = (x^{-1}g^{-1}) \cdot \mathbf{r} = \eta_R(\mathbf{r})(xg) = (g \cdot \eta_R(\mathbf{r}))(x).$$

Let $f^\sharp \in \text{Hom}_H(R \downarrow H, L)$. The map $f^\sharp \uparrow G$ acts from $(R \downarrow H) \uparrow G$ to $L \uparrow G$ by

$$(f^\sharp \uparrow G)\mathbf{s}(g) = (f^\sharp \mathbf{s})(g), \qquad \mathbf{s}(g) \in (R \downarrow H) \uparrow G.$$

Define the map $f_\flat \colon R \to L \uparrow G$ by

$$f_\flat = (f^\sharp \uparrow G) \circ \eta_R,$$

that is, first the map η_R maps R to $(R \downarrow H) \uparrow G$, then the map $f^\sharp \uparrow G$ maps $(R \downarrow H) \uparrow G$ to $L \uparrow G$. We have

$$f_\flat \mathbf{r} = f^\sharp(x^{-1} \cdot \mathbf{r}).$$

Next, we construct an inverse isomorphism. Let $\mathbf{s}(x)$ be a square-integrable section of the fibre bundle $(E, \pi_L, G/H)$ with fibre L. Define an element of the linear space L by

$$\varepsilon_L(\mathbf{s}(x)) = \mathbf{s}(e),$$

where e is the identity of H.

The map ε_L intertwines the representation $(L \uparrow G) \downarrow H$ of H with the representation L. Indeed,

$$\varepsilon_L(h \cdot \mathbf{s}(x)) = (h \cdot \mathbf{s})(e) = \mathbf{s}(he) = h \cdot \mathbf{s}(e) = h \cdot \varepsilon_L(\mathbf{s}).$$

Let $f_\flat \in \mathrm{Hom}_G(R, L \uparrow G)$. The map $f_\flat \downarrow H$ acts from $R \downarrow H$ to $(L \uparrow G) \downarrow H$ by

$$(f_\flat \downarrow H)\mathbf{r} = f_\flat\mathbf{r}.$$

Define the map $f^\sharp \colon (R \downarrow H) \to L$ by

$$f^\sharp = \varepsilon_L \circ (f_\flat \downarrow H),$$

that is, first the map $f_\flat \downarrow H$ maps $R \downarrow H$ to $(L \uparrow G) \downarrow H$, then the map ε_L maps $(L \uparrow G) \downarrow H$ to L. We have

$$f^\sharp\mathbf{r} = (f_\flat\mathbf{r})(e).$$

Theorem 13 (Frobenius reciprocity). *The isomorphisms $(f^\sharp \uparrow G) \circ \eta_R$ and $\varepsilon_L \circ (f_\flat \downarrow H)$ are each other's inverses.*

Remark 9. If $\mathbb{K} = \mathbb{C}$, then all linear spaces of intertwining operators between the copies of L are one-dimensional, and we obtain the classical formulation of the Frobenius reciprocity for complex representations: the multiplicity of R in $L \uparrow G$ is equal to the multiplicity of L in $R \downarrow H$.

Remark 10. A little bit strange notation L and R can be explained as follows. The functor Res that restricts a representation of G to that of H, is left-adjoint to the functor Ind that induces the representation of H to that of G. Dually, the functor Ind is right adjoint to Res. The map η_R (resp., ε_L) is the unit (resp., counit) of the Frobenius reciprocity.

A proof of Frobenius reciprocity is sketched in Example 111. The examples will be given in Chapter 3.

1.11 Probability

1.11.1 Random Elements

It is supposed that the possible states of any stochastic system form a set, traditionally denoted by the symbol Ω. Some subsets of Ω are more important than other subsets, they should form a σ-field denoted by $\mathfrak{F}$. The elements of $\mathfrak{F}$ are called *events*. There is a measure P of total mass 1 defined on $\mathfrak{F}$. For any event $A \in \mathfrak{F}$ the real number $\mathsf{P}(A)$ is called the *probability* of that event. The triple $(\Omega, \mathfrak{F}, \mathsf{P})$ is called a *probability space*.

All inner products on a right finite-dimensional linear space L over a (skew) field $\mathbb{K}$ generate the same collection of open subsets of L. The minimal σ-field that contains all of them, is called the σ-field of *Borel sets* and is denoted by $\mathfrak{B}(L)$. The maps $\mathbf{X}\colon \Omega \to L$ that are measurable with respect to the σ-fields $\mathfrak{F}$ and $\mathfrak{B}(L)$, have special names formed by addition of the adjective 'random' to the name of an element of the space L, for example, random vector, random tensor, random spinor, etc. The only exception from this rule is the case of $\dim_{\mathbb{K}} L = 1$; in that case the map X is called a *random variable* instead of 'a random scalar' (for a random variable we use normal Roman font). We will use the term *random element* for such a $\mathbf{X}$ if L is not specified.

Two events A, $B \in \mathfrak{F}$ are called *independent* if $\mathsf{P}(A \cap B) = \mathsf{P}(A)\mathsf{P}(B)$. This notion is not a part of measure theory, therefore probability is not a part of measure theory as well.

Two random elements $\mathbf{X}$ and $\mathbf{Y}$ taking values in the same space L are called *independent* if for all A, $B \in \mathfrak{B}(L)$, the events $\{\mathbf{X} \in A\}$ and $\{\mathbf{Y} \in B\}$ are independent. Here, we used a shortcut: $\{\mathbf{X} \in A\} = \{\omega \in \Omega\colon \mathbf{X}(\omega) \in A\}$.

Fix an inner product $(\cdot, \cdot)_L$ in L and the corresponding norm $\|\cdot\|_L$.

By definition, the *expected value* of a random element $\mathbf{X}$ is the point $\mathsf{E}[\mathbf{X}] \in L$ given by the following Lebesgue integral:

$$\mathsf{E}[\mathbf{X}] = \int_{\Omega} \mathbf{X}(\omega) \, \mathrm{d}\mathsf{P}(\omega).$$

According to the properties of the Lebesgue integral, the expected value of a random *element* $\mathbf{X}$ exists if and only if the expected value of the random *variable* $\|\mathbf{X}\|_L$ exists.

In what follows, we consider only such random elements $\mathbf{X}$, for which the expected value of the random variable $\|\mathbf{X}\|_L^2$ exists. Consider the function $K(\mathbf{x}, \mathbf{y})\colon L \times L \to \mathbb{K}$ given by

$$K(\mathbf{x}, \mathbf{y}) = \mathsf{E}[(\mathbf{x}, \mathbf{X}(p) - \langle \mathbf{X}(p)\rangle)_L (\mathbf{X}(q) - \langle \mathbf{X}(q)\rangle, \mathbf{y})_L].$$

It is easy to see that $K(\mathbf{x}, \mathbf{y})$ is a bi-additive form. Moreover, it is

- Real symmetric bilinear of type (1.1) if $\mathbb{K} = \mathbb{R}$.

- Complex Hermitian bi-additive of type (1.4) if $\mathbb{K} = \mathbb{C}$.

- Quaternionic Hermitian bi-additive of type (1.5) if $\mathbb{K} = \mathbb{H}$.

Lemma 2. *In each of the three cases above, there exists a unique linear operator $K \in \mathrm{Hom}_{\mathbb{K}}(L)$ such that*

$$K(\mathbf{x}, \mathbf{y}) = (\mathbf{x}, K\mathbf{y})_L. \tag{1.61}$$

This result is well-known when $\mathbb{K} \in \{\mathbb{R}, \mathbb{C}\}$ and is proved in (Vakhaniya 1998, Proposition 1) when $\mathbb{K} = \mathbb{H}$.

We define the *mutual correlation operator* $K_{\mathbf{XY}}$ of two random elements $\mathbf{X}$ and $\mathbf{Y}$ as the linear operator in L satisfying (1.61). We introduce the shortcut $K_{\mathbf{X}} = K_{\mathbf{XX}}$ and call $K_{\mathbf{X}}$ the *correlation operator* of the random vector $\mathbf{X}$.

Remark 11. When $\mathbb{K} \in \{\mathbb{R}, \mathbb{C}\}$, one can define the mutual correlation operator of two random elements in terms of tensor products. This does not work for the remaining case of $\mathbb{K} = \mathbb{H}$, because the tensor product of two quaternionic linear spaces is real rather than quaternionic one.

Let X and Y be two $\mathbb{K}^1$-valued random *variables*. In Equation (1.61), put $\mathbf{x} = \alpha \in \mathbb{K}^1$, $\mathbf{y} = 1 \in \mathbb{K}^1$. We obtain

$$K_{XY}\alpha = \mathsf{E}[\alpha \overline{(X - \mathsf{E}[X])}(Y - \mathsf{E}[Y])] = \alpha\mathsf{E}[\overline{(X - \mathsf{E}[X])}(Y - \mathsf{E}[Y])],$$

that is, K_{XY} is the operator of multiplication by the number

$$\langle X, Y \rangle = \mathsf{E}[\overline{(X - \mathsf{E}[X])}(Y - \mathsf{E}[Y])]$$

from the right. The above number is called the *correlation* of X and Y. The *variance* of X is defined as $\mathsf{V}[X] = \langle X, X \rangle$.

It is easy to prove the following: if two random variables X and Y are independent, then they are *uncorrelated*, that is, $K_{XY} = 0$. The converse statement is wrong.

However, there is an important particular case when the converse statement holds true. Define the *characteristic function* of a $\mathbb{R}^1$-valued random variable X by

$$\varphi_X(t) = \mathsf{E}[\exp(\mathrm{i}tX)], \qquad t \in \mathbb{R}.$$

A random variable X is called *normal* or *Gaussian* if its characteristic function has the form

$$\varphi_X(t) = \exp\left(\mathrm{i}t\mu - \frac{1}{2}\sigma^2 t^2\right), \qquad \mu \in \mathbb{R}, \quad \sigma \geq 0.$$

In this case, we have: $\mathsf{E}[X] = \mu$, $\mathsf{V}[X] = \sigma^2$. An important difference between the cases of $\sigma > 0$ and $\sigma = 0$ is as follows. In the former case, there exists a *probability density function* for X, such a function $f_X(x)$ that for any $A \in \mathfrak{B}(\mathbb{R}^1)$ we have

$$\mathsf{P}\{X \in A\} = \int_A f_X(x)\,\mathrm{d}x.$$

Moreover, we have

$$f_X(x) = \frac{1}{\sqrt{2\pi}\sigma} \exp\left(-\frac{(x - \mu)^2}{2\sigma^2}\right).$$

In the latter case, a probability density function of X does not exists, and we have $\mathsf{P}\{X = \mu\} = 1$.

Observe!

Mathematicians call a normal random variable *standard* if $\mu = 0$ and $\sigma = 1$. Physicists require $\mu = 0$ and $\sigma = \frac{1}{\sqrt{2}}$.

Definition 92. A U-valued random element $\mathbf{X}$ is called *normal* if for any $\mathbf{x}^* \in U^*$, the $\mathbb{R}^1$-valued random variable $\mathbf{x}^*(\mathbf{X})$ is normal.

We have the following result.

Theorem 14. *The components* $\mathbf{X}$ *and* $\mathbf{Y}$ *of a* $U \oplus U$*-valued normal random vector* $\mathbf{X} \oplus \mathbf{Y}$ *are independent if and only if they are uncorrelated.*

A $\mathbb{C}^1$-valued random variable X is called *normal* if $\operatorname{Re} X$ and $-\operatorname{Re}(iX)$ are independent $\mathbb{R}^1$-valued normal random variables with equal variances. In order to define V-valued normal random vectors, replace U by V and $\mathbb{R}^1$ by $\mathbb{C}^1$ in Definition 92. Vakhaniya and Kandelaki (1996) prove that Theorem 14 holds true for such a definition if one replaces U with V.

Similarly, a $\mathbb{H}^1$-valued random variable X is called *normal* if $\operatorname{Re} X$, $-\operatorname{Re}(iX)$, $-\operatorname{Re}(jX)$, and $-\operatorname{Re}(kX)$ are *pairwise* independent $\mathbb{R}^1$-valued normal random variables with equal variances. In order to define W-valued normal random vectors, replace U by W and $\mathbb{R}^1$ by $\mathbb{H}^1$ in Definition 92. Vakhaniya (1998) proves that Theorem 14 holds true for such a definition if one replaces U with W.

1.11.2   Random Cross-Sections

Let L be a finite-dimensional linear space over a (skew) field $\mathbb{K}$, let M be a smooth manifold, and let (E, π, M) be a fibre bundle over M with fibre L. How to define a random cross-section $\mathbf{X}(p)$ of the bundle (E, π, M)? One idea is to use a general scheme invented in probability. Let $(\Omega, \mathfrak{F}, \mathsf{P})$ be a probability space.

Definition 93. A function $\mathbf{X}$ defined on the Cartesian product $M \times \Omega$ is called a *random cross-section* of the fibre bundle (E, π, M) if for any $p_0 \in M$, the function $\mathbf{X}(p_0, \omega)$ of the variable $\omega \in \Omega$ is a $\pi^{-1}(p_0)$-valued random element.

Let n be a positive integer, and let $p_1, \ldots, p_n$ be n distinct points in M. The random elements $\mathbf{X}(p_1) \oplus \cdots \oplus \mathbf{X}(p_n) \in \pi^{-1}(p_1) \oplus \cdots \oplus \pi^{-1}(p_n)$ are called the *finite-dimensional distributions* of the random cross-section $\mathbf{X}(p)$.

Definition 94. A random cross-section $\mathbf{X}(p)$ of a fibre bundle (E, π, M) is called *Gaussian* if all its finite-dimensional distributions are normal random elements.

How to construct random cross-sections? We consider a particular case which will be enough for our purposes.

Consider a principal bundle $\xi = (G, \pi, G/H, H)$ of Example 37. Assume that G is a compact Lie group, and H acts in L by a representation. Let $L^2(E)$ be the Hilbert space of all square-integrable cross-sections of the fibre bundle $\xi[L]$ associated with the principal bundle ξ, and let $\{\,\mathbf{s}_i(p) \colon i \in I\,\}$ be an orthonormal basis in $L^2(E)$. We assume that the set I is finite or countable. Let $\mathbb{K}_i^1$ be copies of the right linear space $\mathbb{K}^1$ indexed by I. Finally, let X_i be a *centred* normal $\mathbb{K}_i^1$-valued random variable, that is, $\mathsf{E}[X_i] = 0$, and let P_i be the measure defined on the Borel σ-field $\mathfrak{B}(\mathbb{K}_i^1)$ by

$$\mathsf{P}_i(B) = \mathsf{P}\{X_i \in B\}, \qquad B \in \mathfrak{B}(\mathbb{K}_i^1).$$

Define Ω as the Cartesian product of the linear spaces $\mathbb{K}_i^1$ over all $i \in I$. A point $\omega \in \Omega$ is a sequence $\omega = \{\,\alpha_i \colon i \in I\,\}$. Let n be a positive integer, and let $i_1, \ldots, i_n$ be n distinct points in I. Define the map

$$\pi_{i_1, \ldots, i_n} \colon \Omega \to \mathbb{K}_{i_1}^1 \times \cdots \times \mathbb{K}_{i_n}^1$$

by $\pi_{i_1, \ldots, i_n}(\{\,\alpha_i \colon i \in I\,\}) = (\alpha_{i_1}, \ldots, \alpha_{i_n})$. A *cylindrical set* is a subset of Ω of the form $\pi_{i_1, \ldots, i_n}^{-1}(B)$, where B is a Borel set in $\mathbb{K}_{i_1}^1 \times \cdots \times \mathbb{K}_{i_n}^1$. Define $\mathfrak{F}$ as the minimal σ-field that contains all cylindrical sets and P as the product over all $i \in I$ of the measures P_i.

Assume that $\sum_{i \in I} \mathsf{V}[X_i] < \infty$, and define

$$\mathbf{X}(p) = \sum_{i \in I} X_i \mathbf{s}_i(p).$$

This series converges in $L^2(E)$ with probability 1 by (Vakhania, Tarieladze, and Chobanyan 1987, p. 331). The properties of this Gaussian random section will be further investigated in Chapter 3.

1.12   Bibliographical Remarks

For further information about (skew) fields as well as linear and multi-linear algebra, see (Bourbaki 2003; Lang 2002; Waerden 1991a, 1991b). Sir William Rowan Hamilton invented quaternions in W. R. Hamilton (1844).

More information about the abstract index notation may be found in Penrose and Rindler (1987) and Wald (1984). The term 'tensor' was introduced by W. R. Hamilton (1854) and appeared in the modern meaning in Voigt (1898).

Observe!

Different authors follow various conventions about Greek and Latin indices. See a table in (Misner, Thorne, and Wheeler 2017).

Tensor products that involve quaternionic spaces, are discussed in Grimus and Urbantke (1997)

William Kingdon Clifford invented Clifford algebras in Clifford (1873). More information about Clifford algebras may be found in (Abłamowicz and

Lounesto 1995; Abłamowicz et al. 2004; Abłamowicz and Fauser 2000; Anglès 2008; Benn and Tucker 1989; Budinich and Trautman 1988; Cartan 1981; Chevalley 1997; Garling 2011; Harvey 1990; Lawson and Michelsohn 1989; Lounesto 2001; Porteous 1995; Vaz and Rocha 2019). See also the papers (Budinich 2019; Floerchinger 2021; Trautman 1997).

Observe!

Many of the above authors omit the minus sign in Definition 17.

The spinors were discovered by Élie Cartan in Cartan (1913). See the historical survey by Diek and Kantowski (1995) and the references therein. We would like to mention (Benn and Tucker 1989; Budinich and Trautman 1988; Carmeli and Malin 2000; Chevalley 1997; Cartan 1981; Dąbrowski 1988; Harvey 1990; Lounesto 2001; Vaz and Rocha 2019).

The modern notion of a manifold goes back to the doctoral thesis by Riemann (1851) and his habilitation lecture 'On the hypotheses which lie at the bases of geometry' (1854), see Riemann (2016). In Henri Poincaré's *Analysis Situs* (1895), a definition of a manifold as a level set of a differentiable function is given, see Poincaré (2010). The modern definition through atlases appeared in Whitney (1936). See the historical account in (Scholz 1980, 1999).

For books in modern differential geometry, see (Berger 2003, 2009; Besse 2008; Chern 1995; Greub, Halperin, and Vanstone 1972, 1973, 1976; Husemöller et al. 2008; Huybrechts 2005; Kobayashi 1995, 2014; Kobayashi and Nomizu 1996a, 1996b; Michor 2008; Morrow and Kodaira 2006; Munteanu 2004; Nakahara 2003; Okonek, Schneider, and Spindler 2011; Robbin and Salamon 2022; Rudolph and Schmidt 2013, 2017; Sharpe 1997; Trautman 1984; Taubes 2011; Yano 1965).

Variational calculus and Euler–Lagrange equations are discussed in (Giaquinta and Hildebrandt 1996a, 1996b; Morrey 2008)

The theory of Lie groups was created by Sophus M. Lie, see the English translation Lie (2015) of his book. The structure theory of Lie groups was developed in the series of papers by Wilhelm Killing (Killing 1888a, 1888b, 1889a, 1889b, 1890). Some gaps in Killing's proofs were corrected in the thesis by Cartan (1894). For recent monographs and textbooks on this subject, see (Duistermaat and Kolk 2000; Gallier and Quaintance 2020a, 2020b; Godement 2017; Hall 2015; Hsiang 2017; Tapp 2016).

It is well-known that some of the low-dimensional so-called simple Lie groups that belong to different series, are isomorphic. The explicit form of isomorphisms can be found in (Gilmore 2012; Yokota and Miyashita 1990).

Group actions are investigated in Bredon (1972).

Gauge theory lies in the very heart of physics. For texts in it, see (M. J. D. Hamilton 2017; Naber 2011) among others.

For textbooks and monographs in variational calculus, see Rindler (2018).

For early history of representation theory, see Curtis (1999). The definition of the group representation appeared in Frobenius (1897). The Schur Lemma appeared in Schur (1906) and the Frobenius–Schur indicator in Frobenius and Schur (1906).

Books in representation theory include (Altmann 1986; Bröcker and Dieck 1995; Jeevanjee 2015; Fulton and Harris 1991; Weyl 1997). See also the paper by Baez and Huerta (2010) among others.

Harmonic analysis constitutes a separate part of representation theory. See (Folland 2016; Wallach 1973)

Spin structures on manifolds are described in Bourguignon et al. (2015). See also the papers (Francis and Kosowsky 2005; Friedrich and Trautman 2000; Trautman 1998, 2008).

The *Infeld–van der Waerden symbols* were introduced by Infeld and Waerden (1933), see also (Cardoso 2005; Harish-Chandra 1946).

The Newman–Penrose formalism was introduced by Newman and Penrose (1962). See also (Campbell and Wainwright 1977; Ko, Newman, and Penrose 1977; Penrose 1967).

Induced representations of groups were invented by Frobenius (1898), where he also proved his reciprocity theorem. For locally compact topological groups, the theory of induced unitary representations was developed by Mackey (1949, 1951, 1952, 1953) and Blattner (1961). See also Kaniuth and Taylor (2013).

In our treatment of random elements in complex and quaternionic finite-dimensional linear spaces, we follow (Vakhaniya 1998; Vakhaniya and Kandelaki 1996). For more deep studies in modern measure-theoretical probability, we recommend (Brémaud 2020; Durrett 2019; Gentili 2020; Grimmett and Stirzaker 2020; Kallenberg 2021; Klenke 2020; Le Gall 2022; Song, Park, and Yoon 2022).

2

Space-Times and Spacetimes

We start introductory notes to this chapter by the two fundamental observations. Trautman (1965, p. 101) formulates the first one as follows.

> ... space and time can be represented by a 4-dimensional differentiable manifold.

We denote that manifold by M and suppose that it is *smooth*. An arbitrary point $p \in M$ is called an *event*.

According to Trautman (1965, p. 101), the second fundamental observation is:

> ... the differentiable manifold of space and time is endowed with an affine connection whose geodesics form a privileged set of world-lines in space-time.

In our terms, the Trautman's 'affine connection' is a Koszul connection, we denote it by ∇_X in the coordinate-free notation, by Γ_{ij}^k in the abstract index notation, and by $\Gamma_{\alpha\beta}^\gamma$ in a given chart $(\mathcal{U}, \mathbf{x})$ of M. We consider only symmetric connections. By definition, a *world-line* is the image $\gamma(\mathbb{R}^1) \subset M$ of a smooth map $\gamma \colon \mathbb{R}^1 \to M$.

Remark 12. In some cases, only a subset of the set of all geodesic line of a given connection forms a privileged set. An example of such a situation is given in Section 2.1.

Penrose (1968) defines five types of space-times and spacetimes: Aristotelian, Galilean, Newtonian, Minkowskian, and Einsteinian. We describe them in Sections 2.1–2.5. We dub the first three as *space-times*, where time is clearly separated from space. The last two are *spacetimes*, the unified structures.

Observe!

To avoid a kind of a 'religious war' between the followers of different sign conventions, we use both of them. Specifically, when the Minkowskian spacetime is used for the purposes of quantum physics, we use the mostly minuses convention (otherwise, the relationship between mass and four momentum becomes $m^2 = -p^\mu p_\mu$) which is strange. On the other hand,

DOI: 10.1201/9781003344353-2

when the Minkowskian spacetime becomes the tangent space to the Einsteinian one, we use the mostly pluses convention. Otherwise, the spatial part of the metric becomes negative, which is also strange.

See also a detailed discussion in Burgess and Moore (2007, Appendix E).

Following Penrose (1968), we model the Aristotelian space-time as the Cartesian product $M = E^3 \times E^1$ of two affine spaces E^3 and E^1 of dimensions 3 and 1, where the points of M are events. The manifold M is partitioned into the family of copies of E^3 indexed by time moments belonging to E^1. Two events belonging to the same copy are simultaneous. An observer is a one-dimensional affine subspace of M that is not contained in any copy, that is, somebody who is present at each instant of time. For a fixed space point $N \in E^3$, the observer $\{(N, P): P \in E^1\}$ is at rest.

A Galilean space-time is a fibre bundle over a one-dimensional affine space whose points are time instants. The fibres of the bundle are three-dimensional affine subspaces of the total space E of the bundle. The affine transformations of E that preserve the bundle structure, form a group, and the Galilean relativity principle state that the Euler–Lagrange equations of a mechanical system are invariant with respect to the above group.

One needs to modify a Galilean space-time because the celebrated Maxwell equations do not satisfy the Galilean relativity principle. In a little bit more sophisticated scheme, we introduce a smooth function Φ defined on a Galilean space time and call it a *scalar field*. A gravitational field with the prescribed Lagrangian density given by Equation (2.2). With such a field and some additional mathematical structures described in Definition 98, we obtain what is called a Newtonian space-time. The unique features of that one are the *absolute time*, the *absolute space*, and the *gravitational force*. The weak equivalence principle states the equality of the two masses of any physical object: gravitational and inertial.

A trouble with Newtonian space-times is that the Maxwell equations can only be proved by introducing a special vector field called the *ether*. However, the existence of the ether has been disproved experimentally.

To overcome this difficulty, Einstein and his teacher of mathematics, Minkowski, introduced the *Minkowskian spacetime*, where the *Poincaré group* acts and mixes time with space. The modified relativity principle due to Einstein states that the Euler–Lagrange equations of a physical system are invariant with respect to the above group.

The neutrino field on the Minkowski spacetime is described by certain relativistic wave Equations (2.6) and (2.7). The electromagnetic field is described by the Maxwell equations. What about the gravitational one? Subsection 2.4.6 describes the trouble connected with introducing gravitation to the Minkowskian spacetime.

That problem was solved by Einstein. He replaced the weak equivalence principle with the strong one, supposed that equations of the gravitational field must be invariant in form under the action of spacetime diffeomorphisms (in physical language, 'covariant'), and that the source of a gravitational field is a metric g_{ij} defined on the spacetime. In Section 2.5, we demonstrate mathematically how these assumptions lead to the Einstein equations of general relativity.

Moreover, the Einstein equations together with the principle of isotropy of the Universe on large scales can be applied to the whole Universe. As a result, we obtain the Copernican principle: there is nothing special about the Earth's location, it is not the centre of the Universe as well as the celebrated Robertson–Walker metric and the Friedman equations.

The relativistic wave equations that describe the three physical fields constituting the cosmic backgrounds, are introduced twice: first on the Minkowskian spacetime using the theory of induced representations, and second, on Einstein spacetimes using the spinorial approach by Penrose and Rindler (1987). Under very mild physical conditions, the so-called geometrical optics approximation states that the waves are propagated along null geodesics.

Finally, we introduce the *gauge-equivalent classes* and show how this leads to the unique Stokes parameter in the case of the Cosmic Neutrino Background and to the four parameters in the case of the two remaining backgrounds.

2.1 The Aristotelian Space-Time

2.1.1 The Description by Penrose

To describe the Aristotelian space-time, we use affine spaces of Example 32. Consider two affine spaces: $(E^3, U^3, +)$ and $(E^1, U^1, +)$ with $\dim U^3 = 3$ and $\dim U^1 = 1$ and assume that U^3 and U^1 carry inner products. Set $M = E^3 \times E^1$.

Call a nonempty subset $F \subseteq E$ an *affine subspace* of E if the set

$$U_F = \{\, A - B \colon A, B \in F \,\}$$

is a linear subspace of U and the translation of F by any vector $\mathbf{x} \in U_F$ is a subset of F. By definition, the dimension of F is equal to that of U_F.

Choose a point $O \in M$ and consider the chart $(\mathcal{U}, \mathbf{x})$ with $\mathcal{U} = M$ and $\mathbf{x}(N) = N - O \in \mathbb{R}^4$, $N \in M$. In this particular chart, define the connection coefficients by $\Gamma^\gamma_{\alpha\beta} = 0$. The geodesic Equation (1.42) takes the form

$$\frac{\mathrm{d}^2 \gamma^\mu(t)}{\mathrm{d}t^2} = 0.$$

The solutions to this equation are one-dimensional affine subspaces of M.

On the other hand, the manifold M is partitioned by a family

$$\{\, E^3 \times \{N\} \colon N \in E^1 \,\}$$

of copies of the space E^3. Two events are simultaneous if and only if they belong to the same copy of E^3. A one-dimensional affine subspace is called an *observer* if it is not contained in any copy. Non-formally, an observer is somebody who is present at each instant of time.

Consider a tangent vector $X \in T_{(N_1, N_2)}M$ with $N_1 \in E^3$, $N_2 \in E^1$. X is called *timelike* if it is orthogonal to the affine subspace $E^3 \times \{N_2\}$. A geodesic line is timelike if all its tangent vectors are timelike. It is easy to see that the timelike geodesics are the family of one-dimensional affine subspaces of M parameterised by a point $N_1 \in E^3$. Each of them has the form

$$\{\, (N_1, N_2) \colon N_2 \in E^1 \,\}.$$

The observers of the above family are *observers at rest*. They are characterised by the fact that the spatial separation between any two points on the corresponding geodesic is equal to 0. Moreover, one can speak about the spatial separation between two events (N_1, N_2) and (P_1, P_2) even if they are not simultaneous, that is, $N_2 \neq P_2$. The above distance is equal to the distance between N_1 and P_1 in E^3.

2.1.2   The Modification by Ptolemy

In fact, Ptolemy modified the Aristotelian space-time in the following way. The space-time M is the Cartesian product $\mathbb{R}^3 \times (0, \infty)$. In the chart $(\mathcal{U}, \mathbf{x})$ with $\mathcal{U} = M$ and $\mathbf{x} \colon M \to \mathbb{R}^4$ with $\mathbf{x}(\mathbf{y}, t) = (\mathbf{y}, t)$, the observers at rest are the family of geodesics $\{\, (\mathbf{y}, t) \colon t \in (0, \infty) \,\}$ parameterised by $\mathbf{y} \in \mathbb{R}^3$.

What we obtain is the *Ptolemaic system*: there exists a special vector $\mathbf{0} \in \mathbb{R}^3$, the 'centre of the world', where the Earth is located at any time instance. Moreover, there exists a special time $t = 0$, the 'moment of creation'.

The only transformations that respect this structure are spatial rotations. That is, the physical laws should be same at any point on the globe. However, nothing may prevent them to depend on the altitude or even the time.

2.2   The Galilean Space-Time

2.2.1   Coordinate-Free Definitions

To 'forget' the special vector $\mathbf{0}$, we use affine spaces of Example 32. Consider an affine space $(E, U, +)$ with $\dim U = 4$ and call it a space-time. However, this definition does not reflect the special role of time coordinate.

Let U be a real linear space of dimension 4, and let $\tau \in U^*$ with $\tau \neq \mathbf{0}$. Assume that the linear subspace $\tau^{-1}(0)$ has an inner product $(\cdot, \cdot)$.

Definition 95. A *Galilean space-time* is a fibre bundle (E, π, I), where I is an affine space modelled upon $\mathbb{R}^1$, E is an affine space modelled upon U, and for any $t \in I$ the fibre $\pi^{-1}(t)$ is an affine subspace of E modelled on $\tau^{-1}(0)$.

The points of E are called *events* while the points of I are called *time instants*. The nonnegative real number $|\pi(A) - \pi(B)|$ is called the *time* between the events A and B. If $|\pi(A) - \pi(B)| = 0$, then the events A and B are called *simultaneous*. In that case, the nonnegative real number $\|B - A\|$ is called the *distance* between the simultaneous events A and B. The total space E is called the *Galilean universe*.

Intuitively, a world line is a history of something moving in the space-time. Formally, a *world line* is a smooth cross-section of a Galilean space-time. The world lines $\mathbf{s}(t)$ for which $\mathbf{s}(I)$ is an affine subspace of A, play a special role. It is easy to see that an affine subspace may serve as an image of a world line if and only if it intersects with each fibre at exactly one point.

A map $f \colon E \to E$ is called an *affine transformation* if there exists a linear operator $Df \colon U \to U$ with $f(A) - f(B) = Df(A - B)$ for all events A and B. We are interested in affine transformations that do not change the time between events and the distance between simultaneous events.

Definition 96. An affine transformation $f \colon E \to E$ is called *Galilean* if for any events A, B in E, we have $|\pi(f(A)) - \pi(f(B))| = |\pi(A) - \pi(B)|$ and for all time instances $t \in I$ and for all simultaneous events A, B in $\pi^{-1}(t)$, we have $\|f(B) - f(A)\| = \|B - A\|$.

It is easy to see that the set of all Galilean transformations is a group with respect to the composition called the *Galilean group*. To give a better description of this group, we introduce a coordinate system in the Galilean universe.

2.2.2   Introducing Coordinates

Choose $U = \mathbb{R}^1 \oplus \tau^{-1}(0)$ and fix a basis $\{\mathbf{e}_1, \mathbf{e}_2, \mathbf{e}_3\}$ in the space $\tau^{-1}(0)$ which is orthonormal with respect to the inner product $(\cdot, \cdot)$. The set $\{1, \mathbf{e}_1, \mathbf{e}_2, \mathbf{e}_3\}$ is a basis in U. Denote by t, x^1, x^2, x^3 the coordinates of a vector $\mathbf{x} \in U$ in the introduced basis. Let $E = U$ as sets. Then, any event $A \in E$ is a vector $\mathbf{x}_A \in U$, and U acts on E by $A + \mathbf{y} = B$, where $\mathbf{x}_B = \mathbf{x}_A + \mathbf{y}$. Identify I with $\mathbb{R}^1$ by the same procedure. Put $\pi(t, x^1, x^2, x^3) = t$. We constructed the Galilean space-time $(\mathbb{R}^1 \oplus \tau^{-1}(0), \pi, \mathbb{R}^1)$ with coordinates (t, x^1, x^2, x^3).

Let (E, π, I) be a Galilean space-time without coordinates, where E is modelled on U, and let O be an observer. Let U_O be the linear subspace of U upon which O is modelled. There exists a unique vector $\mathbf{v}_O \in U_O$ with $\tau(\mathbf{v}_O) = 1$ (the *velocity* of the observer O).

Define the map $h_O \colon E \to \mathbb{R}^1 \oplus \tau^{-1}(0)$ by $h_O(A) = (\tau(\mathbf{x}_A), \mathbf{x}_A - \tau(\mathbf{x}_A)\mathbf{v}_O)$. This is a special chart of the manifold E called an *inertial frame* for the observer O.

It turns out that for any observers O_1 and O_2 the map

$$h_{O_2} \circ h_{O_1}^{-1} \colon \mathbb{R}^1 \oplus \tau^{-1}(0) \to \mathbb{R}^1 \oplus \tau^{-1}(0)$$

is a Galilean transformation f which can be uniquely represented as the composition $f_1 \circ f_2 \circ f_3$, where:

- f_1 maps a vector $(t, \mathbf{x})^\top \in \mathbb{R}^1 \oplus \tau^{-1}(0)$ to the vector $(t, \mathbf{x} + \mathbf{v}t)^\top$, where $\mathbf{v} \in \tau^{-1}(0)$. This transformation is the *uniform motion* with velocity $\mathbf{v}$.

- f_2 maps a vector $(t, \mathbf{x})^\top \in \mathbb{R}^1 \oplus \tau^{-1}(0)$ to the vector $(t, R\mathbf{x})$, where $R \in \mathrm{O}(\tau^{-1}(0))$. This transformation is a *rotation* if R lies in the connected component of the identity element of the group $\mathrm{O}(\tau^{-1}(0))$, and a *rotation with reflection* otherwise.

- f_3 maps a vector $(t, \mathbf{x})^\top \in \mathbb{R}^1 \oplus \tau^{-1}(0)$ to the vector $(t + u, \mathbf{x} + \mathbf{y})^\top$. This transformation is a translation of the origin.

2.2.3   The Galilean Relativity Principle

The *Galilean relativity principle* states that the laws of physics are the same in all inertial frames. Experiments show that the above principle holds true with the great accuracy in classical mechanics. How to give the exact sense to this formulation?

Let (E, π, I) be a Galilean space-time, where E is modelled upon U. Let a and b with $a < b$ be two real numbers. We introduce a smooth function $L \colon E \times E \times [a, b] \to \mathbb{R}^1$ and call it a *Lagrangian* of a mechanical system.

Fix an observer O and the corresponding inertial frame (t, x^1, x^2, x^3). Let $x^i \colon \mathbb{R}^1 \to E$ be a smooth map expressed in the above frame. Denote $A = x^i(a)$, $B = x^i(b)$, $\dot{x}^i = \mathrm{d}x^i(t)/\mathrm{d}t$, $\gamma = x^i([a, b]) \subset E$, that is, γ is a smooth curve joining A and B.

Definition 97. The functional

$$S[\gamma] = \int_\gamma L(x^i(t), \dot{x}^i(t), t)\, \mathrm{d}t$$

is called the *action* of a mechanical system with Lagrangian L that corresponds to the *motion* γ.

The *Hamilton principle of least action* states that a mechanical system is completely determined by its Lagrangian L, and the motions of the system coincide with the extremals of its action defined by the Euler–Lagrange equations

$$\frac{\partial L}{\partial x^i} - \frac{\mathrm{d}}{\mathrm{d}t}\frac{\partial L}{\partial \dot{x}^i} = 0.$$

Another useful form for the above equations is $\dot{p}_i = f_i$, where

$$p_i = \frac{\partial L}{\partial \dot{x}^i}, \qquad f_i = \frac{\partial L}{\partial x^i},$$

and where f_i is the *force*, while p_i is the *impulse*. The function

$$E(x^i(t), \dot{x}^i(t), t) = L(x^i(t), \dot{x}^i(t), t) - \dot{x}^i f_i$$

is called the *energy* of the mechanical system.

The Galilean relativity principle is as follows: the Euler–Lagrange equations of a mechanical system are invariant with respect to the Galilean group.

Example 67. The Lagrangian of a free particle of *inertial mass* m is

$$L(x^i(t), \dot{x}^i(t), t) = \frac{m}{2}((\dot{x}^1)^2 + (\dot{x}^2)^2) + (\dot{x}^3)^2).$$

We have $f_i = 0$, $p_i = m\dot{x}^i$, and the Euler–Lagrange equations take the form $m\ddot{x}^i = 0$, which is the first Newton law. The motions are straight lines, and the energy is the kinetic energy $E = \frac{1}{2}m\|\mathbf{v}\|^2$, where $\mathbf{v} = (\dot{x}^1, \dot{x}^2, \dot{x}^3)^\top$ is the velocity of the particle.

Example 68. Define a *potential* as a smooth function $U(t, x^i)$ defined on an open subset of the Galilean space-time. The Lagrangian of a free particle of inertial mass m is defined by

$$L(x^i(t), \dot{x}^i(t), t) = \frac{m}{2}((\dot{x}^1)^2 + (\dot{x}^2)^2) + (\dot{x}^3)^2) - U(t, x^1, x^2, x^3).$$

Then we have $f_i = -\frac{\partial U}{\partial x_i}$, $p_i = m\dot{x}^i$. The Euler–Lagrange equations take the form $m\ddot{x}^i = f_i$, which is the second Newton law.

2.2.4 A Trouble with the Galilean Space-Time

The Maxwell equations in the Galilean space-time are as follows:

$$\frac{1}{c}\frac{\partial \mathbf{B}}{\partial t} + \nabla \times \mathbf{E} = 0, \quad \frac{1}{c}\frac{\partial \mathbf{E}}{\partial t} - \nabla \times \mathbf{B} = 0, \tag{2.1}$$
$$\nabla \cdot \mathbf{B} = 0, \qquad\qquad \nabla \cdot \mathbf{E} = 0,$$

where $\mathbf{E}$ (resp., $\mathbf{B}$) is the electric (resp., magnetic) field, and where c is the speed of light in vacuum. We prove that they are *not* invariant with respect to the Galilean group.

Indeed, assume they are invariant. Write the first equation in the form

$$\nabla \times \mathbf{E} = -\frac{1}{c}\frac{\partial \mathbf{B}}{\partial t},$$

and multiply both hand sides by ∇ from the left:

$$\nabla \times (\nabla \times \mathbf{E}) = -\frac{1}{c}\nabla \times \frac{\partial \mathbf{B}}{\partial t}.$$

On the left hand side, apply the vector identity $\nabla \times (\nabla \times \mathbf{E}) = \nabla(\nabla \cdot \mathbf{E}) - \nabla^2 \mathbf{E}$. We obtain

$$\nabla(\nabla \cdot \mathbf{E}) - \nabla^2 \mathbf{E} = -\frac{1}{c}\frac{\partial}{\partial t}(\nabla \times \mathbf{B}).$$

The remaining Maxwell equations produce the classical wave equation

$$\frac{1}{c^2}\frac{\partial^2 \mathbf{E}}{\partial t^2} - \nabla^2 \mathbf{E} = 0,$$

or

$$\frac{1}{c^2}\frac{\partial^2 E_i}{\partial t^2} - \frac{\partial^2 E_i}{\partial (x^1)^2} - \frac{\partial^2 E_i}{\partial (x^2)^2} - \frac{\partial^2 E_i}{\partial (x^3)^2} = 0.$$

A simple exercise in the Chain Rule shows that the last three terms are invariant under the Galilean transformation $x^{1'} = x^1 + vt$, while the first term transforms as follows:

$$\frac{\partial E_i}{\partial t} = \frac{\partial E_i}{\partial x^{1'}}\frac{\partial x^{1'}}{\partial t} + \frac{\partial E_i}{\partial t'}\frac{\partial t'}{\partial t} = \frac{\partial E_i}{\partial t'} + v\frac{\partial E_i}{\partial x^{1'}},$$

$$\frac{\partial^2 E_i}{\partial t^2} = \frac{\partial E_i}{\partial x^{1'}}\left(\frac{\partial E_i}{\partial t'} + v\frac{\partial E_i}{\partial x^{1'}}\right)\frac{\partial x^{1'}}{\partial t} + \frac{\partial E_i}{\partial t'}\left(\frac{\partial E_i}{\partial t'} + v\frac{\partial E_i}{\partial x^{1'}}\right)\frac{\partial t'}{\partial t}$$

$$= \frac{\partial^2 E_i}{\partial (t')^2} + 2v\frac{\partial^2 E_i}{\partial x^{1'}\partial t'} + v^2\frac{\partial^2 E_i}{\partial (x^{1'})^2}.$$

We conclude that *the Maxwell equations do not satisfy the Galilean relativity principle*. In other words, either the Maxwell equations are wrong or the Galilean relativity principle must be modified.

The first alternative is contradicted by the experimental data. The possible solution may be as follows. There exists an algorithm for calculating the symmetry group for any system of either ordinary or partial differential equations developed by Sophus Lie. In the case of Maxwell's equations, this algorithm produces the so-called conformal group. Experiments show that there exist laws of classical physics that are not invariant with respect to that group. Einstein discovered that they are invariant with respect to a special subgroup of the conformal group.

2.3   Newtonian Space-Times

Up to now, the space-time structure was uniquely defined. In contrast, there exist many non-isomorphic Newtonian space-times.

2.3.1   Definitions

Let M be a smooth manifold. Assume the existence of a $\mathbb{R}^4$-valued chart $(\mathcal{U}, x^\alpha)$ with $\mathcal{U} = M$. Equivalently, M is homeomorphic to $\mathbb{R}^4$. Let $\Gamma^\alpha_{\beta\gamma}$ be a symmetric Koszul connection on M, let $R^\alpha{}_{\beta\gamma\delta}$ be its curvature tensor, $R_{\alpha\beta} = R^\mu{}_{\alpha\beta\mu}$ its Ricci tensor. Let $g^{\alpha\beta}$ be a symmetric tensor field on M. Denote

$$R^\alpha{}_\beta{}^\gamma{}_\delta = g^{\gamma\mu} R^\alpha{}_{\beta\mu\delta}.$$

Let $t\colon M \to \mathbb{R}^1$ be a smooth function such that for each $\alpha \in \mathbb{R}^1$ the set $t^{-1}(\alpha) \subset M$ is homeomorphic to $\mathbb{R}^3$, and denote $t_\alpha = \frac{\partial t}{\partial x^\alpha}$. Finally, let $\rho\colon M \to \mathbb{R}^1$ be the density of matter.

Definition 98. A tuple $(M, \Gamma^\alpha_{\beta\gamma}, g^{\alpha\beta}, t, \rho)$ is called a *Newtonian space-time* if the following conditions are satisfied.

- $R^\mu{}_{\mu\alpha\beta} = 0$.

- There exist three linearly independent vector fields X, Y, and Z with

$$X^\alpha t_\alpha = Y^\alpha t_\alpha = Z^\alpha t_\alpha = 0, \qquad \nabla_\alpha X^\beta = \nabla_\alpha Y^\beta = \nabla_\alpha Z^\beta = 0.$$

- $R^\alpha{}_\beta{}^\gamma{}_\delta = R^\gamma{}_\delta{}^\alpha{}_\beta$.

- $\nabla_\gamma g^{\alpha\beta} = 0$.

- The rank of the matrix $g^{\alpha\beta}$ is 3, $g^{\alpha\beta} t_\beta = 0$.

- Free falls follow the geodesics of the connection $\Gamma^\alpha_{\beta\gamma}$.

- The gravitational field equations are $R_{\alpha\beta} = 4\pi G \rho t_\alpha t_\beta$, where G is the Newton gravitational constant.

The last point requires explanation.

2.3.2 Gravitational Field

Let Φ be a smooth real-valued function defined on E. Call it a *scalar field*. Denote a point in E by $(t, \mathbf{x})$ with $\mathbf{x} = (x^1, x^2, x^3)$, and $\nabla\Phi = (\partial_{x_1}\Phi, \partial_{x_2}\Phi, \partial_{x_3}\Phi)$, where we introduced a shortcut $\partial_{x_i}\Phi = \frac{\partial\Phi}{\partial x^i}$, $1 \leq i \leq 3$. Make the following assumption.

There is a function $\mathcal{L}\colon \mathbb{R}^9 \to \mathbb{R}$ called the Lagrangian density *of the field Φ and a closed set $\Omega \subseteq \mathbb{R}^3$ such that the Lagrangian L of the field Φ has the form*

$$L = \int_\Omega \mathcal{L}(\Phi, \dot{\Phi}, \mathbf{x}, t)\, d\mathbf{x},$$

where we put $\dot{\Phi} = (\frac{\partial\Phi}{\partial t}, \nabla\Phi)$.

The *action* of the field Φ becomes

$$S[\Phi] = \int_{t_0}^{t_1} \int_\Omega \mathcal{L}(\Phi, \dot{\Phi}, \mathbf{x}, t)\, d\mathbf{x}\, dt.$$

Definition 99. The *variational derivative* of the action with respect to the field is given by

$$\frac{\delta S}{\delta \Phi} = \partial_t \left(\frac{\partial \mathcal{L}}{\partial(\partial_t \Phi)} \right) + \boldsymbol{\nabla} \cdot \left(\frac{\partial \mathcal{L}}{\partial(\boldsymbol{\nabla}\Phi)} \right) - \frac{\partial \mathcal{L}}{\partial \Phi}.$$

It may be proved that the Euler–Lagrange equation for the field Φ has the form

$$\frac{\delta S}{\delta \Phi} = 0,$$

see Vecchiato (2017).

Assume that the Lagrangian density of a field Φ has the form

$$\mathcal{L} = \frac{1}{2}k(\boldsymbol{\nabla}\Phi \cdot \boldsymbol{\nabla}\Phi) + V(\Phi),$$

where k is a nonzero constant and the *interaction potential energy* $V(\Phi)$ describes a 'self-interaction' of the field Φ with itself. A field is called *free* if $V(\Phi) = 0$. The variational derivative of the action with respect to a free field becomes

$$\frac{\delta S}{\delta \Phi} = \boldsymbol{\nabla} \cdot \left(\frac{\partial \mathcal{L}}{\partial(\boldsymbol{\nabla}\Phi)} \right) = k\nabla^2 \Phi.$$

We obtain: a free field satisfies the Laplace equation $\nabla^2 \Phi = 0$.

Call a field *gravitational* if its Lagrangian density has the form

$$\mathcal{L} = \frac{1}{8\pi G_N}(\boldsymbol{\nabla}\Phi \cdot \boldsymbol{\nabla}\Phi) + \rho(t, \mathbf{x})\Phi, \qquad (2.2)$$

where G_N is the Newton gravitational constant and $\rho(t, \mathbf{x})$ is the density of the *gravitational mass* at the point $(t, \mathbf{x}) \in E$. We say that the gravitational mass is the *source* of the gravitational field.

A simple calculation shows that the Euler–Lagrange equations for the gravitational field take the form

$$\nabla^2 \Phi = 4\pi G_N \rho,$$

that is, the gravitational field satisfies the Poisson equation.

At this point, it is convenient to introduce an axiom.

Axiom 1 (The weak equivalence principle). *The inertial mass of any physical object is equal to its gravitational mass.*

Definition 100. A function $\varphi \colon M \to \mathbb{R}^1$ is called a *gravitational potential* if it satisfies equation

$$g^{\alpha\beta}\nabla_\alpha\nabla_\beta\varphi = 0.$$

Theorem 15. *For an arbitrary gravitational potential φ, there exists a special chart (M, y^α) for which we have*

$$\Gamma^\alpha_{\beta\gamma} - g^{\alpha\delta}\frac{\partial\varphi}{\partial y^\delta}t_\beta t_\gamma = 0, \qquad t_\alpha = \delta^0{}_\alpha, \qquad g = \begin{pmatrix} 0\ 0\ 0\ 0 \\ 0\ 1\ 0\ 0 \\ 0\ 0\ 1\ 0 \\ 0\ 0\ 0\ 1 \end{pmatrix}.$$

The chart in not unique, but any two charts are connected with a Galilean transformation.

The physical interpretation of this mathematical result is as follows. The function t is the *absolute time*. For any time instance $\alpha \in \mathbb{R}^1$, the tensor field g determines a metric in the instance $t^{-1}(\alpha)$ of the *absolute space*. In an *inertial coordinate system* (M, y^α), the gravitational field

$$\tilde{\Gamma}^\alpha_{\beta\gamma} = \Gamma^\alpha_{\beta\gamma} - g^{\alpha\delta}\frac{\partial\varphi}{\partial y^\delta}t_\beta t_\gamma$$

vanishes. The equation of a geodesic line determines the trajectory of free fall and becomes

$$\frac{\mathrm{d}^2 y^\alpha}{\mathrm{d}t^2} = -\frac{\partial\varphi}{\partial y^\alpha}, \qquad 1 \le \alpha \le 3.$$

The right hand side of this equation is the *gravitational force*. If the chart is not inertial, the above equation becomes

$$\frac{\mathrm{d}^2 y^\alpha}{\mathrm{d}t^2} + \tilde{\Gamma}^\alpha_{\beta\gamma}\frac{\mathrm{d}y^\beta}{\mathrm{d}t}\frac{\mathrm{d}y^\gamma}{\mathrm{d}t} = -\frac{\partial\varphi}{\partial y^\alpha}, \qquad 1 \le \alpha \le 3.$$

The second term in the left hand side describes the *inertial forces*, for example, centrifugal, Coriolis, etc. The gravitational field equations become the Poisson equation

$$\Delta\varphi = 4\pi G\rho.$$

The tensor $g^{\alpha\beta}$ does not completely determine the gravitational connection $\Gamma^\alpha_{\beta\gamma}$. The non-uniqueness is controlled by the following observation: if φ^α is a gravitational force, then for arbitrary smooth functions $a(t)$, $b(t)$, and $c(t)$, the vector field

$$\tilde{\varphi}^\alpha = \varphi^\alpha + a(t)X^\alpha + b(t)Y^\alpha + c(t)Z^\alpha$$

is also a gravitational force.

2.3.3 A Trouble with Newtonian Space-Times

To obtain the Maxwell equations, we introduce a vector field u^α satisfying $u^\alpha t_\alpha = 1$ and call it the *ether*. The metric $\eta^{\alpha\beta} = g^{\alpha\beta} - c^{-2}u^\alpha u^\beta$ is non-degenerate. Denote by $F_{\alpha\beta}$ the skew-symmetric tensor of the electromagnetic field, and let us raise its indices with the help of $\eta^{\alpha\beta}$: $F^{\alpha\beta} = \eta^{\alpha\gamma}\eta^{\beta\delta}F_{\gamma\delta}$. The Maxwell equations in the absence of sources are

$$\frac{\partial F_{\alpha\beta}}{\partial x^\gamma} + \frac{\partial F_{\beta\gamma}}{\partial x^\alpha} + \frac{\partial F_{\gamma\alpha}}{\partial x^\beta} = 0, \qquad \nabla_\beta F^{\alpha\beta} = 0.$$

In an inertial coordinate system, we have $u^\alpha = \delta^\alpha{}_0$. Denote

$$\mathbf{E} = c^{-1}(F_{10}, F_{20}, F_{30})^\top, \qquad \mathbf{B} = (F_{23}, F_{31}, F_{12})^\top.$$

We obtain the classical Maxwell Equation (2.1).

The problem is that experiments disproved the existence of any kind of ether.

2.4   The Minkowskian Spacetime

Einstein supposed that the elements $g^{\alpha\beta}$ and t of a Newtonian space-time do not have any physical significance, while the element $\eta^{\alpha\beta}$ does. Indeed, that is what he obtained.

2.4.1   Basic Definitions

Definition 101. A *Minkowski spacetime* is a triple $(E, U, (\cdot, \cdot))$, where $(\cdot, \cdot)$ is a symmetric non-degenerate bilinear form on a real linear space U of dimension 4 with one positive and three negative eigenvalues, and where E is an affine space modelled on V.

The points of E are again called *events*, while the space E is called a *Minkowski universe*.

We identify E and U with the space $\mathbb{R}^{1,3}$. The contravariant coordinates are denoted by $x^0 = ct$, x^1, x^2, and x^3, where c is the speed of light in vacuum. In this chart, an event has coordinates x^μ. The symmetric bilinear form (1.1) becomes $\eta(x^\mu, y^\nu) = \eta_{\mu\nu}x^\mu y^\nu$, where the pseudo-Riemannian metric $\eta_{\mu\nu}$ has the form $\eta_{00} = c^2$, $\eta_{\alpha\alpha} = -1$ for $1 \le \alpha \le 3$, and $\eta_{\mu\nu} = 0$ if $\mu \ne \nu$. This convention is known as *timelike convention, particle physics convention, west coast convention, Landau–Lifshitz convention*, or just *mostly minuses*. This convention was used by Richard Feynmann living on the West Coast of the US and in Landau and Lifshitz (1975). In the above chart, the Minkowski universe $(\mathbb{R}^{1,3}, \eta_{\mu\nu})$ is a pseudo-Riemannian manifold.

We can also write our convention in the following form.

$$\eta^{\mu\nu} = S_1 \begin{pmatrix} -c^2 & 0 & 0 & 0 \\ 0 & 1 & 0 & 0 \\ 0 & 0 & 1 & 0 \\ 0 & 0 & 0 & 1 \end{pmatrix} \tag{2.3}$$

with $S_1 = -1$.

It is also useful to introduce the covariant components with the help of the index lowering: $x_\mu = \eta_{\mu\nu}x^\nu$. Moreover, let $\eta^{\mu\nu}$ be the matrix inverse to $\eta_{\mu\nu}$, that is, $\eta^{\mu\rho}\eta_{\rho\nu} = \delta^\mu_\nu$. We have $\eta^{00} = c^{-2}$, $\eta_{\alpha\alpha} = -1$ for $1 \le \alpha \le 3$, and $\eta_{\mu\nu} = 0$ if $\mu \ne \nu$. The index raising is performed by $x^\mu = \eta^{\mu\nu}x_\nu$. The bilinear form $\eta(x^\mu, y^\nu)$ may be written in four different ways:

$$\eta(x^\mu, y^\nu) = \eta_{\mu\nu}x^\mu y^\nu = \eta^{\mu\nu}x_\mu y_\nu = x^\mu y_\mu = x_\mu y^\mu.$$

The group $O(1,3)$ preserves the metric $\eta_{\mu\nu}$. It contains a subgroup $SO_0(1,3)$. The transformation $x^\mu \mapsto L^\mu_\nu x^\nu$, $x^\mu \in \mathbb{R}^{1,3}$, $L^\mu_\nu \in SO_0(1,3)$, is called the *homogeneous Lorentz transformation*.

The corresponding *affine* transformation of the Minkowski universe has the form $x^\mu \mapsto L^\mu_\nu x^\nu + a^\mu$. The corresponding group of transformations is the Cartesian product $SO_0(1,3) \boxtimes \mathbb{R}^{1,3}$. However, physicists use another group, $G = \text{Spin}_+(1,3) \boxtimes \mathbb{R}^{1,3}$. This group acts on the Minkowski universe by

$$(L, \mathbf{a})\mathbf{x} = \pi(L)\mathbf{x} + \mathbf{a}, \qquad (L, \mathbf{a}) \in G, \quad \mathbf{x} \in \mathbb{R}^{1,3}, \tag{2.4}$$

where π is the two-fold covering (1.39). The group G is called the *proper orthochronous Poincaré group*. The transformation (2.4) is called the *inhomogeneous Lorentz transformation*.

Example 69. We find the homogeneous Lorentz transformation between two Minkowski observers, where the second observer $x^{\mu'}$ moves along the x^1 axis of the first observer x^μ with constant velocity v. The matrix of the transformation may only change $x^0 = ct$ and $x^{0'} = ct'$. The interval $c^2t^2 - (x^1)^2$ remains constant under a hyperbolic rotation by an 'angle' ψ:

$$x^0 = x^{0'}\cosh\psi + x^{1'}\sinh\psi, \qquad x^1 = x^{0'}\sinh\psi + x^{1'}\cosh\psi.$$

For the second observer, we have $x^{1'} = 0$, and her motion in the inertial frame x^μ is described as $x^0 = x^{0'}\cosh\psi$, $x^1 = x^{0'}\sinh\psi$. Then we have

$$\frac{x^1}{x^0} = \frac{x^1}{ct} = \frac{v}{c} = \tanh\psi.$$

This gives $\cosh\psi = \gamma = (1 - v^2/c^2)^{-1/2}$, $\sinh\psi = \gamma v/c$, and the Lorentz transformation takes the form

$$t = \gamma t' + \frac{\gamma v}{c^2}x^{1'}, \qquad x^1 = \gamma(x^{1'} + vt'). \tag{2.5}$$

We observe that $\gamma \to 1$ as $c \to \infty$, and the classical mechanics limit gives the Galilean transformation $t = t'$, $x^1 = x^{1'} + vt'$ and the Galilean spacetime. The motion with $v > c$ is impossible, because the coordinates cannot be imaginary. The inertial system cannot move with the velocity of light, because the coordinates become infinite.

We divide the Minkowski universe into three nonintersecting regions.

Definition 102. A vector x^μ is called *timelike* (resp., *lightlike*, resp., *spacelike*) if $x^\mu x_\mu > 0$ (resp., $x^\mu x_\mu = 0$ and $x^\mu \neq 0$, resp., $x^\mu x_\mu < 0$ or $x^\mu = 0$). A vector x^μ is called *causal* if it is not spacelike.

It is easy to see that the *causal cone*, the set $\mathcal{C}$ of all causal vectors is a non-intersecting union of two sets $\mathcal{C}_+$ and $\mathcal{C}_-$, where

$$\mathcal{C}_+ = \{\, x^\mu \in \mathbb{R}^{1,3} \colon x^\mu x_\mu \geq 0, x^0 > 0 \,\},$$
$$\mathcal{C}_- = \{\, x^\mu \in \mathbb{R}^{1,3} \colon x^\mu x_\mu \geq 0, x^0 < 0 \,\}.$$

Definition 103. The set $\mathcal{C}_+$ (resp., $\mathcal{C}_-$) is called the *future causal cone* (resp., the *past causal cone*). The vectors of $\mathcal{C}_+$ (resp., $\mathcal{C}_-$) are called *future-pointing* (resp., *past-pointing*).

Observe that the future causal cone $\mathcal{C}_+$ is a subset of the Minkowski universe $\mathbb{R}^{1,3}$, while the forward light cone $\hat{\mathcal{C}}_+$ is a subset of another copy, $\hat{\mathbb{R}}^{1,3}$.

Definition 104. A *world line* is an image $\gamma(I)$ of a smooth curve $\gamma \colon I \to M$, where I is an opened interval in $\mathbb{R}^1$, such that for every $s \in I$ the vector $\dot{\gamma}(s)$ is timelike and future-pointing.

Some world lines are more important than others.

Definition 105. A *Minkowski observer* is a one-dimensional affine subspace of $\mathbb{R}^{1,3}$ parallel to the x^0-axis.

The frame x^μ is centred at the Minkowski observer who is the x^0 axis. The *inertial frames* of all other observers are obtained by applying an inhomogeneous Lorentz transformation (2.4).

2.4.2   The Einstein Relativity Principle

Before 1957, this principle was formulated as follows: the Euler–Lagrange equations of a physical system are invariant with respect to the group $\mathrm{Pin}(1,3) \boxtimes \mathbb{R}^{1,3}$.

Consider two 4×4 diagonal matrices: the matrix P with diagonal elements $P^0{}_0 = 1$, $P^1{}_1 = P^2{}_2 = P^3{}_3 = -1$, and $T = -P$. The matrix P is called the *space inversion*, while T is called the *time-reversal*. The group $\mathrm{O}(1,3)$ is the non-intersecting union of four connected components:

$$\mathrm{O}(1,3) = \mathrm{SO}_0(1,3) \cup \{\, gP \colon g \in \mathrm{SO}_0(1,3) \,\} \cup \{\, gT \colon g \in \mathrm{SO}_0(1,3) \,\}$$
$$\cup \{\, gPT \colon g \in \mathrm{SO}_0(1,3) \,\}.$$

The union of the first and fourth term in the right hand side is the group $\mathrm{SO}(1,3)$.

An experiment performed in 1957 showed that there exist physical processes that are not invariant with respect to the space inversion. Later we will discuss some important consequences of this result. Similar result holds true for the time-reversal. As a result, nowadays, the Einstein Relativity Principle is formulated as follows: the Euler–Lagrange equations of a physical system are invariant with respect to the group $\mathrm{SL}(2, \mathbb{C}) \boxtimes \mathbb{R}^{1,3}$.

2.4.3 The Energy-Momentum Tensor

The energy-momentum tensor is a *tensor field*, i.e., a function T defined on $\mathbb{R}^{1,3}$ and taking values in the ten-dimensional linear space of rank 2 symmetric tensors over $\mathbb{R}^{1,3}$. Its components have the following physical sense: $T^{00}(x^\alpha)$ is the energy density, $T^{\mu 0}(x^\alpha)$ and $T^{0\nu}(x^\alpha)$ with $1 \le \mu, \nu \le 3$ is the energy flux density in μ or ν direction, $T^{\mu\nu}(x^\alpha)$ with $\mu \ne 0$ and $\nu \ne 0$ is the μ-momentum flux density in ν-direction.

Example 70. The energy-momentum tensor of a perfect fluid with density ρ, pressure p, and velocity u^μ is

$$T^{\mu\nu} = (\rho + p)u^\mu u^\nu + p\eta^{\mu\nu}.$$

2.4.4 Neutrino Field

Let S (resp., tS) be the two-dimensional complex linear space of left-handed (resp., right-handed) Weyl spinors.

Definition 106. A *left-handed* (resp., *right-handed*) *Weyl field* is a map $\psi_A \colon U \to S$ (resp., $\psi_{\dot A} \colon U \to tS$) such that

$$\psi_A(g\mathbf{x}) = g \cdot \psi_A(\mathbf{x}), \qquad \psi_{\dot A}(g\mathbf{x}) = g \cdot \psi_{\dot A}(\mathbf{x}),$$

where the group $\mathrm{SL}(2, \mathbb{C})$ acts by matrix-vector multiplication in the left hand side of both equations, and by the representation $D^{(1/2,0)}$ (resp., $D^{(0,1/2)}$) in the right hand side of the first (resp., the second) equation.

Introduce the following notation: $\overline{\sigma}^\mu = (1, -\sigma^1, -\sigma^2, -\sigma^3)$. Consider the following Lagrangian density:

$$\mathcal{L}_L(x^\mu) = \mathrm{i}c\psi_A(x^\mu)^\top \overline{\sigma}^\mu \frac{\partial \psi_A}{\partial x^\mu}.$$

In order to obtain the Euler–Lagrange equations for the Weyl field, we vary the Lagrangian density with respect to $c\psi_A$, considering $c\psi_A$ and ψ_A as independent fields. We obtain $\frac{\partial \mathcal{L}}{\partial c\psi_A} = 0$, which gives

$$\frac{\partial \psi_A}{\partial x^0} - \sum_{\mu=1}^{3} \sigma^\mu \frac{\partial \psi_A}{\partial x^\mu} = 0. \tag{2.6}$$

This equation describes the *neutrino field*.

For a right-handed Weyl field, similar calculation with the Lagrangian density

$$\mathcal{L}_R(x^\mu) = \mathrm{i}c\psi_{A'}(x^\mu)^\top \overline{\sigma}^\mu \frac{\partial \psi_{A'}}{\partial x^\mu}$$

gives the equation

$$\frac{\partial \psi_{\dot A}}{\partial x^0} + \sum_{\mu=1}^{3} \sigma^\mu \frac{\partial \psi_{\dot A}}{\partial x^\mu} = 0. \tag{2.7}$$

There are three families of neutrino labelled ν_e, ν_μ, and ν_τ. At least two of them have small but nonzero masses. In what follows, we consider only massless neutrino, because the cosmic background of massive ones should be described by different methods.

The energy-momentum tensor of a left-handed neutrino field is

$$T^{\mu\nu} = i\psi_A^* \overline{\sigma}^\mu \frac{\partial \psi_A}{\partial x^\nu},$$

while that of a right-handed neutrino field is

$$T^{\mu\nu} = i\psi_{\dot{A}}^* \overline{\sigma}^\mu \frac{\partial \psi_{\dot{A}}}{\partial x^\nu},$$

where $*$ denotes the Hermitian conjugation.

2.4.5   Electromagnetic Field

An electromagnetic field is described by a vector field A^α in the Minkowski space called the *electromagnetic four-potential*. We define the *electromagnetic tensor* by

$$F^{\alpha\beta} = \frac{\partial A^\beta}{\partial x^\alpha} - \frac{\partial A^\alpha}{\partial x^\beta},$$

and the action of the electromagnetic field by

$$S[F] = -\frac{1}{16\pi c} \int_N F_{\alpha\beta} F^{\alpha\beta} \, d\mathbf{x},$$

where the integral is taken over the region

$$N = \{ \mathbf{x} \in \mathbb{R}^{1,3} \colon t_1 \leq t \leq t_2 \}$$

of the Minkowski space. We suppose that the electromagnetic four-potential tends to 0 for any fixed moment of time as the space coordinates tend to infinity.

Calculating the variation of S, we obtain

$$\begin{aligned}
\delta S &= -\frac{1}{16\pi c} \int_N \delta(F_{\alpha\beta} F^{\alpha\beta}) \, d\mathbf{x} = -\frac{1}{8\pi c} \int_N \delta(F_{\alpha\beta}) F^{\alpha\beta} \, d\mathbf{x} \\
&= -\frac{1}{8\pi c} \int_N \delta \left(\frac{\partial A^\beta}{\partial x^\alpha} - \frac{\partial A^\alpha}{\partial x^\beta} \right) F^{\alpha\beta} \, d\mathbf{x} \\
&= -\frac{1}{8\pi c} \int_N \left(\frac{\partial(\delta A^\beta)}{\partial x^\alpha} - \frac{\partial(\delta A^\alpha)}{\partial x^\beta} \right) F^{\alpha\beta} \, d\mathbf{x} \\
&= \frac{1}{4\pi c} \int_N \frac{\partial(\delta A^\beta)}{\partial x^\alpha} F^{\alpha\beta} \, d\mathbf{x},
\end{aligned}$$

where in the last equality we used the fact that the tensor $F_{\alpha\beta}$ is skew-symmetric. Integration by part gives

$$\delta S = -\frac{1}{4\pi c}\int_N \delta A^\alpha \frac{\partial F^{\alpha\beta}}{\partial x^\beta}\,\mathrm{d}\mathbf{x} - \frac{1}{4\pi c}\int_{\partial N} F^{\alpha\beta}\delta A_\alpha\,\mathrm{d}S_\beta.$$

In the second integral, the tensor $F^{\alpha\beta}$ vanishes on the space part of the boundary ∂N of the region N by our assumptions, while the variations of the electromagnetic four-potential δA_α vanish in the time part of the boundary because the time moments are fixed. Thus, the second integral vanishes, and the Euler–Lagrange equations become

$$\frac{\partial F^{\alpha\beta}}{\partial x^\beta} = 0.$$

To obtain the Maxwell Equations (2.1), define $\mathbf{A} = (A^1, A^2, A^3)^\top$ and

$$\mathbf{E} = -\frac{1}{c}\frac{\partial \mathbf{A}}{\partial t} - \nabla A^0, \qquad \mathbf{B} = \nabla \times \mathbf{A}.$$

The zeroth component of the Euler–Lagrange equations becomes $\nabla \cdot \mathbf{E} = 0$, while the remaining components give $\nabla \times \mathbf{B} = \frac{1}{c}\frac{\partial \mathbf{E}}{\partial t}$. The remaining Maxwell equations are satisfied identically. Indeed, we have

$$\nabla \times \mathbf{E} = -\frac{1}{c}\nabla \times \frac{\partial \mathbf{A}}{\partial t} - \nabla \times (\nabla A^0) = -\frac{1}{c}\frac{\partial \nabla \times \mathbf{A}}{\partial t} = -\frac{1}{c}\frac{\partial \mathbf{B}}{\partial t}$$

and

$$\nabla \cdot \mathbf{B} = \nabla \cdot (\nabla \times \mathbf{A}) = 0.$$

The energy-momentum tensor of an electromagnetic field is given by

$$T^{\mu\nu} = \frac{1}{4\pi}\left(F^{\mu\alpha}F^\nu{}_\alpha - \frac{1}{4}\eta^{\mu\nu}F^{\alpha\beta}F_{\alpha\beta}\right).$$

2.4.6 A Trouble with Minkowskian Spacetime

Consider two Minkowski observers

$$x^\mu = (ct, x^1, x^2, x^3)^\top, \qquad x^{\mu'} = (ct', x^{1'}, x^{2'}, x^{3'})^\top.$$

Assume that x^μ is at rest, while $x^{\mu'}$ is in free fall in a constant gravitational field accelerated along x^3-axis with the constant acceleration $a^\mu = a(0, 0, 0, 1)^\top$.

The observer x^μ emits a monochromatic light signal of wavelength Δt. Consider the following two events: 'a signal is emitting in the direction of x_3' and 'the above photon is reaching the point with $x^1 = x^2 = 0$, $x^3 = h > 0$'. The coordinates of the first (resp., the second) event are $\mathbf{0}$ (resp., $(h, 0, 0, h)^\top$). What is the signal's wavelength $\Delta't$ at the second event measured by the observer $x^{\mu'}$?

On the one hand, we have the following result.

Theorem 16.

$$\Delta t' = \exp(-ah)\Delta t.$$

See proof in Boskoff and Capozziello (2020, Theorem 8.6.6).

On the other hand, assume that a photon is emitted by the observer $x^{\mu'}$ at the origin O along the t'-axis. The next photon will be emitted at the point A with coordinates $(c\Delta t', 0, 0, 0)$. The first photon reaches the hyperplane $x^{3'} = h$ at a point M, the second at a point N. The trajectories of both photons are identical, therefore $OA = MN = c\Delta t'$. But MN is equal to $c\Delta t$ corresponding to the wavelength measured by the observer m^{μ}. The resulting equality $\Delta t = \Delta t'$ contradicts Theorem 16.

2.5 Einsteinian Spacetimes

To overcome the difficulty described in Subsection 2.4.6, Einstein strengthened the weak equivalence principle (Axiom 1) and gave it the following form.

Axiom 2 (Strong equivalence principle). *Inertial effects are locally indistinguishable from gravitational ones.*

In other words, in a local experiment, it is impossible to distinguish between a uniform acceleration and an external gravitational field, independently of when and where it is performed.

Moreover, Einstein supposed that equations of the gravitational field must be invariant in form under the action of spacetime diffeomorphisms (in physical language, 'covariant'). Moreover, each point of spacetime admits a valid notion of past, present, and future (causality principle). Finally, the source of a gravitational field is a metric g_{ij} defined on the spacetime.

2.5.1 Notation

The symbol M denotes a manifold of dimension 4, a spacetime. The points of M are called *events* and are denoted by $p, q, \ldots$. The tangent space at a point $p \in M$ is denoted by T_pM. It is a real linear space of dimension 4 together with a symmetric and non-degenerate bilinear form g_p that has one negative and three positive eigenvalues. This convention is called *spacelike convention* or *relativity convention* or *east coast convention* or just *mostly pluses*. This convention was used by Julian Schwinger living at the East Coast of the US. In other words, in the context of general relativity we choose the value $S_1 = 1$ in Equation (2.3).

We denote by π the projection $TM \to M$ of the tangent bundle (TM, π, M). At each point $p \in M$, the map $T_pM \to T_p^*M$, $X \mapsto X^*$ with

$X^*(Y) = g(X, Y)$ is an isomorphism of Minkowski spacetimes. In coordinates, it is given by $X_\mu = g_{\mu\nu} X^\nu$, the index lowering. The inverse matrix, $g^{\mu\nu}$, may rise indices: $X^\mu = g^{\mu\nu} X_\nu$, and similarly for the space $T_l^k(T_p M)$.

With our convention, a vector X^μ is *timelike* (resp., *lightlike*, resp., *spacelike*) if $g_{\mu\nu} X^\mu X^\nu < 0$ (resp., $g_{\mu\nu} X^\mu X^\nu = 0$ and $X^\mu \neq 0$, resp., $g_{\mu\nu} X^\mu X^\nu > 0$ or $X^\mu = 0$).

Let $\mathcal{T} \subset TM$ be the union over $p \in M$ of the sets $\mathcal{T}_p$ of all timelike vectors in $T_p M$. To satisfy the causality principle, we suppose that the spacetime M is connected and *time orientable*, that is, the set $\mathcal{T}$ has two connected components. Fix a component in $\mathcal{T}$, denote it by $\mathcal{T}^+$ and call it the *future*. Denote the set $\mathcal{T} \setminus \mathcal{T}^+$ by $\mathcal{T}^-$ and call it the *past*. With such a choice, the spacetime M is called *time oriented*.

The set $\mathcal{L}_p^+ = \mathcal{T}^+ \cap T_p M$ is called the *future light cone*. A tangent vector $X \in T_p M$ is called *future pointing* if $X \in \mathcal{L}_p^+ \cup \mathcal{T}_p^+$. A vector field $X(p)$ defined on a domain $\mathcal{U}$ of a chart in M is called *future pointing* if $X(p)$ is future pointing for all $p \in \mathcal{U}$. The definition of past pointing vectors and vector fields may be left to the reader.

Finally, we assume that M is *orientable*, that is, reducible to the subgroup $SO_0(1, 3) \subset O(1, 3)$.

As an example, let $M = \mathbb{R}^{3,1}$ be the Minkowski space. In this special case, we denote by $\eta_{\mu\nu}$ the metric with $\eta_{00} = -c^2$, $\eta_{11} = \eta_{22} = \eta_{33} = 1$, and $\eta_{\mu\nu} = 0$ for $\nu \neq \mu$.

2.5.2 Gravitational Field

Two more results should be mentioned before we continue. The first one is called the *Ostrogradsky instability* after Ostrogradsky (1850). It says that a physical system is subject to a linear instability provided that its Lagrangian depends on time derivative higher than first order. By this reason, an action of a gravitational field should lead to the second-order field equations.

Secondly, we have the following result due to Lovelock (1969).

Theorem 17. *All covariant Lagrangian densities on a four-dimensional Einstein spacetime that lead to the second-order field equations, have the form*

$$\mathcal{L} = (aR + b)\sqrt{-g},$$

where a and b are constants, and where g is the determinant of the metric $g_{\mu\nu}$.

By the above reasoning, we choose the following *Einstein–Hilbert action* for the gravitational field with the source $g_{\mu\nu}$ in vacuum:

$$S[g_{\mu\nu}] = \int_\Omega R\sqrt{-g}\, \mathrm{d}^4 x,$$

where Ω is a compact subset of M with nonempty interior and smooth boundary, and where $\sqrt{-g}\, \mathrm{d}^4 x$ is the Riemannian measure on M.

We calculate the first variation of S.

$$\delta S = \delta \int_\Omega R\sqrt{-g}\, \mathrm{d}^4 x = \delta \int_\Omega g^{\mu\nu} R_{\mu\nu} \sqrt{-g}\, \mathrm{d}^4 x = \int_\Omega \delta(g^{\mu\nu} R_{\mu\nu} \sqrt{-g})\, \mathrm{d}^4 x$$

$$= \int_\Omega (\delta g^{\mu\nu}) R_{\mu\nu} \sqrt{-g}\, \mathrm{d}^4 x + \int_\Omega g^{\mu\nu} (\delta R_{\mu\nu}) \sqrt{-g}\, \mathrm{d}^4 x + \int_\Omega R(\delta\sqrt{-g})\, \mathrm{d}^4 x.$$

$$(2.8)$$

First, we calculate $\delta\sqrt{-g}$. For arbitrary j, $0 \le j \le 3$, the determinant g of the matrix g_{ij} can be calculated by

$$g = \sum_{i=0}^{3} (-1)^{\mu+\nu} g_{\mu\nu} M_{\mu\nu},$$

where $M_{\mu\nu}$ is the determinant of the matrix that results when the μth row and the νth column are eliminated.

On the one hand,

$$\delta\sqrt{-g} = \frac{\partial}{\partial g_{\mu\nu}}(\sqrt{-g})\delta g_{\mu\nu} = -\frac{1}{2\sqrt{-g}}\frac{\partial}{\partial g_{ij}}\delta g_{\mu\nu} = -\frac{1}{2\sqrt{-g}}(-1)^{\mu+\nu} M_{\mu\nu}\delta g_{\mu\nu}$$

$$= -\frac{1}{2\sqrt{-g}}gg^{\mu\nu}\delta g_{\mu\nu} = \frac{1}{2}\sqrt{-g}g^{\mu\nu}\delta g_{\mu\nu}.$$

On the other hand, we have $g^{\kappa\mu} g_{\mu\lambda} = \delta^\kappa_\lambda$, therefore $\delta g^{\kappa\mu} g_{\mu\lambda} + g^{\kappa\mu}\delta g_{\mu\lambda} = 0$, that is, $g^{\kappa\mu}\delta g_{\mu\lambda} = -\delta g^{\kappa\mu} g_{\mu\lambda}$. Multiply both hand sides by $g_{\nu\kappa}$. We obtain

$$g_{\nu\kappa}g^{\kappa\mu}\delta g_{\mu\lambda} = -g_{\nu\kappa}g_{\mu\lambda}\delta g^{\kappa\mu}.$$

In this equation, put $\nu = \mu = \alpha$, $\lambda = \beta$. We have

$$\delta g_{\alpha\beta} = -g_{\alpha\kappa}g_{\alpha\beta}\delta g^{\alpha\kappa}.$$

Then

$$\delta\sqrt{-g} = -\frac{1}{2}\sqrt{-g}g^{\alpha\beta}g_{\alpha\kappa}g_{\beta\alpha}\delta g^{\alpha\kappa} = -\frac{1}{2}\sqrt{-g}g_{\alpha\kappa}\delta g^{\alpha\kappa}.$$

Substitute this value to the rightmost side of Equation (2.8). We obtain

$$\delta S = \int_\Omega \delta g^{\alpha\beta} R_{\alpha\beta} \sqrt{-g}\, \mathrm{d}^4 x + \int_\Omega g^{\alpha\beta}\delta R_{\alpha\beta}\sqrt{-g}\, \mathrm{d}^4 x - \frac{1}{2}\int_\Omega R\sqrt{-g}g_{\alpha\beta}\delta g^{\alpha\beta}\, \mathrm{d}^4 x$$

$$= \int_\Omega \left(R_{\alpha\beta} - \frac{1}{2}Rg_{\alpha\beta}\right)\sqrt{-g}\delta g^{\alpha\beta}\, \mathrm{d}^4 x + \int_\Omega g^{\alpha\beta}\delta R_{\alpha\beta}\sqrt{-g}\, \mathrm{d}^4 x.$$

$$(2.9)$$

In order to calculate the second integral, we use the following result.

Theorem 18 (Palatini's formula).

$$\delta R_{\alpha\beta} = \nabla_\kappa(\delta\Gamma^\kappa_{\alpha\beta}) - \nabla_\beta(\delta\Gamma^\kappa_{\alpha\kappa}).$$

Proof. On the one hand, by definition of the Ricci tensor $R_{\alpha\beta}$, we have

$$R_{\alpha\beta} = R^\kappa_{\alpha\kappa\beta} = \frac{\partial \Gamma^\kappa_{\alpha\beta}}{\partial x^\kappa} - \frac{\partial \Gamma^\kappa_{\alpha\kappa}}{\partial x^\beta} + \Gamma^\kappa_{\kappa\lambda}\Gamma^\lambda_{\alpha\beta} - \Gamma^\kappa_{\beta\lambda}\Gamma^\lambda_{\alpha\kappa}.$$

Its variation becomes

$$\delta R_{\alpha\beta} = \frac{\partial(\delta\Gamma^\kappa_{\alpha\beta})}{\partial x^\kappa} - \frac{\partial(\delta\Gamma^\kappa_{\alpha\kappa})}{\partial x^\beta} + \delta\Gamma^\kappa_{\kappa\lambda}\Gamma^\lambda_{\alpha\beta} + \Gamma^\kappa_{\kappa\lambda}\delta\Gamma^\lambda_{\alpha\beta} - \delta\Gamma^\kappa_{\beta\lambda}\Gamma^\lambda_{\alpha\kappa} - \Gamma^\kappa_{\beta\lambda}\delta\Gamma^\lambda_{\alpha\kappa}.$$
$$(2.10)$$

On the other hand, the covariant derivative of the tensor $\delta\Gamma^\kappa_{\alpha\beta}$ is equal to

$$\nabla_\kappa(\delta\Gamma^\kappa_{\alpha\beta}) = \frac{\partial \Gamma^\kappa_{\alpha\beta}}{\partial x^\kappa} + \Gamma^\kappa_{\kappa\lambda}\delta\Gamma^\lambda_{\alpha\beta} - \delta\Gamma^\kappa_{\beta\lambda}\Gamma^\lambda_{\alpha\kappa} - \delta\Gamma^\kappa_{\alpha\lambda}\Gamma^\lambda_{\beta\kappa}.$$

Similarly,

$$\nabla_\beta(\delta\Gamma^\kappa_{\alpha\kappa}) = \frac{\partial \Gamma^\kappa_{\alpha\kappa}}{\partial x^\beta} + \delta\Gamma^\lambda_{\alpha\kappa}\Gamma^\kappa_{\beta\lambda} - \delta\Gamma^\kappa_{\kappa\lambda}\Gamma^\lambda_{\alpha\beta} - \delta\Gamma^\kappa_{\alpha\lambda}\Gamma^\lambda_{\beta\kappa}.$$

We observe that the difference between the right hand side of the first equation and that of the second one is equal to the right hand side of Equation (2.10). The difference between the left hand sides is equal to the left hand side of Equation (2.10), *quod erat demonstrandum.* □

The second integral in the rightmost part of Equation (2.9) becomes

$$\int_\Omega g^{\alpha\beta}\delta R_{\alpha\beta}\sqrt{-g}\,\mathrm{d}^4x = \int_\Omega g^{\alpha\beta}(\nabla_\kappa[\delta\Gamma^\kappa_{\alpha\beta}] - \nabla_\beta(\delta\Gamma^\kappa_{\alpha\kappa})]\sqrt{-g}\,\mathrm{d}^4x$$

$$= \int_\Omega [\nabla_\kappa(g^{\alpha\beta}\delta\Gamma^\kappa_{\alpha\beta}) - \nabla_\beta(g^{\alpha\beta}\delta\Gamma^\kappa_{\alpha\kappa})]\sqrt{-g}\,\mathrm{d}^4x.$$

To prove the last equality, we need to show that the tensor $g^{\alpha\beta}$ is *covariantly constant*: $\nabla_\kappa g^{\alpha\beta} = 0$. First, we prove that $g_{\alpha\beta}$ has this property. Indeed,

$$\nabla_\kappa g_{\alpha\beta} = \frac{\partial g_{\alpha\beta}}{\partial x^\kappa} - g_{\lambda\beta}\Gamma^\lambda_{\alpha\kappa} - g_{\lambda\alpha}\Gamma^\lambda_{\beta\kappa}$$

$$= \frac{\partial g_{\alpha\beta}}{\partial x^\kappa} - \frac{1}{2}g_{\lambda\beta}g^{\lambda\mu}\left(\frac{\partial g_{\alpha\mu}}{\partial x^\kappa} + \frac{\partial g_{\kappa\mu}}{\partial x^\alpha} - \frac{\partial g_{\alpha\kappa}}{\partial x^\mu}\right)$$

$$- \frac{1}{2}g_{\lambda\alpha}g^{\lambda\mu}\left(\frac{\partial g_{\beta\mu}}{\partial x^\kappa} + \frac{\partial g_{\kappa\mu}}{\partial x^\beta} - \frac{\partial g_{\beta\kappa}}{\partial x^\mu}\right)$$

$$= \frac{\partial g_{\alpha\beta}}{\partial x^\kappa} - \frac{1}{2}\frac{\partial g_{\alpha\beta}}{\partial x^\kappa} - \frac{1}{2}\frac{\partial g_{\kappa\beta}}{\partial x^\alpha} + \frac{1}{2}\frac{\partial g_{\alpha\kappa}}{\partial x^\beta} - \frac{1}{2}\frac{\partial g_{\beta\alpha}}{\partial x^\kappa} - \frac{1}{2}\frac{\partial g_{\kappa\alpha}}{\partial x^\beta} + \frac{1}{2}\frac{\partial g_{\beta\kappa}}{\partial x^\alpha}$$

$$= 0.$$

Now, consider the equality $g_{\alpha\lambda}g^{\lambda\beta} = \delta^\beta_\alpha$ and calculate the covariant derivative of both hand sides. On the left hand side, we have

$$\nabla_\kappa(g_{\alpha\lambda}g^{\lambda\beta}) = (\nabla_\kappa(g_{\alpha\lambda})g^{\lambda\beta} + g_{\alpha\lambda}(\nabla_\kappa g^{\lambda\beta}) = g_{\alpha\lambda}(\nabla_\kappa g^{\lambda\beta}).$$

On the right hand side, $\nabla_\kappa \delta^\beta_\alpha = 0$, that is, $\nabla_\kappa g^{\lambda\beta} = 0$.

We continue to calculate the second integral:

$$\int_\Omega g^{\alpha\beta}\delta R_{\alpha\beta}\sqrt{-g}\,\mathrm{d}^4x = \int_\Omega [\nabla_\kappa(g^{\alpha\beta}\delta\Gamma^\kappa_{\alpha\beta}) - \nabla_\kappa(g^{\alpha\kappa}\delta\Gamma^\beta_{\alpha\beta})]\sqrt{-g}\,\mathrm{d}^4x$$
$$= \int_\Omega \nabla_\kappa(g^{\alpha\beta}\delta\Gamma^\kappa_{\alpha\beta} - g^{\alpha\kappa}\delta\Gamma^\beta_{\alpha\beta})\sqrt{-g}\,\mathrm{d}^4x.$$

Recall that we suppose that $\partial\Omega$, the boundary of Ω, is smooth. By this reason, at each point of $\partial\Omega$, there exists a normal outward vector n_κ, and a Riemannian measure $\sqrt{-h}\,\mathrm{d}^3x$ such that the *divergence theorem* holds true:

$$\int_\Omega \nabla_\kappa(g^{\alpha\beta}\delta\Gamma^\kappa_{\alpha\beta} - g^{\alpha\kappa}\delta\Gamma^\beta_{\alpha\beta})\sqrt{-g}\,\mathrm{d}^4x = \int_{\partial\Omega} (g^{\alpha\beta}\delta\Gamma^\kappa_{\alpha\beta} - g^{\alpha\kappa}\delta\Gamma^\beta_{\alpha\beta})n_\kappa\sqrt{-h}\,\mathrm{d}^3x.$$

The vector $g^{\alpha\beta}\delta\Gamma^\kappa_{\alpha\beta} - g^{\alpha\kappa}\delta\Gamma^\beta_{\alpha\beta}$ vanishes on $\partial\Omega$. It follows that second integral in the rightmost part of Equation (2.9) is equal to 0. The above equation takes the form

$$\delta S = \int_\Omega \left(R_{\alpha\beta} - \frac{1}{2}Rg_{\alpha\beta}\right)\sqrt{-g}\delta g^{\alpha\beta}\,\mathrm{d}^4x.$$

The condition $\delta S = 0$ for arbitrary $g^{\alpha\beta}$ leads to

$$R_{\alpha\beta} - \frac{1}{2}Rg_{\alpha\beta} = 0. \tag{2.11}$$

We proved the following result.

Theorem 19 (The Einstein field equations in vacuum). *If the variations $\delta\Gamma^\alpha_{\beta\kappa}$ vanish on a smooth boundary $\partial\Omega$ of a compact region Ω with nonempty interior without matter and energy inside it, then the Einstein field equations in vacuum (2.11) hold true.*

Following the description of Newton's gravity in Subsection 2.3.2, we allow the presence of matter and energy. As a result, we have

Theorem 20 (The Einstein field equations). *If the variations $\delta\Gamma^i_{jk}$ vanish on a smooth boundary $\partial\Omega$ of a compact region Ω with nonempty interior, then there exists a constant K and a tensor T_{ij} such that*

$$R_{\alpha\beta} - \frac{1}{2}Rg_{\alpha\beta} = KT_{\alpha\beta}. \tag{2.12}$$

Proof. We replace the Einstein–Hilbert Lagrangian density $R\sqrt{-g}$ with a new one, $(kR+\mathcal{L}_M)\sqrt{-g}$, where $\mathcal{L}_M$ is the Lagrangian density of all kinds of matter and energy inside Ω. The new action becomes

$$S = \int_\Omega (kR + \mathcal{L}_M)\sqrt{-g}\,\mathrm{d}^4x.$$

The first variation of this action is

$$\delta S = k \int_\Omega \left(R_{\alpha\beta} - \frac{1}{2} R g_{\alpha\beta} \right) \sqrt{-g} \delta g^{\alpha\beta} \, \mathrm{d}^4 x + \int_\Omega \frac{\delta \mathcal{L}_M}{\delta g^{\alpha\beta}} \sqrt{-g} \delta g^{\alpha\beta} \, \mathrm{d}^4 x.$$

If it vanishes, then Equation (2.12) holds true with

$$T_{\alpha\beta} = -\frac{\delta \mathcal{L}_M}{\delta g^{\alpha\beta}}, \qquad K = k^{-1}.$$

$\square$

The tensor $T_{\alpha\beta}$ is called the covariant *energy-momentum tensor*. It remains to find the value of K.

To do that, we assume that the interior of Ω is filled by particles with impulse $P^i = (mc, p^1, p^2, p^3)^\top$ and number of particles per unit volume $N^i = (nc, nv^1, nv^2, nv^3)$. The *contravariant* energy-momentum tensor becomes

$$T^{\alpha\beta} = P^\alpha N^\beta.$$

On the one hand, $T^{00} = P^0 N^0 = mcnc = \rho c^2$, where $\rho = mn$ is the density of mass. We choose a coordinate system with $p^1 = p^2 = p^3 = 0$ and see that

$$T_{00} = T^{00} g_{00} g_{00} = \rho c^2.$$

On the other hand, we multiply both hand sides of the Einstein field Equation (2.12) by $g^{\mu\alpha}$ and obtain

$$g^{\mu\alpha} R_{\alpha\beta} - \frac{1}{2} R g^{\mu\alpha} g_{\alpha\beta} = K g^{\mu\alpha} T_{\alpha\beta},$$

or

$$R^\mu_\mu \frac{1}{2} \delta^\mu_\mu R = K T^\mu_\mu.$$

Denote the *Laue scalar* T^μ_μ by T and taking into account that in the Einstein summation convention we have $\delta^\mu_\mu = \dim M = 4$, we obtain

$$R_{\alpha\beta} = K \left(T_{\alpha\beta} - \frac{1}{2} T g_{\alpha\beta} \right).$$

In particular, the Einstein field equation with $\alpha = \beta = 0$ becomes

$$R_{00} = K \left(T_{00} - \frac{1}{2} T_{00} g_{00} \right) = \frac{1}{2} K \rho c^2,$$

and the Poisson equation gives

$$R_{00} = \frac{1}{2} \sum_{\alpha=1}^{3} \frac{\partial^2 g_{00}}{\partial (x^\alpha)^2} = \frac{1}{c^2} \nabla^2 \Phi = \frac{1}{c^2} 4\pi G_N \rho.$$

Then, we have

$$\frac{1}{c^2}4\pi G_N \rho = \frac{1}{2}K\rho c^2,$$

or

$$R_{\alpha\beta} - \frac{1}{2}Rg_{\alpha\beta} = \frac{8\pi G}{c^4}T_{\alpha\beta}. \tag{2.13}$$

In fact, Jonh Archibald Wheeler said in Wheeler (1998):

Spacetime tells matter how to move; matter tells spacetime how to curve.

Observe that the final form of the equations of the gravitational field is an equality between the tensors and does not depend on the choice of local charts. With our sign conventions, we may write Equation (2.13) in the following form:

$$R_{ij} - \frac{1}{2}Rg_{ij} = S_3\frac{8\pi G}{c^4}T_{ij}.$$

For the reader's convenience, we repeat all our sign conventions again:

$$\eta^{\mu\nu} = S_1\begin{pmatrix} -c^2 & 0 & 0 & 0 \\ 0 & 1 & 0 & 0 \\ 0 & 0 & 1 & 0 \\ 0 & 0 & 0 & 1 \end{pmatrix},$$

$$R^{\varepsilon}_{\alpha\beta\gamma} = S_2\left(\Gamma^{\delta}_{\alpha\gamma}\Gamma^{\varepsilon}_{\delta\beta} - \Gamma^{\delta}_{\alpha\beta}\Gamma^{\varepsilon}_{\delta\gamma} + \frac{\partial\Gamma^{\varepsilon}_{\alpha\gamma}}{\partial x^{\beta}} - \frac{\partial\Gamma^{\varepsilon}_{\alpha\beta}}{\partial x^{\gamma}}\right),$$

$$R_{\alpha\gamma} = S_2 S_3 R^{\varepsilon}_{\alpha\varepsilon\gamma},$$

$$R_{ij} - \frac{1}{2}Rg_{ij} = S_3\frac{8\pi G}{c^4}T_{ij},$$

The conventions used is several books are summarised in Table 2.1.

In what follows, we will use the following property of the energy-momentum tensor.

Theorem 21.

$$\nabla_{\kappa}T_{\alpha\beta} = 0.$$

Proof. By the second Bianchi identity

$$\nabla_{\lambda}R^{\mu}_{\alpha\beta\kappa} + \nabla_{\beta}R^{\mu}_{\alpha\kappa\lambda} + \nabla_{\kappa}R^{\mu}_{\alpha\lambda\beta} = 0.$$

Contracting the indices β and μ, we obtain

$$\nabla_{\lambda}R^{\mu}_{\alpha\mu\kappa} + \nabla_{\mu}R^{\mu}_{\alpha\kappa\lambda} + \nabla_{\kappa}R^{\mu}_{\alpha\lambda\mu} = 0.$$

The two terms of the left hand side are the covariant derivatives of the Ricci tensor as follows:

$$\nabla_{\lambda}R^{\mu}_{\alpha\mu\kappa} = \nabla_{\lambda}R_{i\alpha\kappa}, \qquad \nabla_{\kappa}R^{\mu}_{\alpha\lambda\mu} = -\nabla_{\kappa}R^{\mu}_{\alpha\mu\lambda} = -\nabla_{\kappa}R_{\alpha\lambda}.$$

TABLE 2.1
Sign Conventions

Book	S_1	S_2	S_3
This book	$\pm 1^1$	+1	+1
Adler (2021)	−1	−1	−1
Bambi (2018)	+1	+1	+1
Boskoff and Capozziello (2020)	−1	+1	+1
Carlip (2019)	−1	−1	+1
Carroll (2004)	+1	+1	+1
Cheng (2010)	+1	+1	−1
Chruściel (2019)	+1	+1	+1
Deruelle and Uzan (2018)	+1	+1	+1
Dhurandhar and Mitra (2022)	−1	+1	+1
Dodelson and Schmidt (2021)	+1	+1	+1
Durrer (2020)	+1	+1	+1
Ferrari, Gualtieri, and Pani (2020)	+1	+1	+1
Foster and Nightingale (2006)	−1	+1	−1
Frankel (1979)	+1	+1	+1
Gourgoulhon (2012)	+1	+1	+1
Griffiths (1991)	−1	+1	−1
Griffiths and Podolský (2009)	+1	−1	+1
Grøn and Hervik (2007)	+1	+1	+1
Guidry (2019)	+1	+1	+1
Hartle (2013)	+1	+1	+1
Hawking and Ellis (2023)	+1	+1	+1
Hobson, Efstathiou, and Lasenby (2006)	−1	+1	+1
Jetzer (2022)	−1	+1	−1
Kopczyński and Trautman (1992)	−1	−1	+1
Lambourne (2010)	−1	+1	−1
Misner, Thorne, and Wheeler (2017)	+1	+1	+1
Narlikar (2010)	−1	−1	−1
O'Donnell (2003)	−1	+1	+1
Ohanian and Ruffini (2013)	−1	+1	−1
Penrose and Rindler (1987)	−1	−1	−1
Peter and Uzan (2009)	+1	+1	+1
Poisson and Will (2014)	+1	+1	+1
Rahaman (2021)	−1	+1	−1
Ryder (2009)	+1	+1	+1
Shibata (2015)	+1	+1	+1
Soffel and Han (2019)	+1	+1	+1
Stephani (1982)	+1	+1	+1
Stephani et al. (2003)	+1	+1	+1
Stewart (1990)	−1	−1	−1
Straumann (2013)	+1	+1	+1
Thorne and Blandford (2017)	+1	+1	+1
Vecchiato (2017)	+1	+1	+1
Wald (1984)	$\pm 1^2$	+1	+1
Weinberg (1972)	+1	−1	−1
Weinberg (2008)	+1	−1	−1

[1] +1 if the Minkowskian spacetime is tangent to an Einsteinian one.
[2] −1 in Chapter 13.

Our equation becomes

$$\nabla_\lambda R_{\alpha\kappa} + \nabla_\mu R^\mu_{\alpha\kappa\lambda} - \nabla_\kappa R_{\alpha\lambda} = 0.$$

We multiply it by a covariantly constant term $g^{\alpha\nu}$ and obtain

$$\nabla_\lambda(g^{\alpha\nu} R_{\alpha\kappa}) + \nabla_\mu(g^{\alpha\nu} R^\mu_{\alpha\kappa\lambda}) - \nabla_\kappa(g^{\alpha\nu} R_{\alpha\lambda}) = 0,$$

which is the same as

$$\nabla_\lambda R^\nu_\kappa + \nabla_\mu(g^{\alpha\nu} R^\mu_{\alpha\kappa\lambda}) - \nabla_\kappa R^\nu_\lambda = 0.$$

We contract the indices ν and λ and see that

$$\nabla_\lambda R^\lambda_\kappa + \nabla_\mu(g^{\alpha\lambda} R^\mu_{\alpha\kappa\lambda}) - \nabla_\kappa R^\lambda_\lambda = 0.$$

The second term in the left hand side is

$$\nabla_\mu(g^{\alpha\lambda} R^\mu_{\alpha\kappa\lambda}) = \nabla_\mu(g^{\alpha\lambda} g^{\mu\pi} R_{\pi\alpha\kappa\nu}) = \nabla_\mu(g^{\mu\pi} g^{\alpha\lambda} R_{\pi\alpha\kappa\nu})$$
$$= \nabla_\mu(g^{\mu\pi} R^\lambda_{\pi\lambda\nu}) = \nabla_\mu R^\mu_\kappa.$$

Then, we have

$$\nabla_\lambda R^\lambda_\kappa + \nabla_\mu R^\mu_\kappa - \nabla_\kappa R = 0, \quad \text{or} \quad 2\nabla_\lambda R^\lambda_\kappa - \nabla_\kappa R = 0.$$

Write this equation in the form

$$\nabla_\lambda\left(R^\lambda_\kappa - \frac{1}{2}\delta^\lambda_\kappa R\right) = 0$$

and multiply both hand sides by a covariant constant $g_{\mu\lambda}$. We obtain

$$\nabla_\lambda\left(g_{\mu\lambda} R^\lambda_\kappa - \frac{1}{2}\delta^\lambda_\kappa g_{\mu\lambda} R\right) = 0,$$

or

$$\nabla_\lambda\left(R_{\mu\kappa} - \frac{1}{2} R g_{\mu\kappa}\right) = 0.$$

We proved that the covariant derivative of the left hand side of the Einstein field Equation (2.13) is null. The same holds true for their right hand side, which is proportional to $T_{\alpha\beta}$. □

2.5.3 The Robertson–Walker Metric

It is practically impossible to construct a detailed model of the Universe as a manifold. Usually, one considers a Riemannian manifold of dimension 5 which has a chart with coordinate, say l. The level sets N_l of that coordinate are submanifolds of dimension 4. The submanifold N_1 models our Universe, while $M = N_0$ is its idealised approximation.

Definition 107. A *reference frame* on M is a smooth future pointing vector field $X^\mu(p)$ such that $g_{\mu\nu}X^\mu X^\nu(p) = -c^2$.

A reference frame represents an average velocity of the 'cosmic flow' of all types of energy in the Universe.

Observations show that our Universe is isotropic at big scales: the galaxies are located equally dense in all directions. We formalise this observation.

On the one hand, let $p \in M$, and let $H_p = \{\, Y^\mu \in T_pM \colon g_{\mu\nu}(p)X^\mu(p)Y^\nu = 0 \,\}$. The real linear space H_p has dimension 3, and the restriction of g to H_p is positive-definite. That is: H_p is the linear space of spacelike vectors. Actually, this is the reason for our current sign convention. Denote by $\mathrm{SO}(H_p)$ the connected component of identity in the group of orthogonal linear operators on H_p.

On the other hand, let $\psi \colon M \to M$ be an isometry with $\psi(p) = p$. The push forward ψ_* acts on vectors $Y^\mu \in T^pM$, and this action is linear. Assume that $\psi_*X^\mu(p) = X^\mu(p)$, i.,e., ψ_* fixes the reference frame. Then the restriction of ψ_* to H_p is a linear map of H_p to itself.

Definition 108. A manifold M with a reference frame $X(p)$ is called *isotropic* at a point p if any element of the group $\mathrm{SO}(H_p)$ is a restriction to H_p of the push forward of some isometry $\psi \colon M \to M$ with $\psi(p) = p$.

In other words, a manifold M with a reference frame $X(p)$ is isotropic if it has a big amount of isometries.

Let $I \subseteq \mathbb{R}^1$ be an open interval, and let $k \colon I \to M$ be a smooth map. The curve k is called *timelike* if for all $t \in I$ the vector $\dot{k}(t) \in T_{k(t)}M$ is timelike. A submanifold $N \subset M$ is called spacelike if $T_pN \subset H_p$ for all $p \in N$.

Definition 109. An *observer* is a timelike curve k. A *comoving observer* is a timelike geodesic line $k(t)$ such that $\dot{k}(t)$ belongs to the reference frame. A comoving observer consists of *instant comoving observers*: the pairs $(k(t), \dot{k}(t))$, $t \in I$.

In other words, comoving observers follow the cosmic flow. Another name for them is *free falling observers*.

Theorem 22. *Let M be a manifold with reference frame $X(p)$ which is isotropic at all points. Then there is an open interval $I \subseteq \mathbb{R}^1$, a smooth function $a \colon I \to (0, \infty)$ and real numbers K and R such that*

- *M can be decomposed into a family $\{\, M_t \colon t \in I \,\}$ of slices: 3-dimensional spacelike submanifolds such that the tangent space of each of them at a point p is equal to H_p;*

- *for each $t \in I$, the map ψ_t that maps an event $p \in M_t$ to the unique comoving observer $k(u)$ with $k(t) = p$ is a one-to-one correspondence between M_t and the set of all comoving observers;*

- *for all t, $u \in I$, the map $\psi_u^{-1}\psi_t \colon M_t \to M_u$ is one-to-one and smooth with the smooth inverse, the image of the metric g restricted to M_t under the push forward $(\psi_u^{-1}\psi_t)_*$ is proportional to the metric g restricted to M_u;*

- *for every instant comoving observer $(p, X(p))$, there is a chart $(\mathcal{U}, \mathbf{x})$ of M with*

$$p \mapsto \mathbf{x}(p) = (t, r, \varphi, \theta)^\top \in I \times (0, R) \times (0, 2\pi) \times (0, \pi)$$

such that in this chart the metric g has the form

$$ds^2 = -c^2\,dt^2 + a^2(t)\left[\frac{dr^2}{1 - Kr^2} + r^2(d\theta^2 + \sin^2\theta\,d\varphi^2)\right]. \qquad (2.14)$$

The metric (2.14) is the celebrated *Robertson–Walker metric*. The parameter K has units of length to the power -2, r has units of length, $a(t)$ is unit-less. The coordinate t is the *cosmic time*. We follow the standard convention: the cosmic time t with $a(t) = 1$ is the current moment of time. As $r \downarrow 0$, a point q with the coordinate r goes to the location p of the instant comoving observer $(p, X(p))$. The submanifolds M_t are the level sets of cosmic time.

The comoving distance r is not a physical observable, the physical coordinate is $d = ra(t)$. Let a galaxy has a trajectory $\mathbf{r}(t)$ in comoving coordinates and $\mathbf{d}(t)$ in physical coordinates. The physical velocity of the galaxy is

$$\mathbf{v}(t) = \frac{d\mathbf{d}(t)}{dt} = \dot{a}(t)\mathbf{r}(t) + a(t)\dot{\mathbf{r}}(t),$$

where an overdot denotes the time derivative.

Definition 110. The *Hubble parameter* is

$$H(t) = \frac{\dot{a}(t)}{a(t)}.$$

We see that

$$\mathbf{v}(t) = H(t)\mathbf{d}(t) + \mathbf{v}_p(t).$$

The first term in this equation is called the *Hubble flow*. This is the velocity of the galaxy due to the dynamics of the Universe. The second term is the *peculiar velocity* of the galaxy measured by a comoving observer, who follows the Hubble flow.

All instant comoving observers located at the points $p \in M_t$, observe the same. In other words, the universe is not just isotropic, but also homogeneous. This assertion is known as the *Copernican principle*: there is nothing special about the Earth's location, it is not the centre of the Universe. As we see, isotropy implies the Copernican principle.

It remains to find the function $a(t)$, which determines the dynamics of our Universe.

2.5.4   The Friedman Equations

To consider how the spacetime is evolving, we substitute the Robertson–Walker metric to the left hand side of the Einstein Equation (2.13). First, we calculate the metric tensor. By Equation (2.14), the only nonzero entries of g_{ij} are

$$g_{tt} = -c^2, \qquad g_{rr} = \frac{a^2(t)}{1 - Kr^2}, \qquad g_{\theta\theta} = a^2(t)r^2, \qquad g_{\varphi\varphi} = a^2(t)r^2 \sin^2\theta.$$

Next, we calculate the connection coefficients, using Equation (1.41), and the entries of the Ricci tensor Equation (1.47). Nowadays, such calculations are usually performed using a suitable computer algebra system. We write down the results.

Among the 40 connection coefficients, the following 13 are non-zero:

$$\Gamma^t_{rr} = \frac{a(t)\dot{a}(t)}{c^2(1 - Kr^2)}, \qquad \Gamma^t_{\theta\theta} = \frac{r^2 a(t)\dot{a}(t)}{c^2}, \quad \Gamma^t_{\varphi\varphi} = \frac{r^2 \sin^2\theta \, a(t)\dot{a}(t)}{c^2},$$

$$\Gamma^r_{tr} = \frac{\dot{a}(t)}{a(t)}, \qquad \Gamma^r_{rr} = \frac{Kr}{1 - Kr^2}, \qquad \Gamma^r_{\theta\theta} = -r(1 - Kr^2),$$

$$\Gamma^r_{\varphi\varphi} = -r(1 - Kr^2)\sin^2\theta, \quad \Gamma^\theta_{t\theta} = \frac{\dot{a}(t)}{a(t)}, \qquad \Gamma^\theta_{r\theta} = \frac{1}{r},$$

$$\Gamma^\theta_{\varphi\varphi} = -\sin\theta\cos\theta, \qquad \Gamma^\varphi_{t\varphi} = \frac{\dot{a}(t)}{a(t)}, \qquad \Gamma^\varphi_{r\varphi} = \frac{1}{r},$$

$$\Gamma^\varphi_{\theta\varphi} = \cot\theta.$$

The nonzero entries of the Ricci tensor are:

$$R_{tt} = -\frac{3}{c^2}\frac{\ddot{a}(t)}{a(t)},$$

$$R_{rr} = \frac{a^2(t)}{c^2(1 - Kr^2)}\left[\frac{\ddot{a}(t)}{a(t)} + 2H^2(t) + 2\frac{Kc^2}{a^2(t)}\right],$$

$$R_{\theta\theta} = \frac{a^2(t)r^2}{c^2}\left[\frac{\ddot{a}(t)}{a(t)} + 2H^2(t) + 2\frac{Kc^2}{a^2(t)}\right],$$

$$R_{\varphi\varphi} = \frac{a^2(t)r^2 \sin^2\theta}{c^2}\left[\frac{\ddot{a}(t)}{a(t)} + 2H^2(t) + 2\frac{Kc^2}{a^2(t)}\right],$$

and the Ricci scalar is

$$R = \frac{6}{c^2}\left[\frac{\ddot{a}(t)}{a(t)} + H^2(t) + \frac{Kc^2}{a^2(t)}\right].$$

The nonzero entries of the *Einstein tensor*

$$G^m_n = g^{lm}\left(R_{ln} - \frac{1}{2}g_{ln}R\right)$$

become

$$G_t^t = -\frac{3}{c^2}\left[H^2(t) + \frac{Kc^2}{a^2(t)}\right],$$

$$G_r^r = G_\theta^\theta = G_\varphi^\varphi = -\frac{1}{c^2}\left[2\frac{\ddot{a}(t)}{a(t)} + H^2(t) + \frac{Kc^2}{a^2(t)}\right].$$

Following Einstein (1917), in cosmology all local irregularities (stars, planets, people, etc.) are smoothing out into a *cosmic fluid*. Any such fluid is described by its density $\rho(t)$, pressure $P(t)$, which do not depend on the space coordinates, and an *equation of state* that connects the two. The nonzero components of the energy-momentum tensor T_j^i of such a fluid in the frame of a comoving observer have the form

$$T_t^t = -\rho(t)c^2, \qquad T_r^r = T_\theta^\theta = T_\varphi^\varphi = P(t). \qquad (2.15)$$

The time component of the Einstein equation gives

$$H^2(t) = \frac{8\pi G_N}{3}\rho(t) - \frac{Kc^2}{a^2(t)},$$

and is called the *Friedmann equation*. Each space component gives

$$\frac{\ddot{a}(t)}{a(t)} = -\frac{4\pi G_N}{3}\left(\rho(t) + \frac{3P(t)}{c^2}\right).$$

This equation is called either the second Friedmann equation or the *Raychaudhuri equation*.

The cosmological fluids have a constant equation of state: $P(t) = w\rho(t)c^2$ with $w \in \mathbb{R}$. In particular, a fluid called *matter* has zero pressure: $w = 0$.

Soon after Einstein proved Equation (2.13), he realised that his equation can be applied to the whole Universe. At this moment, Edwin Hubble still did not discover his flow. For that reason, Einstein assumed that the Universe is static (that is $a(t) = 1$), spatially closed and the matter is distributed uniformly within it. Then the Raychaudhuri equation gives

$$0 = -\frac{4\pi G_N}{3}\rho,$$

which may happen only if the Universe is empty. This contradicts the observations.

To overcome this difficulty, Einstein used Theorem 21: the right hand side of Equation (2.13) is covariantly constant. He added a covariantly constant term Λg_{ij} to the left hand side of the above equation and obtained

$$G_{ij} + \Lambda g_{ij} = \frac{8\pi G_N}{c^4}T_{ij}.$$

The right hand side of this equation still satisfies the law of conservation of the energy-momentum tensor (Theorem 21). The positive real number Λ is called the *cosmological constant.*

The Raychaudhuri equation takes the form

$$\frac{\ddot{a}(t)}{a(t)} = -\frac{4\pi G_N}{3}\left(\rho(t) + \frac{3P(t)}{c^2}\right) + \frac{\Lambda c^2}{3}.$$

The data $a(t) = 1$, $P(t) = 0$, $\rho(t) = \frac{\Lambda c^2}{4\pi G_N}$ solve this equation. The Friedmann equation becomes

$$0 = \frac{8\pi G_N}{3}\frac{\Lambda c^2}{4\pi G_N} - Kc^2 + \frac{\Lambda c^2}{3},$$

and gives $K = \Lambda$. This means that the model is a four-dimensional ball of radius $r = \Lambda^{-1/2}$, volume $V = 2\pi^2 r^3 = 2\pi^2 \Lambda^{-3/2}$, and mass $M = V\rho = \frac{\pi c^2}{2G_N\sqrt{\Lambda}}$. This model is called the *Einstein 1917 static model of the Universe* after Einstein (1917).

In 1929, Hubble published the first evidence of the existence of the Hubble flow. Subsequently, Einstein published a *dynamical* model of the Universe in Einstein (1931). In particular, he wrote

> Nachdem nun aber durch HUBBELS Resultate klar geworgen ist, daß die außer-galaktischen Nebel gleichmäßig über den Raum verteilt und in einer Dilatationsbewegung begriffen sind (wenigstens sofern man deren systematische Rotverschiebungen alls Dopplereffekte zu deuten hat), hat die Annahme (2) von der statischen Natur des Raumes keine Berechtigung mehr...

Or in translation by O'Raifeartaigh et al. (2017):

> Now that it has become clear from Hubble's results that the extra-galactic nebulae are uniformly distributed throughout space and are in dilatory motion (at least if their systematic redshifts are to be interpreted as Doppler effects), assumption (2) concerning the static nature of space has no longer any justification...

At the sunset of the past millennium, the cosmological constant returned, but this time to the *right* hand side of Equation (2.13). In fact, it became a contribution to the energy-momentum tensor

$$T^i_j(\Lambda) = -\frac{\Lambda c^4}{8\pi G_N}\delta^i_j.$$

Compare this contribution with Equation (2.15). We obtain $P(t) = -\rho(t)c^2$, that is, $w = -1$. This fluid is called the *dark energy.*

Besides these two fluids, there exists only one more called the *radiation* for which $w = \frac{1}{3}$. The density of the matter (resp., dark matter, resp., radiation) is denoted by $\rho_m(t)$ (resp., ρ_Λ, resp., $\rho_r(t)$). We have to explain why ρ_Λ does not depend on t.

Recall the law of conservation of the energy-momentum tensor: $\nabla_i T^i_j = 0$. By definition of the covariant derivative, we have

$$\frac{\partial T^i_j}{\partial x^i} + \Gamma^j_{ik} T^k_j - \Gamma^k_{ij} T^i_k = 0.$$

The evolution of the energy density is determined by the component with $j = 0$:

$$\frac{\partial T^i_0}{\partial x^i} + \Gamma^0_{ik} T^k_0 - \Gamma^k_{i0} T^i_k = 0.$$

We have $T^i_0 = 0$ if $i \neq 0$, that is,

$$\frac{1}{c} \dot{\rho}(t) c^2 + \Gamma^i_{i0} \rho(t) c^2 - \Gamma^k_{i0} T^i_k = 0.$$

The term Γ^k_{i0} does not vanish if and only if $i = k > 0$, and in this case $\Gamma^i_{i0}(t) = \frac{3}{c} H(t)$. We obtain the energy conservation law:

$$\dot{\rho}(t) + 3H(t) \left(\rho(t) + \frac{P(t)}{c^2} \right) = 0.$$

The constant equation of state $P(t) = w\rho(t)c^2$ gives

$$\frac{\dot{\rho}(t)}{\rho(t)} = -3(1 + w) \frac{\dot{a}(t)}{a(t)}.$$

It follows that $\rho(t)$ is proportional to $a^{-3(1+w)}(t)$. In particular, ρ_Λ does not depend on t, ρ_m is proportional to $a^{-3}(t)$ as expected, because the volume of a region in space is proportional to $a^3(t)$ while the amount of matter inside the region stays constant. Below we will explain why $\rho_r(t)$ is proportional to $a^{-4}(t)$.

Denote $H = H(0)$ and call this number the *Hubble constant*. The Friedmann equation becomes

$$\frac{H^2(t)}{H_0^2} = \frac{\rho_{m0}}{\rho_0} a^{-3}(t) + \frac{\rho_{r0}}{\rho_0} a^{-4}(t) + \frac{\rho_{\Lambda 0}}{\rho_0},$$

where $\rho_0 = \frac{3c^2 H_0^2}{8\pi G_N}$ is the density of the Universe today. It is called the *critical density*, because it makes Universe flat. Introducing the notation

$$\Omega_i = \frac{\rho_{i0}}{\rho_0}, \qquad i \in \{m, r, \Lambda\},$$

we write down the Friedmann equation as follows:

$$\frac{H^2(t)}{H_0^2} = \Omega_m a^{-3}(t) + \Omega_r a^{-4}(t) + \Omega_\Lambda. \tag{2.16}$$

The determination of the values of H_0 and the *density parameters* Ω_m, Ω_r, and Ω_Λ is one of the main tasks of cosmology.

The function $a(t)$ in Equation (2.16) is not observable. To solve this issue, assume that an electromagnetic wave leaves a distant galaxy and reaches an observer. The spread of electromagnetic waves in M is described by a suitable modification of the Maxwell equations without sources:

$$\nabla_n F^{mn} = 0, \qquad \nabla_s F_{mn} + \nabla_m F_{ns} + \nabla_n F_{sm} = 0.$$

Under very mild conditions, the approximation of *geometric optics* holds true. In particular, light rays are null geodesics. See a detailed discussion in Misner, Thorne, and Wheeler (2017, Chapter 22).

Let a distant galaxy is located at the time-independent comoving distance $r = r_1$, and the observer at $r = 0$. A wave crest leave galaxy at the time t_1 and reaches the observer at the time t_0. Then, we have

$$r_1 = \int_{t_1}^{t_0} \frac{dt}{a(t)}.$$

The next crest leaves the galaxy at the time $t_1 + \Delta t_1$ and arrives at $t_0 + \Delta t_0$:

$$r_1 = \int_{t_1 + \Delta t_1}^{t_0 + \Delta t_0} \frac{dt}{a(t)}.$$

Over the period given by the frequency of light, the scale factor $a(t)$ does not change significantly. This gives $\frac{\Delta t_0}{a(t_0)} = \frac{\Delta t_1}{a(t_1)}$. Then, the observed wave length λ_0 is related to the emitted wave length λ_1 by

$$\frac{\lambda_0}{\lambda_1} = \frac{a(t_0)}{a(t_1)}.$$

Observing a wave of length λ_0, astronomers identify it with a known wave of length λ_1 and denote $z = \frac{\lambda_0 - \lambda_1}{\lambda_1}$. The observable z is called the *redshift*. We have

$$\frac{a(t_0)}{a(t_1)} = 1 + z,$$

that is, at the redshift z the Universe was $(1+z)$ times less, while we are living at $z = 0$. The redshift was discovered by Hubble in 1920s, and the model of static Universe was abandoned. In the proportionality between the radiation density $\rho_m(t)$ and $a^{-4}(t)$, three units appear because of expansion and one unit by the redshifting of the radiation energy.

According to our agreement, $a(t_0) = 1$, so $a(t) = (1+z)^{-1}$. The Friedmann Equation (2.16) becomes

$$H^2(z) = H_0^2 [\Omega_m (1 + z)^3 + \Omega_r (1 + z)^4 + \Omega_\Lambda].$$

The values of constants are usually obtained by parameter fits. The results are considered as normal random variables and are usually quoted as $m(n)$, where m is the expectation of the corresponding random variable, while n is a multiple of its standard deviation, σ. The properties of normal distribution imply that if $n = \sigma$ (resp $m = 1.96\sigma$), then $(m - n, m + n)$ is the 68 % (resp., 95 %) confidence interval. For example, Aghanim (2020b) gives the 95 % confidence interval $H_0 = 67.70(81)\,\mathrm{km\,s}^{-1}\,\mathrm{Mpc}^{-1}$. Recall that the parsec (pc) is the distance from which the diameter of the Earth orbit has angle size $1''$ and is equal to 3.26 light years.

Similarly, Aghanim (2020b) gives the 95 % confidence intervals $\Omega_m = 0.3106(110)$, $\Omega_\Lambda = 0.6894(56)$. The radiation density of *photons* is known very precisely from the measurements of the temperature of the Cosmic Microwave Background in units h^2, where $H_0 = 100h\,\mathrm{km\,s}^{-1}\,\mathrm{Mpc}^{-1}$, and is equal to $\Omega_r h^2 = 4.183\,43 \times 10^{-5}$. The radiation density of *neutrinos* can be calculated with the help of the standard model of particle physics.

We have $a(t) \to 0$ as $z \to \infty$. The time moment t with $a(t) = 0$ is a singular point of the Friedmann equations known to the general public as the *Big Bang*. We have

$$H(t) = a^{-1}\frac{\mathrm{d}a}{\mathrm{d}t} = (1+z)\frac{\mathrm{d}a}{\mathrm{d}z}\dot{z} = -(1+z)^{-1}\dot{z}.$$

The Friedmann equation takes the form

$$\mathrm{d}t = -\frac{1}{H_0}\frac{\mathrm{d}z}{(1+z)\sqrt{\Omega_m(1+z)^3 + \Omega_r(1+z)^4 + \Omega_\Lambda}}.$$

Then an electromagnetic wave that we observe now from a distant galaxy at redshift z, left the galaxy

$$t(z) = \frac{1}{H_0}\int_0^z \frac{\mathrm{d}u}{(1+u)\sqrt{\Omega_m(1+u)^3 + \Omega_r(1+u)^4 + \Omega_\Lambda}}$$

time units ago. As $z \to \infty$, $t(z)$ tends to t_0, the age of the Universe. Aghanim (2020b) gives $t_0 = 13.801(24)$ billion years at the confidence level 68 %.

Measuring distances in the expanding Universe is a tricky business. How to calculate the *comoving distance* (the distance r that participates in the Robertson–Walker metric) to a galaxy with the redshift z? We have

$$r = \int_0^r \mathrm{d}r = c\int_0^t \frac{\mathrm{d}u}{a(u)}.$$

We change integration variable from t to $z = a^{-1}(t) - 1$, take the Friedmann equation into account and obtain

$$r = \frac{c}{H_0}\int_0^z \frac{\mathrm{d}u}{\sqrt{\Omega_m(1+u)^3 + \Omega_r(1+u)^4 + \Omega_\Lambda}}.$$

If we could connect us with the galaxy by a ruler, it would show this distance.

Imagine a galaxy of known physical diameter d at the redshift z. Choose a chart in such a way that the galaxy is extended in the θ direction and occupies the angle $\Delta\theta$. Then we have $d = a(t)r\Delta\theta$. The *angular diameter distance* is $D_A = d/\Delta\theta$, and we obtain

$$D_A = \frac{c}{H_0(1+z)} \int_0^z \frac{du}{\sqrt{\Omega_m(1+u)^3 + \Omega_r(1+u)^4 + \Omega_\Lambda}}.$$

A ruler between us and the galaxy could show this distance at the moment of light emitting.

On the other hand, assume that the galaxy has known luminosity L. The flux per unit area in the stationary universe would be $f = L/(4\pi r^2)$. The galaxy is emitting a given amount of energy per second at time moment t, but we receive it in a time that is stretched by $a^{-1}(t) = 1 + z$. Moreover, each photon is redshifted and has its frequency reduced by another $1 + z$. The observer at $r = 0$ receives the flux $f = L/(4\pi r^2(1+z)^2)$ and sees the galaxy at if it were at the *luminosity distance* $D_L = (1+z)^2 D_A$. This gives

$$D_L = \frac{c(1+z)}{H_0} \int_0^z \frac{du}{\sqrt{\Omega_m(1+u)^3 + \Omega_r(1+u)^4 + \Omega_\Lambda}}$$

and explains why it is so hard to observe distant galaxies.

Finally, the *lookback distance* D_l is the distance travelled by the electromagnetic waves from the galaxy at the redshift z to the observer. We have $D_l = ct(z)$, which gives

$$D_l = \frac{c}{H_0} \int_0^z \frac{du}{(1+u)\sqrt{\Omega_m(1+u)^3 + \Omega_r(1+u)^4 + \Omega_\Lambda}}.$$

For example, 3C273, one of the first discovered quasars, is located at $z = 0.1583$. At the moment, when the observable flux has been emitted, the distance between the Earth and the quasar (its angular diameter distance) was equal to 1.853 billion light years. The electromagnetic wave travelled during 1.994 billion years. The comoving distance to 3C273, that could be measured by a ruler now, is equal to 2.147 billion light years, and the luminosity distance is 2.486 billion light years.

2.6 Relativistic Wave Equations on the Minkowskian Spacetime

Consider the principal bundle $(SL(2,\mathbb{C}) \boxtimes \mathbb{R}^{1,3}, \pi, \hat{C}_+, E(2) \boxtimes \mathbb{R}^{1,3})$ of Example 38. It has no continuous sections. However, there exists a continuous section over the dense subset $\hat{\mathcal{C}}_0^\circ = \hat{\mathcal{C}}_0 \setminus \{\, \mathbf{k} \in \hat{\mathcal{C}}_0 \colon k_3 = -k_0 \,\}$. Choose it in the form $s(\mathbf{k}) = (H_{\mathbf{k}}, 0)$, where

$$H_{\mathbf{k}} = \frac{1}{2k_0(k_0+k_3)} \begin{pmatrix} -k_0^{1/2}(k_0+k_3) & k_0^{-1/2}(k_1-\mathrm{i}k_2) \\ -k_0^{1/2}(k_1+\mathrm{i}k_2) & -k_0^{-1/2}(k_0+k_3) \end{pmatrix}.$$

Consider the representation of the group H given by

$$(L, \mathbf{a}) \mapsto \exp(-\mathrm{i}(\mathbf{k}_0, \mathbf{a})) D_{\ell_1, \ell_2}(L), \qquad L \in \mathcal{E}, \quad \mathbf{a} \in \mathbb{R}^{1,3},$$

where $D_{\ell_1, \ell_2}(L)$ is the restriction of the complex irreducible representation V_{ℓ_1, ℓ_2} of the group $\mathrm{SL}(2, \mathbb{C})$ constructed in Example 50, to the subgroup $\mathcal{E}$. Denote by E_{ℓ_1, ℓ_2} the fibre bundle associated with the principal bundle $(G, \pi, \hat{\mathcal{C}}_0)$ by the above representation. This representation induces the representation of the group G in the complex Hilbert space $\tilde{\mathcal{V}} = L^2(E_{\ell_1, \ell_2})$ given by

$$s(\mathbf{k}) \mapsto \exp(-\mathrm{i}(\mathbf{k}, \mathbf{a})) D_{\ell_1, \ell_2}(H_{\mathbf{k}}^{-1} A H_{\mathbf{q}}) s(\mathbf{q}),$$

where $s(\mathbf{k}) \in \tilde{\mathcal{V}}$, $A \in \mathrm{SL}(2, \mathbb{C})$, $\mathbf{a} \in \mathbb{R}^{1,3}$, and $\mathbf{q} = (\tilde{\pi}(A))^{-1}\mathbf{k}$. This representation is neither unitary nor irreducible.

Definition 111. A linear operator $\pi \colon V_{\ell_1, \ell_2} \to V_{\ell_1, \ell_2}$ is called an *invariant projection* if $\pi^2 = \pi$, π intertwines the representation D_{ℓ_1, ℓ_2} with itself, and

$$\pi D_{\ell_1, \ell_2}^*(L) D_{\ell_1, \ell_2}(L) \pi = \pi, \qquad L \in H.$$

Theorem 23. *If π is an invariant projection, then the complex Hilbert space*

$$\tilde{\mathcal{H}} = \{\, \tilde{s} \in \tilde{\mathcal{V}} \colon \pi \tilde{s}(\mathbf{k}) = \tilde{s}(\mathbf{k}) \,\}$$

is a closed invariant subspace for the representation $\tilde{V}$. The restriction of the representation $\tilde{V}$ to $\tilde{\mathcal{H}}$ is unitary. If the restriction of D_{ℓ_1, ℓ_2} to $\pi V_{\ell_1, \ell_2}$ is irreducible, then the representation $\tilde{\mathcal{H}}$ is irreducible.

In physical applications, the representation $\tilde{\mathcal{H}}$ with $\ell_2 = 0$ describes a massless relativistic field (or a particle) of spin ℓ_1, while that with $\ell_2 = 0$ describes the same object with the opposite helicity. In particular, the value of $\ell_1 = 1/2$ corresponds to a neutrino, 1 to a photon, and 2 to a gravitational wave (or to a graviton, if it exists).

The condition $\pi \tilde{s}(\mathbf{k}) = \tilde{s}(\mathbf{k})$, which defines the Hilbert space $\tilde{\mathcal{H}}$, is a hidden form of a relativistic wave equation. We will recover it in a few steps.

2.6.1   A Transition to an Equivalent Representation

Observe that if $\mathbf{k} \in \hat{\mathcal{C}}_+^\circ$, then the linear operator

$$f(\mathbf{k}) = D_{\ell_1, \ell_2}(H_{\mathbf{k}}^{-1})^* D_{\ell_1, \ell_2}(H_{\mathbf{k}}^{-1})$$

is positive-definite. For simplicity of notation, fix the values of ℓ_1 and ℓ_2, and denote the linear space V_{ℓ_1, ℓ_2} just by V. Fix a $SL(2, \mathbb{C})$-invariant measure m on the Borel σ-field of $\hat{C}_+$. Define the Hilbert space $\mathcal{H}$ by

$$\mathcal{H} = \{\, \mathbf{s} \in \Gamma E_{\ell_1, \ell_2} \colon \|\mathbf{s}\|^2 < \infty \,\},$$

where

$$(\mathbf{s}_1, \mathbf{s}_2) = \int_{\hat{C}_+} (\mathbf{s}_1(\mathbf{k}), f(\mathbf{k}) \mathbf{s}_2(\mathbf{k}))_V \, dm(\mathbf{k}).$$

Under the isometry $L^2(E_{\ell_1, \ell_2}) \to \mathcal{H}$, $\mathbf{s}(\mathbf{k}) \mapsto D(H_{\mathbf{k}}) \mathbf{s}(\mathbf{k})$, the induced representation $\tilde{\mathcal{V}}$ becomes

$$\mathbf{s}(\mathbf{k}) \mapsto \exp(-\mathrm{i}(\mathbf{k}, \mathbf{a})) D(A) \mathbf{s}(\mathbf{q}),$$

where $g = (A, \mathbf{a}) \in G$ and $\mathbf{q} = (\tilde{\pi}(A))^{-1} \mathbf{k}$. The condition $\pi \tilde{\mathbf{s}}(\mathbf{k}) = \tilde{\mathbf{s}}(\mathbf{k})$ becomes $\pi(\mathbf{k}) \mathbf{s}(\mathbf{k}) = \mathbf{s}(\mathbf{k})$, where $\pi(\mathbf{k}) = D(H_{\mathbf{k}}) \pi D(H_{\mathbf{k}})^{-1}$ and $\mathbf{s}(\mathbf{k}) = D(H_{\mathbf{k}}) \tilde{\mathbf{s}}(\mathbf{k})$.

2.6.2 Relativistic Wave Equations in the Frequency-Wave-Number Domain

For a positive integer ℓ, put $\ell_1 = \ell/2$, $\ell_2 = 0$, and $\pi = \left(\begin{smallmatrix} 1 & 0 \\ 0 & 1 \end{smallmatrix}\right)^{\otimes \ell}$. Let $\tilde{\mathbf{s}} \in L^2(E_{\ell, 0})$ and assume that $\tilde{\mathbf{s}}$ satisfies the condition $\pi \tilde{\mathbf{s}}(\mathbf{k}) = \tilde{\mathbf{s}}(\mathbf{k})$. Define $\mathbf{s}(\mathbf{k}) = D_{(\ell/2, 0)}(H_{\mathbf{k}}) \tilde{\mathbf{s}}(\mathbf{k})$. Denote

$$K = \frac{1}{2} \left(k_0 \sigma_0 - \sum_{i=1}^{3} k_i \sigma_i \right),$$

where σ_0 is the 2×2 identity matrix, and σ_i are the classical Pauli matrices. Then, we have

$$\sum_{A_1 = 0}^{1} P_{B A_1} s^{A_1 \cdots A_\ell}(\mathbf{k}) = \sum_{A_1 = 0}^{1} P_{B A_1} \prod_{k=1}^{\ell} (H_{\mathbf{k}})_0^{A_k} \tilde{s}^{0 \cdots 0}(\mathbf{k}) = 0$$

for all $B, A_2, \ldots, A_\ell \in \{0, 1\}$.

Conversely, assume that

$$\sum_{A_1 = 0}^{1} P_{B A_1} s^{A_1 \cdots A_\ell}(\mathbf{k}) = 0, \qquad B \in \{0, 1\}, \quad (A_2, \ldots, A_\ell) \in \{0, 1\}^{\ell - 1}. \quad (2.17)$$

We have

$$\tilde{s}^{B_1 \cdots B_\ell}(\mathbf{k}) = \sum_{(A_1, \ldots, A_\ell) \in \{0, 1\}^\ell} \prod_{k=1}^{\ell} (H_{\mathbf{k}}^{-1})_{A_k}^{B_k} s^{A_1 \cdots A_\ell}(\mathbf{k}).$$

It follows that $\tilde{s}^{1 B_2 \cdots B_\ell}(\mathbf{k}) = 0$ for all $(B_2, \ldots, B_\ell) \in \{0, 1\}^{\ell - 1}$. Therefore $\pi \tilde{\mathbf{s}}(\mathbf{k}) = \tilde{\mathbf{s}}(\mathbf{k})$.

2.6.3   A Transition to the Space Domain

Write the coordinates in the frequency–wavenumber domain as $(\hbar\omega, p_1, p_2, p_3)$, where $\hbar$ is the Planck constant. Assume that Equation (2.17) holds true. Under the inverse Fourier transform it becomes

$$\partial_{BA_1}\psi^{A_1\cdots A_\ell}(\mathbf{x}) = 0, \qquad B \in \{0,1\}, \quad (A_2,\ldots,A_\ell) \in \{0,1\}^{\ell-1}.$$

According to Example 63, in the Minkowski space ∂_{BA_1} is the first-order differential operator acting by

$$\partial_{BA_1}\psi^{A_1\cdots A_\ell}(\mathbf{x}) = \sum_{m=0}^{3} \sigma^m_{BA_1}\frac{\partial\psi^{A_1\cdots A_\ell}(\mathbf{x})}{\partial x^m}.$$

This is the reason why we need to introduce spinor fields.

2.7   Neutrino, Electromagnetic, and Gravitational Waves

2.7.1   Relativistic Wave Equations on Einstein Spacetimes

Let (E, π, M) be a spinor bundle over a smooth pseudo-Riemannian manifold M of dimension 4 such that the metric tensor g_{ij} has one positive and three negative eigenvalues. Assume that the fibres of that bundle are copies of the space $V_{\ell/2,0}$, where ℓ is a positive integer. It turns out that the relativistic wave equations have the same form as before:

$$\nabla_{A_1\dot{A}}\varphi^{A_1\cdots A_\ell} = 0, \tag{2.18}$$

where $\nabla_{A\dot{A}}$ is the spinor covariant derivative constructed in Example 63, and where $\varphi^{A_1\cdots A_\ell}$ is a totally symmetric spinor field in (E, π, M).

We will consider three particular cases: $\ell = 1$, $\ell = 2$, and $\ell = 4$.

In the case of $\ell = 1$, Equation (2.18) has the form

$$\nabla_{A\dot{A}}\varphi^{A} = 0. \tag{2.19}$$

It is known as the *Weyl neutrino equation* and describes the neutrino waves.

To prove that, introduce the tensor

$$F_{ij} = \sigma_i{}^{A\dot{A}}\sigma_j{}^{B\dot{B}}\varphi_A\varphi_B\varepsilon_{\dot{A}\dot{B}}.$$

Penrose and Rindler (1987, Section 4.4) proves that the above spinor equation is equivalent to the non-linear tensor equation

$$F_{ij}\nabla_l F_k{}^l + F_{il}\nabla_k F_j{}^l = 0.$$

To obtain an equivalent linear equation, we apply the operator $\nabla^{A\dot{A}}$ to both hand sides of Equation (2.19) and obtain

$$\Box\varphi^A = 0, \tag{2.20}$$

where $\Box = \nabla^{A\dot{A}}\nabla_{A\dot{A}}$ is the *d'Alembert operator* in M.

In the case of $\ell = 2$, Equation (2.18) has the form

$$\nabla_{A\dot{A}}\varphi^{AB} = 0. \tag{2.21}$$

Let φ^{AB} be a solution to this equation, and define

$$F_{ij} = \sigma_i{}^{A\dot{A}}\sigma_j{}^{B\dot{B}}(\varphi_{AB}\varepsilon_{\dot{A}\dot{B}} + \varphi_{\dot{A}\dot{B}}\varepsilon_{AB}).$$

Then the tensor field F_{ij} satisfies the Maxwell equations on M:

$$\nabla_i F^{ij} = 0, \qquad \nabla_{[i}F_{ik]} = 0. \tag{2.22}$$

Conversely, let a tensor field F_{ij} satisfies the Maxwell equations on M. Define

$$F_{AB\dot{A}\dot{B}} = \sigma^i{}_{A\dot{A}}\sigma^j{}_{B\dot{B}}F_{ij}, \qquad \varphi_{AB} = \frac{1}{2}F_{AB\dot{A}}{}^{\dot{A}}.$$

Then φ^{AB} satisfies Equation (2.21). For proofs, see Penrose and Rindler (1987, Section 5.1).

We adapt the approach by Dolan (2018). Specifically, take a derivative of the first equation in (2.22), re-order derivatives and apply the second equation. We obtain

$$\Box F_{ij} + 2R_{ikjl}F^{kl} + R_i{}^k F_{jk} - R_j{}^k F_{ik} = 0, \tag{2.23}$$

see Dolan (2018, Equation (8)).

Similarly, in the case of $\ell = 4$, write Equation (2.18) in the form

$$\nabla_{A\dot{A}}\varphi^{ABCD} = 0. \tag{2.24}$$

Let φ^{ABCD} be a solution to this equation, and define

$$K_{abcd} = \sigma_a{}^{A\dot{A}}\sigma_b{}^{B\dot{B}}\sigma_c{}^{C\dot{C}}\sigma_d{}^{D\dot{D}}(\varphi_{ABCD}\varepsilon_{\dot{A}\dot{B}}\varepsilon_{\dot{C}\dot{D}} + \varphi_{\dot{A}\dot{B}\dot{C}\dot{D}}\varepsilon_{AB}\varepsilon_{CD}).$$

Then, the tensor K_{abcd} satisfies the equation

$$\nabla_{[a}K_{bc]de} = 0 \tag{2.25}$$

and symmetries

$$K_{abcd} = K_{[cd][ab]}, \qquad K_{[abc]d} = 0, \qquad K_{abc}{}^b = 0. \tag{2.26}$$

Conversely, let a tensor K_{abcd} satisfies the above equation and symmetries. Define

$$K_{ABCD\dot{A}\dot{B}\dot{C}\dot{D}} = \sigma_a{}^{A\dot{A}}\sigma_b{}^{B\dot{B}}\sigma^c{}_{C\dot{C}}\sigma^d{}_{D\dot{D}}K_{abcd},$$
$$\varphi_{ABCD} = \frac{1}{4}K_{(ABCD)\dot{A}\dot{B}\dot{C}\dot{D}}\varepsilon^{\dot{A}\dot{B}}\varepsilon^{\dot{C}\dot{D}}.$$

The spinor φ_{ABCD} satisfies Equation (2.24). For a proof, see (Penrose and Rindler 1987), Section 5.7.

What is the physical sense of the tensor field K_{abcd}? If this field satisfies Equation (2.25), then there is a tensor field h_{ij} such that

$$K_{ijkl} = \frac{1}{2}(\nabla_j\nabla_k h_{il} + \nabla_i\nabla_l h_{jk} - \nabla_j\nabla_l h_{ik} - \nabla_i\nabla_k h_{jl}).$$

Introduce the 'trace-reversed' symbol $\overline{h}_{ij}$ by

$$\overline{h}_{ij} = h_{ij} - \frac{1}{2}g_{ij}h_k{}^k,$$

and substitute it to the last equation in (2.26). We obtain

$$\Box\overline{h}_{ij} - \nabla_i\nabla^k\overline{h}_{jk} - \nabla_j\nabla_k\overline{h}_{ik} + g_{ij}\nabla^k\nabla^l\overline{h}_{kl} = 0, \qquad (2.27)$$

where $\Box = g^{ij}\nabla_i\nabla_j$ is the *d'Alembert operator* in M.

Observe that Equation (2.27) has the following property. Let X_i be a smooth covariant vector field. If $\overline{h}_{ij}$ is a solution to Equation (2.27), then

$$\overline{h}'_{ij} = \overline{h}_{ij} + \nabla_j X_i + \nabla_i X_j - g_{ij}\nabla_i X^i$$

is another solution. This can be checked by direct calculations. It follows that the two above solutions are physically equivalent. In other words, a gauge field $\nabla_j X_i + \nabla_i X_j - g_{ij}\nabla_i X^i$ may be added to a solution.

By direct calculation of $\nabla^j\overline{h}'_{ij}$ we obtain

$$\nabla^j\overline{h}'_{ij} = \nabla^j\overline{h}_{ij} - \Box X_i.$$

Assume that the initial configuration of the field h_{ij} is such that $\nabla^j\overline{h}_{ij} = f_i$. The d'Alembert operator $\Box$ is invertible. Choose a gauge field for which $\Box X_i = f_i$, which is the same as

$$\nabla^j\overline{h}_{ij} = 0. \qquad (2.28)$$

In this gauge, the last three terms in Equation (2.27) vanish and it becomes the wave equation $\Box\overline{h}_{ij} = 0$ and describes gravitational waves. The gauge (2.28) is called the *Lorenz gauge*, or the *Hilbert gauge*, or the *harmonic gauge*, or the *de Donder gauge*.

In Subsection 2.7.3, we will use an additional gauge. Observe that the Lorenz gauge is not spoiled by the coordinate transformation $x^i \to x^i + \zeta^i$ with $\Box\zeta^i = 0$. Under this transformation, $\overline{h}_{ij} \to \overline{h}_{ij} - \zeta_{ij}$, where

$$\zeta_{ij} = \frac{\partial\zeta_j}{\partial x^i} + \frac{\partial\zeta_i}{\partial x^j} - \eta_{ij}\frac{\partial\eta_k}{\partial x^k}.$$

Then also $\Box\zeta_{ij} = 0$, because the *Minkowski space* d'Alembert operator commutes with partial derivatives. It follows that one can subtract the functions

ζ_{ij} from $\overline{h}_{ij}$, and the new $\overline{h}_{ij}$ still satisfy the wave equation $\Box \overline{h}_{ij} = 0$. The above functions depend on four arbitrary functions ζ^i, and we can choose them to impose four conditions on h_{ij}.

We choose ζ^0 in such a way that $h = 0$, and ζ^i, $1 \le i \le 3$ in such a way that $h^{0i} = 0$. The Lorenz gauge (2.28) with $j = 0$ becomes

$$\frac{\partial h^{00}}{\partial x^0} = 0.$$

In particular, h^{00} is time-independent and therefore describes the static part of the gravitational interaction, which can be put equal to 0 for the gravitational wave. The condition $h = 0$ becomes $h_i^i = 0$, and both conditions together with the Lorenz gauge become:

$$h^{0i} = 0, \qquad h_i^i = 0, \qquad \frac{\partial h_{ij}}{\partial x^j} = 0.$$

These conditions determine the *transverse-traceless gauge*. The symmetric and traceless matrix h^{ij} becomes

$$h_{ij}^{TT} = \begin{pmatrix} 0 & 0 & 0 & 0 \\ 0 & h_+ & h_\times & 0 \\ 0 & h_\times & -h_+ & 0 \\ 0 & 0 & 0 & 0 \end{pmatrix}.$$

The matrix entry h_+ (resp., $h_\times$) is called the *amplitude of the* plus (resp., cross) *polarisation*.

Again, we adapt the approach by (Dolan 2018) and write down an equivalent wave equation

$$\Box \overline{h}_{ij} + 2R^k{}_i{}^l{}_j \overline{h}_{kl} = 0. \tag{2.29}$$

2.7.2 The Geometrical Optics Approximation

We assume that the wavelength (resp., inverse frequency) is short in comparison with all involved length scales (resp. time scales).

For Equation (2.20), we use the *plane wave* ansatz (substitution)

$$\varphi^A = \mathcal{A}(x)\Phi^A(x)\exp(i\omega\Psi(x)),$$

for Equation (2.23), the ansatz

$$F_{ij} = \mathcal{A}(x)f_{ij}(x)\exp(i\omega\Psi(x)),$$

for Equation (2.29), the ansatz

$$\overline{h}_{ij} = \operatorname{Re}\mathcal{A}(x)H_{ij}(x)\exp(i\omega\Psi(x)).$$

In all the three plane waves above, ω is the order-counting parameter, and the real-valued field $\mathcal{A}(x)$ (resp., $\Psi(x)$) is called the *amplitude* (resp., the *phase*).

The functions $\Phi^A(x)$, $f_{ij}(x)$, and H_{ij} are called *polarisations*. As we will see later, ω is *not* a frequency!

The above three substitutions give the following results:

$$\exp(i\omega\Psi)(-\omega^2 k^i k_i \mathcal{A}\Phi^A + i\omega[(k^i\nabla_i\mathcal{A} + (\nabla_i k^i)\mathcal{A})\Phi^A) + \mathcal{A}k^i\nabla_i\Phi^A]$$
$$+ O(1) = 0, \tag{2.30a}$$

$$\exp(i\omega\Psi)(-\omega^2 k^i k_i \mathcal{A}f_{ij} + i\omega[(2k^i\nabla_i\mathcal{A} + (\nabla_i k^i)\mathcal{A})f_{ij} + \mathcal{A}k^i\nabla_i f_{ij}]$$
$$+ O(1) = 0, \tag{2.30b}$$

$$\exp(i\omega\Psi)(-\omega^2 k^i k_i \mathcal{A}H_{ij} + i\omega[(2k^i\nabla_i\mathcal{A} + (\nabla_i k^i)\mathcal{A})H_{ij} + \mathcal{A}k^i\nabla_i H_{ij}]$$
$$+ O(1) = 0, \tag{2.30c}$$

where $k_i = \nabla_i\Psi$, and where we denote by $O(1)$ the terms that do not depend on the counting parameter ω. The geometrical optics approach is to examine these equations order-by-order in ω.

At ω^2, all three equations give $k^i k_i = 0$. That is, the gradient vector k^i, tangent to the sub-manifold of constant phase, is null. The gradient satisfies the geodesic equation $k^j\nabla_j k_i = 0$. The solutions to this equation, say $x^i(n)$, are null geodesics lying in the sub-manifold of constant phase. By this reason, the vector k^i is called the *null generator*.

Denote $\theta = \nabla_i k^i$ and call this number the *expansion scalar*. At ω, we obtain the *transport equations* for the amplitude and the polarisation:

$$k^i\nabla_i\mathcal{A} = -\frac{1}{2}\theta\mathcal{A}, \qquad\qquad k^l\nabla_l\psi^A = 0,$$

$$k^l\nabla_l f_{ij} = 0, \qquad\qquad k^l\nabla_l H_{ij} = 0.$$

In other words, the polarisation is parallel-propagated along the null generator.

2.7.3   Polarisation

We represent the plane neutrino wave according to Audretsch and Graf (1970) as

$$\varphi_A = \hat{\varphi}_A \exp(iS), \tag{2.31}$$

the plane electromagnetic and gravitational waves according to Breuer, Tiomno, and Vishveshwara (1975) by

$$F_{ij} = \mathrm{Re}(2k_{[i}m_{j]})\exp(iS), \tag{2.32}$$

and

$$C_{ijpq} = -2\,\mathrm{Re}(k_{[i}m_{j][p}k_{q]}\exp(iS)), \tag{2.33}$$

where the real-valued function S is the *phase*, $k_i = \nabla_i S$ is a real null wave-vector, in Equation (2.32) m_i is a complex-valued vector field, while in Equation (2.33) m_{ij} is a complex-valued rank 2 skew-symmetric tensor field, and C_{ijpq} is the Weyl tensor (1.58a) of the gravitational wave.

In Equation (2.32), the vector fields k_i and m_i satisfy

$$k_i k^i = k_i m^i = \theta = \nabla_i m_i + \theta m_i = 0,$$

where θ is the expansion scalar. In Equation (2.33), the fields k_i and m_{ij} satisfy

$$m_{ij} k^j = m_i{}^i = \nabla_i m_{ij} + \theta m_{ij} = 0.$$

In Equations (2.31)–(2.33), the choice of parameters in not unique: a gauge transformation is possible. In Equations (2.31), this transformation takes the form

$$S \mapsto S - \delta, \qquad \hat{\varphi}_A \mapsto \exp(i\delta)\hat{\varphi}_A, \quad \delta \in \mathbb{R},$$

in Equation (2.32)

$$S \mapsto S - \delta, \qquad m_i \mapsto \exp(i\delta)m_i + \lambda k_i, \quad (\delta, \lambda) \in \mathbb{R}^2,$$

while in Equation (2.33) the gauge transformation is

$$S \mapsto S - \delta, \qquad m_{ij} \mapsto \exp(i\delta)m_{ij} + \lambda_{(i}k_{j)}, \quad (\delta, \lambda_i) \in \mathbb{R}^5.$$

In Equation (2.32) (resp., in Equation (2.33)), at a given point $x \in M$, and for a fixed vector $k_i(x)$, the gauge-equivalent classes of the vectors m_i (resp., skew-symmetric tensors m_{ij}) form a complex linear space H_x^k of dimension 2. Indeed, for the electromagnetic waves, H_x^k is the quotient space of the 3-dimensional complex linear space of the vectors m_i satisfying $m_i k^i = 0$ by the 1-dimensional subspace of the vectors m_i satisfying $m_i = \lambda k_i$. For the gravitational waves, H_x^k is the quotient space of the 6-dimensional complex linear space of the skew-symmetric tensors m_{ij} by the 4-dimensional subspace of m_{ij} satisfying $m_{ij} = \lambda_{(i}k_{j)}$. The inner products in the spaces H_x^k are

$$(m_1, m_2) = m_1{}^i m_{i2}^*, \qquad (m_1, m_2) = \frac{1}{2}m_1{}^{ij}m_{2ij}^*.$$

In Equation (2.31), the gauge-equivalent classes of the scalars $\hat{\varphi}_A$ form a complex linear space H_x^k of dimension 1.

Consider an observer with velocity u^i locating at $x \in M$. By gauge transformation, we can find a representative m_i (resp., m_{ij}) in each gauge-equivalence class with $m_i u^i = 0$ (resp., $m_{ij} u^j = 0$).

In an ensemble of neutrino (resp., electromagnetic, resp., gravitational) waves φ_A (resp., T_{ij}, resp., C_{ijpq}), we define the *polarisation tensor* L_i (resp., L_{ij}, resp., L_{ijpq}) by $L_i = \langle \sigma^i{}_{A\dot{A}}\varphi_A\varphi_{\dot{A}}^* \rangle$ (resp., $L_{ij} = \langle m_i m_j^* \rangle$, resp., $L_{ijpq} = \langle m_{ij}m_{pq}^* \rangle$). The polarisation tensor defines a positive-definite Hermitian linear operator L in the complex linear space H_x^k.

The *energy-momentum tensor* of a neutrino (resp., electromagnetic, resp., gravitational) ensemble is given by

$$T_{ij} = I k_i k_j, \tag{2.34a}$$

resp.,

$$T_{ij} = \frac{1}{2} \operatorname{tr} L k_i k_j, \tag{2.34b}$$

resp.,

$$T_{ij} = \frac{1}{32\pi} \operatorname{tr} L k_i k_j. \tag{2.34c}$$

Equation (2.34a) is proved in Kuchowicz (1974), where I is the intensity of the neutrino field. Equation (2.34b) (resp., (2.34c)) is proved in Breuer, Tiomno, and Vishveshwara (1975) (resp., Isaacson (1968)).

A complex vector t^i with $t^i c t_i = 1$ define an orthonormal basis in the linear space H_x^k as follows. For the case of electromagnetic radiation, the basis is t_i and $c t_i$, for the case of gravitational radiation the basis is

$$e_R = \sqrt{2} t \otimes t, \qquad e_L = \sqrt{2} c t \otimes c t.$$

The *Stokes parameters* of the electromagnetic (resp., gravitational) radiation with respect to the basis $\{t, ct\}$ (resp., $\{e_R, e_L\}$) and an observer with velocity u^i satisfying $u^i t_i = 0$, are defined by

$$I = \omega^2 \operatorname{tr} L, \qquad V = \omega^2 ((t, Lt) - (ct, cLt)), \qquad Q + \mathrm{i}U = 2\omega^2 (ct, Lt),$$

resp.,

$$I = \omega^2 \operatorname{tr} L, \qquad V = \omega^2 ((e_R, Le_R) - (e_L, Le_L)), \qquad Q + \mathrm{i}U = 2\omega^2 (e_L, Le_R),$$

where $\omega = |u_i k^i|$ is the frequency of the wave measured by the observer.

Observe!

For the Weyl neutrino equation, the only Stokes parameter is I.

A rotation of the vector t by an angle δ, $t \mapsto \exp(\mathrm{i}\delta)t$, rotates ct by $-\delta$, e_R by 2δ, and e_R by -2δ. The Stokes parameters T and V are invariant under such a rotation, the electromagnetic (resp., gravitational) combination $Q + \mathrm{i}U$ rotates by 2δ (resp., 4δ).

Next, we introduce the *normalised Stokes parameters* by

$$s_0 = 1, \qquad s_1 = \frac{V}{I}, \qquad s_2 + \mathrm{i}s_3 = \frac{Q + \mathrm{i}U}{I}.$$

The *degree of linear* (resp., *circular*) *polarisation* d_L (resp., d_C) is

$$d_L = \sqrt{s_1^2 + s_2^2}, \qquad d_C = |s_3|.$$

The reader may see circularly polarised waves at Fig. 2.1.[1]

---

1. Source: https://tikz.net/cyclic-polarization/.

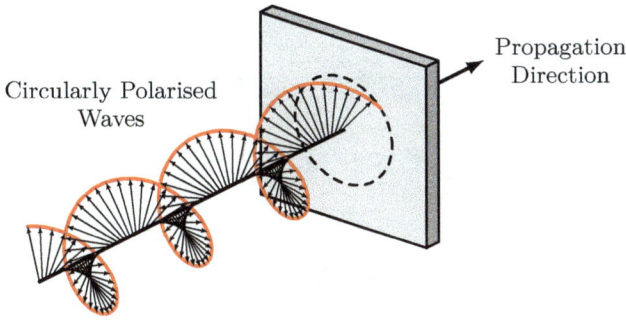

FIGURE 2.1
Circularly polarised waves.

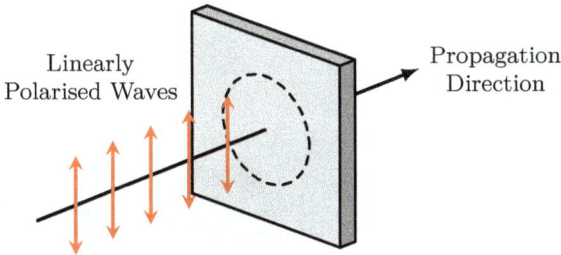

FIGURE 2.2
Linearly polarised waves.

The linearly polarised waves are depicted at Fig. 2.2.[2]

The *degree of polarisation*, d, is the ratio of the completely polarised power to the total power,

$$d = \sqrt{s_1^2 + s_2^2 + s_3^2}.$$

If the wave is completely polarised, then $d = 1$.

By definition, the *tilt angle* τ of the polarisation ellipse is the angle between its major axis and the $e_{(1)}$ axis, see Fig. 2.3.[3] The horizontal (resp., vertical) polarisation corresponds to $\tau = 0$ (resp., $\tau = \pi/2$).

The angle τ is given by

$$\tau = -\frac{i}{4} \ln \frac{s_1^2 + 2is_1s_2 - s_2^2}{s_1^2 + s_2^2}.$$

2. Source: https://tikz.net/optics_polarization/.

3. Source: https://tex.stackexchange.com/questions/474961/ can-i-get-some-help-drawing-a-polarization-ellipse-something-like-this

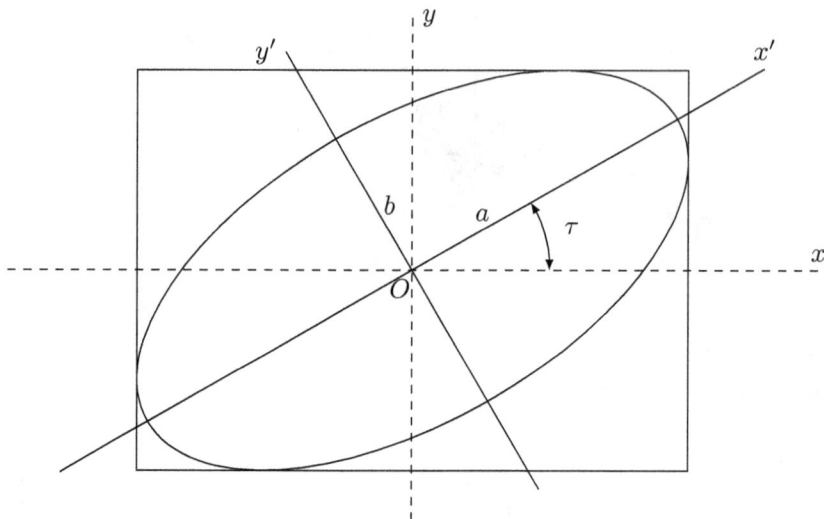

FIGURE 2.3
Polarisation ellipse

The ratio of two axes a and b defines as angle ε by

$$\cot \varepsilon = \pm \frac{a}{b}, \qquad -\frac{\pi}{4} < \varepsilon < \frac{\pi}{4},$$

where the plus (resp., the minus) sign means left-handed (resp., right-handed) polarisation. The case of $\varepsilon = \frac{\pi}{4}$ (resp., $\varepsilon = -\frac{\pi}{4}$, resp., $\varepsilon = 0$) corresponds to the left-handed circular (resp., right-handed circular, resp., pure linear) polarisation.

2.8 Bibliographical Remarks

The classification of space-times and spacetimes into five classes was proposed by Penrose (1968). See also (Earman, Glymour, and Stachel 1977; Earman 1989; Friedman 1983).

Galilean space-time structure is described in (Artz 1981; Fuente, Pelegrín, and Rubio 2021, 2022). The non-invariance of Maxwell equations in Galilean space-times is discussed in Preti, Felice, and Masiero (2009)

For Newtonian space-time structure, see Bernal and Sánchez (2003). In our description, we followed Trautman (1965), Classical mechanics in Newtonian space-times is the subject of Marsden and Ratiu (1999).

Special relativity is the subject of the monographs and textbooks (Naber 1992; Rowe 2001; Sexl and Urbantke 2001). See also the papers (Künzle 1972, 1976)

For theoretical aspects of general relativity, see the books (Bambi 2018; Deruelle and Uzan 2018; Esposito 1995; Felice and Clarke 1990; Ferrari, Gualtieri, and Pani 2020; Foster and Nightingale 2006; Grøn and Hervik 2007; Hartle 2013; Hobson, Efstathiou, and Lasenby 2006; Kopczyński and Trautman 1992; Malament 2012; O'Donnell 2003; Poisson and Will 2014; Rahaman 2021; Ryder 2009; Sachs and Wu 1977b; Stewart 1990; Straumann 2013; Vecchiato 2017; Weinberg 1972) and the papers (Eguchi, Gilkey, and Hanson 1980; Penrose 1960).

For applied and numerical aspects of general relativity, see (Baumgarte and Shapiro 2010; Dhurandhar and Mitra 2022; Griffiths and Podolský 2009; Gourgoulhon 2012; Jetzer 2022; Shibata 2015; Soffel and Han 2019; Stephani et al. 2003).

Literature in cosmology includes (Adler 2021; Baumann 2022; Bergmann and Sabbata 1980; Cheng 2010; Hawking and Ellis 2023; Liddle 2015; Liddle and Lyth 2000; Liebscher 2005; Peter and Uzan 2009; Piattella 2018; Sachs and Wu 1977a). All the above books and many more traditionally include a review of simple cosmological models, where only one or two of the three parameters ρ_m, ρ_r, and ρ_Λ are nonzero.

Theorem 22 has been proved by Straumann (1974).

In our discussion on different distances in the expanding Universe, we do not involve *cosmological horizons*. See Davis and Lineweaver (2004).

Relativistic wave equations and their symmetries are discussed in literature on relativistic quantum theory. See (Baulieu, Iliopoulos, and Sénéor 2017; Başkal, Kim, and Noz 2019; Corinaldesi and Strocchi 1963; Dyson 2011; Faria and Melo 2010; Folland 2008; Gibson and Pollard 1976; Grant 2007; Itzykson and Zuber 1980; Kachelriess 2018; Kim and Noz 1986; Maggiore 2005; Ohnuki 1988; Ryder 1996; Schwartz 2014; Srednicki 2010; Streater and Wightman 2000; Velo and Wightman 1978; Wald 1994; Weinberg 2005a, 2005b, 2005c; Zee 2010). Our exposition is based on Lledó (2004). A sample of papers includes (Exton, Newman, and Penrose 1969; Geroch, Held, and Penrose 1973; Niederer and O'Raifeartaigh 1974a, 1974b; Penrose 1969).

Different kinds of fermions (Dirac, Majorana, Weyl, etc.) are discussed in (Pal 2011; Todorov 2011).

The Einstein–Maxwell equations that describe the electromagnetic field in Einstein spacetimes, are discussed in Torre (2014).

Gravitational waves were predicted by Einstein (1916) and finally discovered in 2015, see Abbott (2016). Their brief history can be found in Taylor (2021, Subsection 2.2.1). A story about their discovery may be found in Cervantes-Cota, Galindo-Uribarri, and Smoot (2016). A preliminary report about the discovery of the Cosmic Gravitational Background has been recently published in Agazie (2023).

A part of the above story is interesting for us. Einstein predicted gravitational waves as a solution to an *approximate* linear partial differential equation. A question about existence of wave solutions to the *exact* Equation (2.13) remained open quite long time. Finally, the existence of such solutions were proved. References include Choquet-Bruhat (1968) and the recent survey paper Roche, Aazami, and Cederbaum (2023) which also describes an approach to definition of plane waves in Einstein spacetimes, alternative to the one chosen in Section 2.7.

Gravitational waves are studied in (Auger and Plagnol 2017; Bethke 2015; Griffiths 1991; MacCallum and Taub 1973; Madore 1972, 1973; Maggiore 2007, 2018; Philippoz 2018; Tiec and Novak 2017).

Concerning the *gauge theory*, see Marathe (2010)

The Stokes parameters have been invented by Stokes (1851). See also Chandrasekhar (1960).

We would like to mention a literature about polarised light: (Degl'Innocenti 2013; Gil Pérez and Ossikovski 2016).

A good account of ray optics in curved manifolds can be found in Perlick (2000).

Papers on Cosmic Microwave Background and its polarisation include (Gluscevic, Kamionkowski, and Cooray 2009; Madison 2021), among others.

Papers about Cosmic Gravitational Background and its polarisation include (Abbott 2018; Afzal et al. 2022; Auclair 2020; Auclair et al. 2020; Bernardo and Ng 2022; Bian et al. 2022; Blanco-Pillado and Olum 2017; Boileau et al. 2022; Buchmüller, Domcke, and Schmitz 2021; Caprini and Figueroa 2018; Chang and Cui 2020; Chen, Wu, and Huang 2022; Cusin, Durrer, and Ferreira 2019; Guedes, Avelino, and Sousa 2018; Inayoshi et al. 2021; Jenkins and Sakellariadou 2018; Kato and Soda 2016; Li and Guo 2022; Liu, Cai, and Guo 2021; Matsui and Kuroyanagi 2019; Ricciardone et al. 2021; Seto and Taruya 2008; Sousa, Avelino, and Guedes 2020; Steane 2017), among others.

The hypothetical Cosmic Neutrino Background is not discovered yet. It is known, however, that at least two of the three neutrino species are moving much slower than light today. As a consequence, they are subject to gravitational clustering, see a review in Ringwald and Wong (2004). The corresponding background must be studied by different methods that lie outside the scope of this book, see Lesgourgues et al. (2013, Subsection 7.2.3).

Concerning the Cosmic Neutrino Background, see (Alvey et al. 2022; Audren et al. 2015; Bauer and Shergold 2023; Bernal et al. 2021; Bernardis et al. 2008; Chen, Upadhye, and Wong 2021; Das, Perez-Gonzalez, and Sen 2022; Dodelson and Vesterinen 2009; Duda, Gelmini, and Nussinov 2001; Follin et al. 2015; Hannestad 2010; Hannestad and Brandbyge 2010; Lin and Holder 2020; Lisanti, Safdi, and Tully 2014; Long, Lunardini, and Sabancilar 2014; Michney and Caldwell 2007; Nussinov and Nussinov 2022; Safdi et al. 2014; Trotta and Melchiorri 2005; Tully and Zhang 2021; Vogel 2015; Zhou 2022).

Monographs on neutrino physics includes (Athar and Singh 2020; Ereditato 2018; Fukugita and Yanagida 2003; Lesgourgues et al. 2013; Mohapatra and

Palash 2004; Valle and Romão 2014; Xing and Zhou 2011; Zuber 2020). See also the papers (Griffiths and Newing 1971; Lesgourgues and Pastor 2006; Lambiase et al. 2005; Wigmans 2000).

3

Random Cross-Sections of Homogeneous Bundles

As a result of our investigations in Chapter 2, we have the following mathematical construction. (M, g) is a pseudo-Riemannian manifold of dimension 4 that represents the Universe, the metric g has one positive and three negative eigenvalues. The manifold M is oriented and time-oriented, its second Stiefel–Whitney characteristic class vanishes. The three cosmic backgrounds, neutrino, microwave, and gravitational ones, spread along the null-geodesics. The background's rays through an observational instrument located at an observer $O \in M$, correspond to null-geodesics through O. Their past directions constitute the field of vision of the instrument and may be represented by the sphere S^2. The observer may imagine herself situated in the centre of the sphere S^2 and mapping onto S^2 all she sees at any instant.

Quantum fluctuations in early Universe imply that the intensity I of the neutrino background is a *random* function on S^2, while the Hermitian matrix

$$P = I\sigma_0 + Q\sigma_1 + U\sigma_2 + V\sigma_3$$

of Stokes parameters of any of the two remaining backgrounds is a random cross-section of a certain spin-tensor bundle over S^2, where σ_0 is the identity matrix and where σ_i are the Pauli matrices (1.14). Further restrictions are induced by the Cosmological Principle. We consider this question in Section 3.1. As a result, we obtain that the backgrounds are described by means of *isotropic* random cross-sections of certain homogeneous fibre bundles over S^2. Specifically, S^2 is represented as a homogeneous space G/H for certain compact topological groups G and H. A representation of the group H acts in a finite-dimensional linear space L over a skew field $\mathbb{K}$ and induces a representation of G in the Hilbert space $L \uparrow G$ of square-integrable sections of the fibre bundle $\xi[L]$ associated to the principal G-bundle (G, π, S^2, H). Theorem 24 gives the spectral expansion of such an isotropic random cross-section for a general bundle.

It turns out that the obtained expansion is a Fourier one with respect to a certain basis in the space $L \uparrow G$. Moreover, the vectors of the basis must be *special* in the following sense. The minimal closed subspace of $L \uparrow G$ that contains the orbit $G \cdot \mathbf{x}$ of a special vector $\mathbf{x}$, must be as small as possible, i.e., carry an *irreducible* representation of G. For the case of a homogeneous fibre bundle over S^2 the elements of these spaces are familiar *spherical harmonics*.

DOI: 10.1201/9781003344353-3

Such vectors can be constructed explicitly, using the Frobenius reciprocity. For that, we first have to construct the orthonormal bases of the linear spaces, where the irreducible representation of G act. This construction is performed in Section 3.2. By applying the Frobenius reciprocity, we construct several special bases in Examples 71–74. Some of them are well known, some are new.

By combining the above constructed special bases with Theorem 24, we obtain several multipole expansions of the intensity, linear polarisation, and circular polarisation of cosmic backgrounds in Section 3.3.

In order to construct such expansions for all Stokes parameters simultaneously, we need to use fibre bundles, where the fibre L carries a finite-dimensional but *reducible* representation of the group H. In that case, the irreducible subspaces of $L \uparrow G$ are constructed with the help of certain numbers called the *Clebsch–Gordan coefficients* of the *complex* representations of the group SU(2). We do not describe the details of the construction, but review it in Section 3.4. The bases constructed there are used in Section 3.5 to build several examples of random multipole expansions of the polarisation matrix.

Two remarks remain. First, in Sections 3.3 and 3.5 we use spherical harmonics with *integer* multipole index ℓ. We believe that the special bases of spherical harmonics with *half-integer* ℓ may find applications elsewhere.

Second, when the representation of the group H in L is *real* and *reducible*, then it is possible to construct the special bases in $L \uparrow G$ using the Clebsch–Gordan coefficients of the *real* representations of the group SU(2). Is it useful to use such bases to study cosmic backgrounds? The question is open, because in that case the helicity basis (3.31) of the *complexified* tangent space of the sky sphere S^2 cannot be defined. By that reason, we do not study such bases.

3.1 Isotropic Random Cross-Sections

Consider the principal fibre bundle $(G, \pi, G/H, H)$ of Example 37. Recall that G is a compact Lie group, H is its closed subgroup, $M = G/H$ is a smooth manifold of left cosets $gH = \{\, gh \colon h \in H \,\}$, and the map $\pi \colon G \to M$ is given by $\pi(g) = gH$. The reason why we weaken the *real-analytic* or *holomorphic* structure of M to the *smooth* one, is as follows. A homogeneous fibre bundle whose associated principal bundle is real-analytic or holomorphic, has very few nonzero holomorphic cross-sections, see Wells (2008).

Let L be a finite-dimensional linear space over a (skew) field $\mathbb{K}$ with inner product $(\cdot, \cdot)_L$, and let a representation of H acts in L and respects the above inner product. In other words, the representation L is orthogonal (resp., unitary, resp., symplectic) if $\mathbb{K} = \mathbb{R}$ (resp., $\mathbb{K} = \mathbb{C}$, resp., $\mathbb{K} = \mathbb{H}$). Let $\xi[L] = (E, \pi_L, M)$ be the homogeneous fibre bundle over M with fibre L

and associated principal fibre bundle (G, π, M, H) where the left action of G is induced by the action $g \cdot (h, \mathbf{1}) = (gh, \mathbf{1})$, g, $h \in G$, $\mathbf{1} \in L$. Finally, let the cross-sections $\{\, \mathbf{s}_i(p) \colon i \in I \,\}$ constitute an orthonormal basis of the Hilbert space $L \uparrow G$ of the square-integrable cross-sections of $\xi[L]$ with respect to the G-invariant probabilistic measure μ on M.

In Subsection 1.11.2, we constructed a random cross-section

$$\mathbf{X}(p) = \sum_{i \in I} X_i \mathbf{s}_i(p) \tag{3.1}$$

of the fibre bundle $\xi[L]$, where X_i constitute a set of independent $\mathbb{K}$-valued random variables, and the series of their variances converges.

Consider the particular case, when the principal fibre bundle is the Hopf one realised as $(\mathrm{SU}(2), \pi, S^2, \mathrm{U}(1))$. For a certain representation of $\mathrm{U}(1)$ in a finite-dimensional linear space L, there is a random section $X(p)$ of the fibre bundle $\xi[L]$ that models some aspect of a cosmic background. By the Cosmological Principle, $X(p)$ must be isotropic. But what does this mean?

Definition 112. A random cross-section $\mathbf{X}(p)$ is called *isotropic* if its finite-dimensional distributions are G-invariant, that is, for any positive integer n, for any set $p_1, \ldots, p_n$ of n distinct points in M, and for any $g \in G$, the random elements $\mathbf{X}(p_1) \oplus \cdots \oplus \mathbf{X}(p_n)$ and $g \cdot \mathbf{X}(g^{-1} \cdot p_1) \oplus \cdots \oplus g \cdot \mathbf{X}(g^{-1} \cdot p_n)$ are identically distributed.

By isotropy, the expected value $\langle \mathbf{X}(p) \rangle = \mathsf{E}[\mathbf{X}(p)] \in L$ of the random cross-section $\mathbf{X}(p)$ must satisfy the following condition:

$$\langle g \cdot \mathbf{X}(g^{-1} \cdot p) \rangle = \langle \mathbf{X}(p) \rangle. \tag{3.2}$$

It follows that $\langle \mathbf{X}(p) \rangle$ does not depend on p, because G acts on M transitively. Moreover, by (Wallach 1973, Lemma 5.2.3), the action of H in $\pi_L^{-1}(p)$ is the representation L. Then we have $h \cdot \langle \mathbf{X}(p) \rangle = \langle \mathbf{X}(p) \rangle$. That is, the vector $\langle \mathbf{X}(p) \rangle$ belongs to the isotypical component of the trivial representation of H is it exists, otherwise $\langle \mathbf{X}(p) \rangle = \mathbf{0}$.

Definition 113. The correlation operator $K(\mathbf{X}(p), \mathbf{X}(q))$ or just $K(p, q)$ of the random cross-section $\mathbf{X}(p)$ is the mutual correlation operator of the L-valued random elements $\mathbf{X}(p) - \langle \mathbf{X}(p) \rangle$ and $\mathbf{X}(q) - \langle \mathbf{X}(q) \rangle$.

In other words, $K(p, q)$ is the linear operator in L satisfying the condition

$$(\mathbf{x}, K(p, q)\mathbf{y})_L = \mathsf{E}[(\mathbf{x}, \mathbf{X}(p) - \langle \mathbf{X}(p) \rangle)_L (\mathbf{X}(q) - \langle \mathbf{X}(q) \rangle, \mathbf{y})_L], \qquad \mathbf{x}, \mathbf{y} \in L.$$

If the random cross-section $\mathbf{X}(p)$ is isotropic, then

$$K(g \cdot \mathbf{X}(g^{-1} \cdot p), g \cdot \mathbf{X}(g^{-1} \cdot q)) = K(\mathbf{X}(p), \mathbf{X}(q)), \qquad g \in G. \tag{3.3}$$

In the remaining part of this section, without lost of generality, we assume that the random cross-section $\mathbf{X}(p)$ is *centred*, that is, $\langle \mathbf{X}(p) \rangle = \mathbf{0}$. Indeed, an isotropic field without this assumption is just $\mathbf{X}(p) + \mathbf{1}$, where $\mathbf{1}$ belongs to the isotypical component of the trivial representation of H if it exists, otherwise $\mathbf{1} = \mathbf{0}$.

Definition 114. A random cross-section $\mathbf{X}(p)$ is called *wide-sense isotropic* if it satisfies (3.2) and (3.3).

Obviously, an isotropic random cross-section $\mathbf{X}(p)$ with $\mathsf{E}[\|\mathbf{X}(p)\|_L^2] < \infty$ is wide-sense isotropic. Conversely, if a random cross-section $\mathbf{X}(p)$ is wide-sense isotropic and Gaussian, then it is isotropic. This follows from Theorem 14.

In what follows, we consider only wide-sense isotropic random cross-sections and call them isotropic.

How the isotropy of the cross-section $\mathbf{X}(p)$ influences Equation (3.1)?

Denote by $L_0^2(\Omega; L)$ the Hilbert space of L-valued random elements $\mathbf{X}$ on a probability space $(\Omega, \mathfrak{F}, \mathsf{P})$ that satisfy the conditions $\mathsf{E}[\|\mathbf{X}\|_L^2] < \infty$ and $\mathsf{E}[\mathbf{X}] = \mathbf{0}$ with inner product

$$(\mathbf{X}, \mathbf{Y}) = \mathsf{E}[(\mathbf{X}, \mathbf{Y})_L].$$

Consider the cross-section $\mathbf{X}(p)$ as the map $M \to L_0^2(\Omega; L)$, $p \mapsto \mathbf{X}(p)$.

Definition 115. The random cross-section $\mathbf{X}(p)$ is called *mean-square continuous* if the above map is continuous.

It is easy to see that in this case, for an arbitrary $j \in I$, the random function $(\mathbf{X}(p), \mathbf{s}_j(p))_L$ determines a continuous map $M \to L_0^2(\Omega; \mathbb{K}^1)$, $p \mapsto (\mathbf{X}(p), \mathbf{s}_j(p))_L$. Integrate this map over M with respect to the measure μ. Using Equation (3.1), we obtain

$$X_j = \int_M (\mathbf{X}(p), \mathbf{s}_j(p))_L \, \mathrm{d}\mu(p), \qquad j \in I.$$

In what follows, we construct a *special basis* in the space $L \uparrow G$.

Let $\mathbf{s} \in L \uparrow G$. Let $H_{\mathbf{s}}$ be the closed linear span of the elements $g \cdot \mathbf{s} \in L \uparrow G$, $g \in G$. The above space carries a representation of the group G.

Definition 116. An element $\mathbf{s} \in L \uparrow G$ is called *special* if the representation $H_{\mathbf{s}}$ is irreducible.

Assume that $H = \{e\}$ and $\mathbb{K}^1$ is the trivial representation of H. For that case, (Hofmann and Morris 2020) call an element $\mathbf{s} \in \mathbb{K}^1 \uparrow G$ *almost invariant* if the space $H_{\mathbf{s}}$ is finite-dimensional. We see that a special element is almost invariant.

We give a construction of a special basis, whose elements are special. First, partition the space L into mutually orthogonal subspaces L_σ, $\sigma \in \hat{H}_{\mathbb{K}}$, the isotypic components of an irreducible representation σ. All of them, except finitely many, are equal to $\mathbf{0} \in L$. Denote $\hat{H}_{\mathbb{K}}(L) = \{\, \sigma \in \hat{H}_{\mathbb{K}} \colon L_\sigma \neq \{\mathbf{0}\} \,\}$.

By Wallach (1973, Lemma 5.2.5), the space $L \uparrow G$ is partitioned into nonzero mutually orthogonal subspaces $L_\sigma \uparrow G$, $\sigma \in \hat{H}_{\mathbb{K}}(L)$, where the representation of G induced by the representation L_σ of H acts. We construct a special basis in each space $L_\sigma \uparrow G$ separately.

Denote by R_λ (resp., $R_\lambda \downarrow H$) the $\mathbb{K}$-linear space where the representation $\lambda \in \hat{G}_{\mathbb{K}}$ (resp., its restriction to H) acts. Denote by $(R_\lambda \downarrow H)_\mu$ the subspace of $R_\lambda \downarrow H$ where the isotypic component of a representation $\mu \in \hat{H}_{\mathbb{K}}$ acts. Introduce the following notation:

$$\hat{G}_{\mathbb{K}}(\mu) = \{\, \lambda \in \hat{G}_{\mathbb{K}} \colon (R_\lambda \downarrow H)_\mu \neq \{\mathbf{0}\} \,\},$$

$$\hat{H}_{\mathbb{K}}(\lambda) = \{\, \mu \in \hat{H}_{\mathbb{K}} \colon (R_\lambda \downarrow H)_\mu \neq \{\mathbf{0}\} \,\},$$

$$n_\sigma = \frac{\dim L_\sigma}{\dim \sigma}, \quad \sigma \in \hat{H}_{\mathbb{K}}(L),$$

$$n_{\lambda\mu} = \frac{\dim(R_\lambda \downarrow H)_\mu}{\dim \mu}, \quad \lambda \in \hat{G}_{\mathbb{K}}(\mu).$$

That is, the linear space L_σ (resp., $(R_\lambda \downarrow H)_\mu$) contains $n_\sigma > 0$ (resp., $n_{\lambda\mu} > 0$) copies of the irreducible representation σ (resp., μ). The partition of the above spaces into irreducible subspaces is not unique and should be performed in a special way in each concrete example. Therefore, assume we already constructed the spaces $(R_\lambda \downarrow H)_{\mu,k}$, $1 \leq k \leq n_{\lambda\mu}$, and $L_{\sigma,i}$, $1 \leq i \leq n_\sigma$, where the copies of the representations μ and σ act.

Now we introduce coordinates. Let the elements

$$\{\, \mathbf{e}_{\lambda\mu n q} \colon \mu \in \hat{H}_{\mathbb{K}}(\lambda), 1 \leq k \leq n_{\lambda\mu}, 1 \leq l \leq \dim \mu \,\}$$

constitute an orthonormal basis of the space $(R_\lambda \downarrow H)$. Similarly, let the elements

$$\{\, \mathbf{f}_{\rho i j} \colon \rho \in \hat{H}_{\mathbb{K}}(L), 1 \leq i \leq n_\rho, 1 \leq j \leq \dim \rho \,\} \tag{3.4}$$

constitute an orthonormal basis of the space L.

The matrix of a linear operator $f^\sharp \in \mathrm{Hom}_{\mathbb{K}H}(R_\lambda \downarrow H, L)$ has $\dim L$ rows and $\dim \lambda$ columns. Enumerate them in such a way that the first $n_\sigma \dim \sigma$ rows correspond to the n_σ copies of the irreducible representation σ inside L, while the first $n_{\lambda\sigma} \dim \sigma$ columns correspond to the $n_{\lambda\sigma}$ copies of σ inside $R_\lambda \downarrow H$. Then, the only potentially nonzero matrix entries with respect to the constructed bases are located at the intersection of the above rows and columns. The above entries are divided into $n_{\lambda\sigma} n_\sigma$ square blocks A_{ik}. Each block has $\dim \sigma$ rows and columns:

$$f^\sharp = \begin{pmatrix} A_{11} & A_{12} & \cdots & A_{1n_{\lambda\sigma}} & 0 \\ A_{21} & A_{22} & \cdots & A_{2n_{\lambda\sigma}} & 0 \\ \cdots & \cdots & \cdots & \cdots & 0 \\ A_{n_\sigma 1} & A_{n_\sigma 2} & \cdots & A_{n_\sigma n_{\lambda\sigma}} & 0 \\ 0 & 0 & 0 & 0 & 0 \end{pmatrix}. \tag{3.5}$$

The structure of each block is determined by Theorems 9, 10, and 11, and depends on $\mathbb{K}$ and the type of the irreducible representation $\sigma \in \hat{H}_{\mathbb{K}}$.

If $\mathbb{K} = \mathbb{C}$ or σ is of real type, then each block A_{ik} is a multiple, say αI, of the identity matrix I. In the remaining four cases, the structure of the block is more sophisticated.

If $\mathbb{K} = \mathbb{R}$ and σ is of complex type, then $\dim \sigma$ is divisible by 2, the diagonal of each block A_{ik} is filled by the 2×2 matrices, and the basis can be chosen in such a way that the matrices have the form $\begin{pmatrix} \alpha & -\beta \\ \beta & \alpha \end{pmatrix}$ with α, $\beta \in \mathbb{R}$, a particular realisation of the unique real representation of the real Clifford algebra $\mathrm{Cl}_{1,0} = \mathbb{C}$. Similarly, if σ is of quaternionic type, then $\dim \sigma$ is divisible by 4, the diagonal of each block A_{ik} is filled by the 4×4 matrices, and the basis can be chosen in such a way that the matrices have the form
$\begin{pmatrix} \alpha & -\delta & -\gamma & -\beta \\ \delta & \alpha & \beta & -\gamma \\ \gamma & -\beta & \alpha & \delta \\ \beta & \gamma & -\delta & \alpha \end{pmatrix}$ with α, β, γ, $\delta \in \mathbb{R}$, a particular realisation of the unique real representation of the real Clifford algebra $\mathrm{Cl}_{2,0} = \mathbb{H}$.

If $\mathbb{K} = \mathbb{H}$ and σ is of complex (resp., quaternionic) type, then each block A_{ik} is again a multiple of the identity matrix, but the diagonal element α belongs to the subset $\mathbb{C} \subset \mathbb{R}$ (resp., $\mathbb{R} \subset \mathbb{H}$).

We proved that for each i, $1 \le i \le n_\sigma$, the set $\{A_{i1}, A_{i2}, \ldots, A_{in_{\lambda_\sigma}}\}$ depends on $\dim_{\mathbb{K}'} \mathrm{Hom}_{\mathbb{K}H}(\sigma, \sigma) n_{\lambda_\sigma}$ $\mathbb{K}'$-valued parameters. The Frobenius reciprocity gives

$$\dim_{\mathbb{K}'} \mathrm{Hom}_{\mathbb{K}H}(\sigma, \sigma) n_{\lambda_\sigma} n_\sigma = \dim_{\mathbb{K}'} \mathrm{Hom}_{\mathbb{K}G}(R_\lambda, L_\sigma \uparrow G).$$

It follows that the representation $L_\sigma \uparrow G$ contains $M n_\sigma$ copies of the irreducible representation λ, where

$$M = M(\sigma, \lambda) = n_{\lambda_\sigma} \frac{\dim_{\mathbb{K}'} \mathrm{Hom}_{\mathbb{K}H}(\sigma, \sigma)}{\dim_{\mathbb{K}'} \mathrm{Hom}_{\mathbb{K}G}(\lambda, \lambda)}. \tag{3.6}$$

Denote by $(L_\sigma \uparrow G)_\lambda$ the isotypic component of the irreducible representation λ in the space $L_\sigma \uparrow G$. Under the Frobenius reciprocity map, an operator $f^\sharp \in \mathrm{Hom}_{\mathbb{K}H}(R_\lambda \downarrow H, L)$ with matrix (3.5) becomes the element $f_\flat \in \mathrm{Hom}_{\mathbb{K}G}(R_\lambda, (L_\sigma \uparrow G)_\lambda)$ given by

$$f_\flat \mathbf{r} = f^\sharp(g^{-1} \cdot \mathbf{r}), \qquad \mathbf{r} \in R_\lambda, \quad g \in G. \tag{3.7}$$

The matrix of the operator $f_\flat$ has $M n_\sigma \dim \lambda$ rows and $\dim \lambda$ columns. We must choose an orthonormal basis $\mathbf{s}^{\lambda \mu \sigma k l m n q}(p)$ of the space $(L_\sigma \uparrow G)_\lambda$. Here the indices k, l, and m with $1 \le k \le n_{\lambda_\sigma}$, $1 \le l \le \dim \sigma$, and $1 \le m \le M$ enumerate the copies of λ inside $L_\sigma \uparrow G$, while the indices μ, n, and q with $\mu \in \hat{H}_{\mathbb{K}}(\lambda)$, $1 \le n \le n_\mu$, and $1 \le q \le \dim \mu$ enumerate the vectors of an orthonormal basis inside each copy. The above matrix consists of the n_σ blocks B_i, each block has $M \dim \lambda$ rows and $\dim \lambda$ columns and depends on $M n_\sigma \dim_{\mathbb{K}'} \mathrm{Hom}_{\mathbb{K}G}(\lambda, \lambda)$ $\mathbb{K}'$-valued parameters. The structure of each square sub-block depends on $\mathbb{K}$ and the type of the irreducible representation $\lambda \in \hat{G}_{\mathbb{K}}$ in the same way as above.

We choose and fix a one-to-one correspondence between the two above described sets of parameters that determine the matrices $f^\sharp$ and $f_\flat$. The chosen correspondence and Equation (3.7) determine the sections $\mathbf{s}^{\lambda\mu\sigma klmnq}(p)$ up to the order.

Observe that the basis created in this way, is special by construction, and each of the L-valued cross-sections $\mathbf{s}^{\lambda\mu\sigma klmnq}(p)$ has components $s_{\rho ij}^{\lambda\mu\sigma klmnq}(p)$, $\rho \in \hat{H}_\mathbb{K}(L)$, $1 \leq i \leq n_\rho$, $1 \leq j \leq \dim\rho$ in the basis (3.4).

We formulate the main result of this section. Denote by $X_{\rho ij}(p)$ the components of the random cross-section $\mathbf{X}(p)$ in the basis (3.4).

Theorem 24. *The centred random cross-section*

$$
X_{\rho ij}(p) = \sum_{\sigma \in \hat{H}_\mathbb{K}(L)} \sum_{\lambda \in \hat{G}_\mathbb{K}(\sigma)} \sum_{\mu \in \hat{H}_\mathbb{K}(\lambda)} \sum_{k=1}^{n_{\lambda\sigma}} \sum_{l=1}^{\dim\sigma} \sum_{m=1}^{M}
$$
$$
\times \sum_{n=1}^{n_\mu} \sum_{q=1}^{\dim\mu} X_{\lambda\mu\sigma klmnq}^{\rho ij} s_{\rho ij}^{\lambda\mu\sigma klmnq}(p)
\tag{3.8}
$$

of a homogeneous vector bundle $\xi[L] = (E, \pi_L, M)$ is isotropic if and only if the centred $\mathbb{K}^1$-valued random variables $X_{\lambda\mu\sigma klmnq}^{\rho ij}$ satisfy the condition

$$
\mathsf{E}[\overline{X_{\lambda\mu\sigma klmnq}^{\rho ij}} X_{\lambda'\mu'\sigma'k'l'm'n'q'}^{\rho'i'j'}] = \delta_{\lambda\lambda'}\delta_{\mu\mu'}\delta_{nn'}\delta_{qq'} b_{\lambda\mu\sigma klmnq\rho'i'j'}^{\rho ij\sigma'k'l'm'},
\tag{3.9}
$$

where for fixed λ_0, μ_0, n_0, and q_0, the matrix $b_{\lambda_0\mu_0\sigma klmn_0 q_0\rho'i'j'}^{\rho ij\sigma'k'l'm'}$ is nonnegative-definite, and

$$
\sum_{\sigma,\sigma' \in \hat{H}_\mathbb{K}(L)} \sum_{\lambda \in \hat{G}_\mathbb{K}(\sigma) \cap \hat{G}_\mathbb{K}(\sigma')} \sum_{\mu \in \hat{H}_\mathbb{K}(\lambda)} \sum_{k=1}^{n_{\lambda\sigma}} \sum_{k'=1}^{n_{\lambda\sigma'}} \sum_{l=1}^{\dim\sigma} \sum_{l'=1}^{\dim\sigma'} \sum_{m=1}^{M(\sigma,\lambda)} \sum_{m'=1}^{M(\sigma',\lambda')}
$$
$$
\times \sum_{n=1}^{n_\mu} \sum_{q=1}^{\dim\mu} b_{\lambda\mu\sigma klmnq\rho'i'j'}^{\rho ij\sigma'k'l'm'} < \infty.
\tag{3.10}
$$

Proof. Assume that the centred random cross-section (3.8) is isotropic. Pick two irreducible components L_λ and $L_{\lambda'}$ of the representation $L \uparrow G$ of the group G. Assume that the representation λ (resp., λ') is located inside $L_\sigma \uparrow G$ (resp., $L_{\sigma'} \uparrow G$) and is enumerated by the indices k, l, and m (resp., k', l', and m'). Let the linear space L_λ (resp., $L_{\lambda'}$) is generated by the $\dim\lambda$ (resp., $\dim\lambda'$) special basis elements $\mathbf{s}^{\lambda\mu\sigma klmnq}(p)$, $\mu \in \hat{H}_\mathbb{K}(\lambda)$, $1 \leq n \leq n_\mu$, $1 \leq q \leq \dim\mu$ (resp., $\mathbf{s}^{\lambda'\mu'\sigma'k'l'm'n'q'}(p')$, $\mu' \in \hat{H}_\mathbb{K}(\lambda')$, $1 \leq n \leq n_{\mu'}$, $1 \leq q \leq \dim\mu'$) with components $s_{\rho ij}^{\lambda\mu\sigma klmpq}(p)$ (resp., $s_{\rho'i'j'}^{\lambda'\mu'k'l'm'p'q'}(p')$). Introduce the random cross-sections $\mathbf{Y}(p) = \mathbf{Y}^{\sigma\lambda klm}(p) \in L_\lambda$ and $\mathbf{Y}'(p') = \mathbf{Y}'^{\sigma'\lambda'k'l'm'}(p') \in L_{\lambda'}$

with components

$$Y_{\rho i j}(p) = \sum_{\mu \in \hat{H}_{\mathbb{K}}(\lambda)} \sum_{n=1}^{n_{mu}} \sum_{q=1}^{\dim \mu} X_{\lambda \mu \sigma k l m n q}^{\rho i j} \mathbf{s}^{\lambda \mu \sigma k l m n q}(p),$$

$$Y'_{\rho' i' j'}(p') = \sum_{\mu' \in \hat{H}_{\mathbb{K}}(\lambda')} \sum_{n'=1}^{n_{mu'}} \sum_{q'=1}^{\dim \mu'} X_{\lambda' \mu' \sigma' k' l' m' n' q'}^{\rho' i' j'} \mathbf{s}^{\lambda' \mu' \sigma' k' l' m' n' q}(p').$$

By definition of an isotropic random cross-section, we have

$$\mathsf{E}[(\mathbf{x}, g \cdot \mathbf{Y}(g^{-1} \cdot p))_{L_\lambda}(g \cdot \mathbf{Y}'(g^{-1} \cdot p'), \mathbf{x}')_{L_{\lambda'}}] = \mathsf{E}[(\mathbf{x}, \mathbf{Y}(p))_{L_\lambda}(\mathbf{Y}'(p'), \mathbf{x}')_{L_{\lambda'}}]$$

for all $\mathbf{x} \in L_\lambda$, $\mathbf{x}' \in L_{\lambda'}$, and for all $g \in G$. We calculate both hand sides of this equation separately.

On the one hand,

$$\mathsf{E}[(\mathbf{x}, g \cdot \mathbf{Y}(g^{-1} \cdot p))_{L_\lambda}(g \cdot \mathbf{Y}'(g^{-1} \cdot p'), \mathbf{x}')_{L_{\lambda'}}]$$
$$= \mathsf{E}[(g^{-1} \cdot \mathbf{x}, \mathbf{Y}(g^{-1} \cdot p))_{L_\lambda}(\mathbf{Y}'(g^{-1} \cdot p'), g^{-1} \cdot \mathbf{x}')_{L_{\lambda'}}]$$
$$= \mathsf{E}[(g^{-1} \cdot \mathbf{x}, \sum_{\mu,n,q} X_{\lambda \mu \sigma k l m n q}^{\rho i j} \mathbf{s}^{\lambda \mu \sigma k l m n q}(g^{-1} \cdot p))_{L_\lambda}$$
$$\times (\sum_{\mu',n',q'} X_{\lambda' \mu' \sigma' k' l' m' n' q'}^{\rho' i' j'} \mathbf{s}^{\lambda' \mu' \sigma' k' l' m' n' q}(p), g^{-1} \cdot \mathbf{x}')_{L_{\lambda'}}]$$
$$= \sum_{\mu,n,q,\mu',n',q'} \mathsf{E}[\overline{X_{\lambda \mu \sigma k l m n q}^{\rho i j}} X_{\lambda' \mu' \sigma' k' l' m' n' q'}^{\rho' i' j'}] (g^{-1} \cdot \mathbf{x}, s_{\sigma i j}^{\lambda \mu k l m}(g^{-1} \cdot p))_{L_\lambda}$$
$$\times (s_{\sigma' i' j'}^{\lambda' \mu' k' l' m'}(g^{-1} \cdot p'), g^{-1} \cdot \mathbf{x}')_{L_{\lambda'}}.$$

On the other hand,

$$\mathsf{E}[(\mathbf{x}, \mathbf{Y}(p))_{L_\lambda}(\mathbf{Y}'(p'), \mathbf{x}')_{L_{\lambda'}}]$$
$$= \mathsf{E}[(\mathbf{x}, \sum_{\mu,n,q} X_{\lambda \mu \sigma k l m n q}^{\rho i j} \mathbf{s}^{\lambda \mu \sigma k l m n q}(p))_{L_\lambda}$$
$$\times (\sum_{\mu',n',q'} X_{\lambda' \mu' \sigma' k' l' m' n' q'}^{\rho' i' j'} \mathbf{s}^{\lambda' \mu' \sigma' k' l' m' n' q}(p'), \mathbf{x}')_{L_{\lambda'}}]$$
$$= \sum_{\mu,i,j,\mu',i',j'} \mathsf{E}[\overline{X_{\lambda \mu \sigma k l m n q}^{\rho i j}} X_{\lambda' \mu' \sigma' k' l' m' n' q'}^{\rho' i' j'}]$$
$$\times [(\mathbf{x}, s_{\sigma i j}^{\lambda \mu k l m n q}(p))_L, (s_{\sigma' i' j'}^{\lambda' \mu' k' l' m' n' q'}(p'), \mathbf{x}')_L].$$

The rightmost parts of the two last displays are equal. This means the following. The linear operator that maps L_λ to $L_{\lambda'}$ and has the matrix entries $\mathsf{E}[\overline{X_{\lambda \mu \sigma k l m n q}^{\rho i j}} X_{\lambda' \mu' \sigma' k' l' m' n' q'}^{\rho' i' j'}]$ with respect to the special basis, *intertwines* the irreducible representations λ and λ'. Schur's Lemma implies the symbol $\delta_{\lambda \lambda'}$ in the right hand side of condition (3.9). When $\lambda = \lambda'$, the same Lemma implies that under transposition the off-diagonal matrix entries are multiplied by -1, while the very definition of the above matrix implies that they are replaced with their conjugates. It follows that all off-diagonal matrix entries are zeroes, the matrix is diagonal with real and nonnegative diagonal entries. We denote them by $b^{\rho i j \sigma' k' l' m'}_{\lambda \mu \sigma k l m n q \rho' i' j'}$ and obtain condition (3.9).

For fixed λ_0, μ_0, n_0, and q_0, the matrix $b^{\rho i j \sigma' k' l' m'}_{\lambda_0 \mu_0 \sigma k l m n_0 q_0 \rho' i' j'}$ is the matrix of the correlation operator of the random vector with components $X^{\rho i j}_{\lambda_0 \mu_0 \sigma k l m n_0 q_0}$, therefore it is nonnegative-definite. Equation (3.10) becomes equivalent to the obvious inequality $\mathsf{E}[\|X_{\rho i j}(p)\|^2] < \infty$.

Conversely, assume that the centred random variables $X^{\rho i j}_{\lambda \mu \sigma k l m n q}$ satisfy Equations (3.9) and (3.10). The latter equation implies that the series (3.8) converges point-wise in mean-square and therefore determines a centred random cross-section of a homogeneous vector bundle $\xi[L] = (E, \pi_L, M)$. We need to prove that the constructed random cross-section is isotropic.

Let g be an arbitrary element of G, and let $\mathbf{x}$ and $\mathbf{x}'$ be arbitrary elements of L. For all $p, p' \in M$ we have

$$
\begin{aligned}
&\mathsf{E}[(\mathbf{x}, g \cdot \mathbf{X}(g^{-1} \cdot p))_L, (g \cdot \mathbf{X}(g^{-1} \cdot p'), \mathbf{x}')_L] \\
&= \sum \mathsf{E}[(\mathbf{x}, g \cdot X^{\rho i j}_{\lambda \mu \sigma k l m n q} s^{\lambda \mu \sigma k l m n q}_{\rho i j}(g^{-1} \cdot p))_L, \\
&\quad (g \cdot X^{\rho' i' j'}_{\lambda' \mu' \sigma' k' l' m' n' q'} s^{\lambda' \mu' \sigma' k' l' m' n' q'}_{\rho' i' j'}(g^{-1} \cdot p'), \mathbf{x}')_L] \\
&= \sum \mathsf{E}[(\mathbf{x}, g^{\nu \xi r}_{\lambda \mu q} X^{\rho i j}_{\lambda \mu \sigma k l m n q} \overline{g^{\nu \xi r}_{\lambda \mu q}} s^{\nu \xi \sigma k l m n r}_{\rho i j}(p))_L, \\
&\quad (g^{\nu' \xi' r'}_{\lambda' \mu' q'} X^{\rho' i' j'}_{\lambda' \mu' \sigma' k' l' m' n' q'} \overline{g^{\nu' \xi' r'}_{\lambda' \mu' q'}} s^{\nu' \xi' \sigma' k' l' m' n' r'}_{\rho' i' j'}(p'), \mathbf{x}')_L] \\
&= \sum \mathsf{E}[(\mathbf{x}, X^{\rho i j}_{\lambda \mu \sigma k l m n q} s^{\nu \xi \sigma k l m n r}_{\rho i j}(p))_L, \\
&\quad (X^{\rho' i' j'}_{\lambda' \mu' \sigma' k' l' m' n' q'} s^{\nu' \xi' \sigma' k' l' m' n' r'}_{\rho' i' j'}(p'), \mathbf{x}')_L] \\
&= \mathsf{E}[(\mathbf{x}, \mathbf{X}(p))_L, (\mathbf{X}(p'), \mathbf{x}')_L],
\end{aligned}
$$

where we used Equation (3.8), the definition of action of $g \in G$ on the elements of a special basis, and the fact that the matrix of a representation of g is either orthogonal, unitary, or symplectic.   □

3.2   Examples of Special Bases: The Irreducible Case

In what follows, we consider special bases in the Hilbert spaces of square-integrable sections of fibre bundles associated to the following three particular principal bundles $(G, \pi, G/H, H)$.

1. A variant of the Hopf bundle, $G = \mathrm{SU}(2)$, $H = \mathrm{U}(1)$. The base G/H is the sphere S^2.

2. $G = \mathrm{SO}(3)$, $H = \mathrm{SO}(2)$.

3. $G = \mathrm{O}(3)$, $H = \mathrm{O}(2)$.

In this section, the representation L of the group H that constructs an associated fibre bundle is *irreducible*, hence the title of the section.

Observe that the map π of item 1 in terms of the Euler angles is given by Equation (1.53). We choose a local spin gauge $\psi(\theta, \varphi) = 0$ of Example 61.

If we identify a quaternion $q = \alpha + \beta\mathrm{i} + \gamma\mathrm{j} + \delta\mathrm{k}$ with unit norm with the matrix given by the rightmost part of Equation (1.40), then, under the map (1.35), the Euler angles in $\mathrm{SO}(3)$ are described by Equation (1.26). In this chart, the subgroup $\mathrm{SO}(2)$ that fixes the vector $(0, 0, 1)^\top$ corresponds to the values $\varphi = \theta = 0$. That is,

$$
\begin{aligned}
\pi_+ &\begin{pmatrix} \cos\varphi\cos\theta\cos\psi - \sin\varphi\sin\psi & -\cos\varphi\cos\theta\sin\psi - \sin\varphi\cos\psi & \cos\varphi\sin\theta \\ \sin\varphi\cos\theta\cos\psi + \cos\varphi\sin\psi & -\sin\varphi\cos\theta\sin\psi + \cos\varphi\cos\psi & \sin\varphi\sin\theta \\ -\sin\theta\cos\psi & \sin\theta\sin\psi & \cos\theta \end{pmatrix} \\
&= \begin{pmatrix} \cos\psi & -\sin\psi & 0 \\ \sin\psi & \cos\psi & 0 \\ 0 & 0 & 1 \end{pmatrix}.
\end{aligned}
\tag{3.11}
$$

The local spin gauge $\psi(\theta, \varphi) = 0$ of Example 61 works here equally good.

The group $\mathrm{O}(3)$ is the Cartesian product $\mathrm{SO}(3) \times \{-I, I\}$. On the connected component $\mathrm{SO}(3) \times \{I\}$, the map π_+ in Euler angles is given by Equation (3.11) and maps its domain $\mathcal{U}_+ \subset \mathrm{SO}(3) \times \{I\}$ to $\mathrm{SO}(2) \subset \mathrm{O}(2)$. For the embedded subgroup $\mathrm{O}(2)$ that fixes the vector $(0, 0, 1)^\top$, we choose the values $\theta = \psi = 180°$ on the connected component $\mathrm{SO}(3) \times \{-I\}$. That is,

$$
\begin{aligned}
\pi_- &\begin{pmatrix} -\cos\varphi\cos\theta\cos\psi + \sin\varphi\sin\psi & \cos\varphi\cos\theta\sin\psi + \sin\varphi\cos\psi & -\cos\varphi\sin\theta \\ -\sin\varphi\cos\theta\cos\psi - \cos\varphi\sin\psi & \sin\varphi\cos\theta\sin\psi - \cos\varphi\cos\psi & -\sin\varphi\sin\theta \\ \sin\theta\cos\psi & -\sin\theta\sin\psi & -\cos\theta \end{pmatrix} \\
&= \begin{pmatrix} -\cos\varphi & -\sin\varphi & 0 \\ -\sin\varphi & \cos\varphi & 0 \\ 0 & 0 & 1 \end{pmatrix}.
\end{aligned}
\tag{3.12}
$$

The map π_- maps its domain $\mathcal{U}_- \subset \mathrm{SO}(3) \times \{-I\}$ to $\mathrm{SO}(2) \subset \mathrm{O}(2)$. Topologically, the group $\mathrm{O}(2)$ is the disconnected union of two circles with charts $\begin{pmatrix} \cos\psi & -\sin\psi \\ \sin\psi & \cos\psi \end{pmatrix}$ and $\begin{pmatrix} -\cos\varphi & -\sin\varphi \\ -\sin\varphi & \cos\varphi \end{pmatrix}$. As a group, it is *not* a Cartesian product of $\mathrm{SO}(2)$ by $\{-I, I\}$.

The matrix in the left hand side of Equation 3.12 defines the Euler angles on the set $\mathrm{SO}(3) \times \{-I\}$. We map the domain $\mathcal{U} \subset S^2$ of the angular spherical coordinates to $\mathcal{U}_+$ (resp., $\mathcal{U}_-$) with the help of the two local gauges

$$
\psi_\pm : \mathcal{U} \to \mathcal{U}_\pm, \qquad \psi_\pm(\theta, \varphi) = (\varphi, \theta, 0) \in \mathcal{U}_\pm.
\tag{3.13}
$$

The irreducible representations of the groups $\mathrm{SU}(2)$ and $\mathrm{SO}(3)$ have been described in Example 52, those of the groups $\mathrm{U}(1)$ and $\mathrm{SO}(2)$ in Example 55. The group $\mathrm{O}(3)$ is isomorphic to the Cartesian product $\mathrm{SO}(3) \times \mathrm{O}(1)$. Therefore, for each irreducible representation L of the group $\mathrm{SO}(3)$, there are two irreducible representations L^g and L^u of the group $\mathrm{O}(3)$. For each $A \in \mathrm{O}(3)$, we have $L^g(-A) = L^g(A)$ and $L^u(-A) = -L^u(A)$. This notation is due to

Mulliken (1955, 1956). The symbol g (resp., u) is the first letter of the German word 'gerade', even (resp., 'ungerade', odd).

Now we classify the complex irreducible representations of the group $O(2)$. Let V be one of them. Restrict V to the subgroup $SO(2)$. Let $\tilde{V}$ be an arbitrary irreducible component of the above restriction. Then, $\tilde{V}$ is isomorphic either to cU_0, or to V_m, or to tV_m, where m is a positive integer. An element $g = \begin{pmatrix} \cos\psi & -\sin\psi \\ \sin\psi & \cos\psi \end{pmatrix} \in SO(2)$ acts on $\tilde{V}$ by multiplication by $\exp(im\psi)$ for some integer m. Define the *reflection* τ by $\tau = \begin{pmatrix} -1 & 0 \\ 0 & 1 \end{pmatrix} \in O(2)$. Consider the linear subspace $\tau \cdot \tilde{V}$ of the space V. For all $\mathbf{x} \in \tilde{V}$ and for all $g \in SO(2)$ we have

$$g \cdot (\tau \cdot \mathbf{x}) = \tau \cdot (g^{-1} \cdot \mathbf{x}) = \exp(-im\psi)(\tau \cdot \mathbf{x}).$$

We conclude that the space $\tau \cdot \tilde{V}$ carries a representation of $SO(2)$ that acts by multiplication by $\exp(-im\psi)$. Moreover, the direct sum $\tilde{V} \oplus (\tau \cdot \tilde{V})$ carries a representation of $O(2)$. Since V is irreducible, we have $V = \tilde{V} \oplus (\tau \cdot \tilde{V})$. If $\tilde{V} = \tau \cdot \tilde{V}$, then $m = 0$ and V is a one-dimensional representation of the group $\{I, \tau\}$. Denote by cU_0^+ (resp., cU_0^-) the representation where the reflection τ acts by multiplication by 1 (resp., -1). Otherwise, V is two-dimensional, and we denote it by cU_m, $m \geq 1$.

In the irreducible real representation U_0^+ (resp., U_0^-), the matrix $g \in O(2)$ acts by multiplication by 1 (resp., by $\det g$). In the irreducible real representation U_m, the matrix $\begin{pmatrix} \cos\psi & -\sin\psi \\ \sin\psi & \cos\psi \end{pmatrix} \in O(2)$ acts by matrix-vector multiplication by $\begin{pmatrix} \cos(m\psi) & -\sin(m\psi) \\ \sin(m\psi) & \cos(m\psi) \end{pmatrix}$, while the matrix $\begin{pmatrix} -\cos\varphi & -\sin\varphi \\ -\sin\varphi & \cos\varphi \end{pmatrix} \in O(2)$ acts by matrix-vector multiplication by $\begin{pmatrix} -\cos(m\varphi) & -\sin(m\varphi) \\ -\sin(m\varphi) & \cos(m\varphi) \end{pmatrix}$. We also have quaternionic irreducible representations qcU_0^+, qcU_0^-, and qcU_m.

Following usual notation, from now on, the symbol σ (resp., λ, resp., μ, resp., $\mathbf{s}$) will be replaced with s (resp., ℓ, resp., m, resp., $\mathbf{Y}$).

Next, for every irreducible representation of the group G, we determine the structure of its restriction to H. To perform that task, we calculate the characters of all representations involved. The results of calculations are shown in Table 3.1.

The results about the characters of irreducible complex representations of G and H are classical. To calculate those of irreducible real and quaternionic representations, we used Definition 83 and the equalities $cr = c'q = 1 + t$ of Remark 8.

A similar table for the case of $G = SO(3)$, $H = SO(2)$ can be obtained from Table 3.1 by deleting all information with half-integers ℓ and m.

A table for the case of $G = O(3)$, $H = O(2)$ follows.

In the third column of Table 3.2, the first value corresponds to the connected component of identity of the group, while the second value corresponds to the second connected component.

We apply Theorem 8 and obtain the following results.

For the pairs $(G, H) = (SU(2), U(1))$ and $(G, H) = (SO(3), SO(2))$, upon restriction to H, the irreducible representations U_0 (resp., cU_0, resp., qcU_0)

TABLE 3.1
The Characters of Irreducible Representations of the Groups SU(2) and U(1)

Group	Representation	Character
$\mathbb{K} = \mathbb{C}$		
SU(2)	$cU_\ell,\ \ell = 0, 1, \ldots$	$\frac{\sin((\ell+1/2)\psi)}{\sin(\psi/2)}$
SU(2)	$c'W_\ell,\ \ell = 1/2, 3/2, \ldots$	$\frac{\sin((\ell+1/2)\psi)}{\sin(\psi/2)}$
U(1)	cU_0	1
U(1)	$V_m,\ m = 1/2, 1, \ldots$	$\exp(im\psi)$
U(1)	$tV_m,\ m = 1/2, 1, \ldots$	$\exp(-im\psi)$
$\mathbb{K} = \mathbb{R}$		
SU(2)	$U_\ell,\ \ell = 0, 1, \ldots$	$\frac{\sin((\ell+1/2)\psi)}{\sin(\psi/2)}$
SU(2)	$rc'W_\ell,\ \ell = 1/2, 3/2, \ldots$	$2\frac{\sin((\ell+1/2)\psi)}{\sin(\psi/2)}$
U(1)	U_0	1
U(1)	$rV_m,\ m = 1/2, 1, \ldots$	$2\cos(m\psi)$
$\mathbb{K} = \mathbb{H}$		
SU(2)	$qcU_\ell,\ \ell = 0, 1, \ldots$	$2\frac{\sin((\ell+1/2)\psi)}{\sin(\psi/2)}$
SU(2)	$W_\ell,\ \ell = 1/2, 3/2, \ldots$	$\frac{\sin((\ell+1/2)\psi)}{\sin(\psi/2)}$
U(1)	qcU_0	2
U(1)	$qV_m,\ m = 1/2, 1, \ldots$	$2\cos(m\psi)$

TABLE 3.2
The Characters of Irreducible Representations of the Groups O(3) and O(2)

Group	Representation	Character
$\mathbb{K} = \mathbb{C}$		
O(3)	$cU_\ell^g,\ \ell = 0, 1, \ldots$	$\frac{\sin((\ell+1/2)\psi)}{\sin(\psi/2)},\ \frac{\sin((\ell+1/2)\psi)}{\sin(\psi/2)}$
O(3)	$cU_\ell^u,\ \ell = 0, 1, \ldots$	$\frac{\sin((\ell+1/2)\psi)}{\sin(\psi/2)},\ -\frac{\sin((\ell+1/2)\psi)}{\sin(\psi/2)}$
O(2)	cU_0^+	$1, 1$
O(2)	cU_0^-	$1, -1$
O(2)	$cU_m,\ m = 1, 2, \ldots$	$2\cos(m\psi),\ 0$
$\mathbb{K} = \mathbb{R}$		
O(3)	$U_\ell^g,\ \ell = 0, 1, \ldots$	$\frac{\sin((\ell+1/2)\psi)}{\sin(\psi/2)},\ \frac{\sin((\ell+1/2)\psi)}{\sin(\psi/2)}$
O(3)	$U_\ell^u,\ \ell = 0, 1, \ldots$	$\frac{\sin((\ell+1/2)\psi)}{\sin(\psi/2)},\ -\frac{\sin((\ell+1/2)\psi)}{\sin(\psi/2)}$
O(2)	U_0^+	$1, 1$
O(2)	U_0^-	$1, -1$
O(2)	$U_m,\ m = 1, 2, \ldots$	$2\cos(m\psi),\ 0$
$\mathbb{K} = \mathbb{H}$		
O(3)	$qcU_\ell^g,\ \ell = 0, 1, \ldots$	$2\frac{\sin((\ell+1/2)\psi)}{\sin(\psi/2)},\ 2\frac{\sin((\ell+1/2)\psi)}{\sin(\psi/2)}$
O(3)	$qcU_\ell^u,\ \ell = 0, 1, \ldots$	$2\frac{\sin((\ell+1/2)\psi)}{\sin(\psi/2)},\ -2\frac{\sin((\ell+1/2)\psi)}{\sin(\psi/2)}$
O(2)	qcU_0^+	$2, 2$
O(2)	qcU_0^-	$2, -2$
O(2)	$qcU_m,\ m = 1, 2, \ldots$	$4\cos(m\psi),\ 0$

of g are equivalent to those oh H denoted by the same symbol. Upon restriction to O(2), the irreducible representations U_0^g, (resp., cU_0^g, resp., qcU_0^g) are equivalent to the representations U_0^+, (resp., cU_0^+, resp., qcU_0^+) of O(2), while the irreducible representations U_0^u, (resp., cU_0^u, resp., qcU_0^u) are equivalent to the representations U_0^-, (resp., cU_0^-, resp., qcU_0^-) of O(2).

To continue, we observe that

$$\frac{\sin((\ell + 1/2)\psi)}{\sin(\psi/2)} = \sum_{m=-\ell}^{\ell} \exp(-im\psi).$$

Moreover, for nonnegative integer ℓ we have

$$\frac{\sin((\ell + 1/2)\psi)}{\sin(\psi/2)} = 1 + \sum_{m=1}^{\ell} (2\cos(m\psi)),$$

while for positive half-integer ℓ,

$$\frac{\sin((\ell + 1/2)\psi)}{\sin(\psi/2)} = \sum_{m=1/2}^{\ell} (2\cos(m\psi)),$$

where we used the geometric sum formula and elementary properties of complex numbers. Our conclusion is summed up in Table 3.3.

Next, we construct the linear spaces where the irreducible representations of the group G act, and fix our choice of orthonormal bases of the above spaces.

For the case of $\mathbb{K} = \mathbb{C}$, the construction is classical. Denote by $R = H_\ell^{\mathbb{C}}$ the complex linear space of homogeneous polynomials of degree 2ℓ in two complex variables ζ_0 and ζ_1, and let $\boldsymbol{\zeta} = (\zeta_0, \zeta_1)^\top$. Introduce the inner product on $H_\ell^{\mathbb{C}}$ by

$$\left(\sum_{m=-\ell}^{\ell} \alpha_m \zeta_0^{\ell+m} \zeta_1^{\ell-m}, \sum_{m=-\ell}^{\ell} \beta_m \zeta_0^{\ell+m} \zeta_1^{\ell-m} \right) = \sum_{m=-\ell}^{\ell} (\ell + m)!(\ell - m)! \overline{\alpha_m} \beta_m.$$

It is well-known that the action of $G = \mathrm{SU}(2)$ on $H_\ell^{\mathbb{C}}$ given by

$$A \cdot f(\boldsymbol{\zeta}) = f(A^{-1}\boldsymbol{\zeta}), \qquad A \in \mathrm{SU}(2), \quad f \in H_\ell^{\mathbb{C}}$$

is the unitary irreducible representation cU_ℓ.

The choice of an orthonormal basis in the space $H_\ell^{\mathbb{C}}$ is definitely far from unique. We choose the following one:

$$f_\ell^m(\zeta_0, \zeta_1) = \frac{\zeta_0^{\ell+m} \zeta_1^{\ell-m}}{\sqrt{(\ell + m)!(\ell - m)!}}, \qquad -\ell \leq m \leq \ell. \tag{3.14}$$

The matrix entries of the representations cU_ℓ in Euler angles have the form of the *Wigner D-functions*:

$$D_{mm'}^\ell(\varphi, \theta, \psi) = \exp(-i(m\varphi + m'\psi))d_{mm'}^\ell(\theta), \tag{3.15}$$

TABLE 3.3

The Structure of the Representation $R \downarrow H$ of the Group G

$G = \mathrm{SU}(2), H = \mathrm{U}(1)$	
R	The structure of $R \downarrow H$
cU_ℓ	$cU_0 \oplus V_1 \oplus \cdots \oplus V_\ell \oplus tV_1 \oplus \cdots \oplus tV_\ell$
$c'W_\ell$	$V_{1/2} \oplus \cdots \oplus V_\ell \oplus tV_{1/2} \oplus \cdots \oplus tV_\ell$
U_ℓ	$U_0 \oplus rV_1 \oplus \cdots \oplus rV_\ell$
$rc'W_\ell$	$2rV_{1/2} \oplus \cdots \oplus 2rV_\ell$
qcU_ℓ	$qcU_0 \oplus 2qV_1 \oplus \cdots \oplus 2qV_\ell$
W_ℓ	$qV_{1/2} \oplus \cdots \oplus qV_\ell$
$G = \mathrm{SO}(3), H = \mathrm{SO}(2)$	
cU_ℓ	$cU_0 \oplus V_1 \oplus \cdots \oplus V_\ell \oplus tV_1 \oplus \cdots \oplus tV_\ell$
U_ℓ	$U_0 \oplus rV_1 \oplus \cdots \oplus rV_\ell$
qcU_ℓ	$qcU_0 \oplus 2qV_1 \oplus \cdots \oplus 2qV_\ell$
$G = \mathrm{O}(3), H = \mathrm{O}(2)$	
cU_ℓ^g	$cU_0^{(-1)^\ell} \oplus cU_1 \oplus \cdots \oplus cU_\ell$
cU_ℓ^u	$cU_0^{(-1)^{\ell+1}} \oplus cU_1 \oplus \cdots \oplus cU_\ell$
U_ℓ^g	$U_0^{(-1)^\ell} \oplus U_1 \oplus \cdots \oplus U_\ell$
U_ℓ^u	$U_0^{(-1)^{\ell+1}} \oplus U_1 \oplus \cdots \oplus U_\ell$
qcU_ℓ^g	$qcU_0^{(-1)^\ell} \oplus qcU_1 \oplus \cdots \oplus qcU_\ell$
qcU_ℓ^u	$qcU_0^{(-1)^{\ell+1}} \oplus qcU_1 \oplus \cdots \oplus qcU_\ell$

where

$$d_{mm'}^\ell(\theta) = (-1)^{\ell+m'} \left(\frac{(\ell+m')!(\ell-m')!}{(\ell+m)!(\ell-m)!} \right)^{1/2} \sin^{2\ell}(\theta/2)$$

$$\times \sum_{k=\max\{0,m+m'\}}^{\min\{\ell+m,\ell+m'\}} (-1)^k \binom{\ell+m}{k} \binom{\ell-m}{k-m-m'} \cot^{2k-m-m'}(\theta/2).$$

$$(3.16)$$

Equations (1.53), (3.15), and (3.16) show the following. The complex irreducible representation cU_0 (resp., V_m, resp., tV_m) of the group $H = \mathrm{U}(1)$ acts in the linear span of the vector $f_\ell^0(\zeta_0, \zeta_1)$ (resp., $f_\ell^{-m}(\zeta_0, \zeta_1)$, resp., $f_\ell^m(\zeta_0, \zeta_1)$). The complex irreducible representation V_m (resp., tV_m) of the group H acts in the linear span of the vector $f_\ell^{-m}(\zeta_0, \zeta_1)$ (resp., $f_\ell^m(\zeta_0, \zeta_1)$).

To construct the real and quaternionic counterparts of the space $H_\ell^{\mathbb{C}}$, we use a result by Itzkowitz, Rothman, and Strassberg (1991). For a $P \in H_\ell^{\mathbb{C}}$, let $\overline{P}$ be P with conjugate coefficients.

Theorem 25 (Itzkowitz, Rothman, and Strassberg (1991)). *For an arbitrary nonnegative integer (resp., positive half-integer) ℓ, equation*

$$(jP)(\zeta_0, \zeta_1) = \overline{P}(\zeta\zeta_1, -\zeta\zeta_0) \tag{3.17}$$

establishes a one-to one correspondence between the set $S^1 = \{\,\zeta \in \mathbb{C}\colon |\zeta| = 1\,\}$ and the set of all real (resp., quaternionic) structures on $H_\ell^{\mathbb{C}}$.

Take $\zeta = 1$. An easy calculation shows that

$$(jf_\ell^m)(\zeta_0, \zeta_1) = (-1)^{\ell-m} f_\ell^{-m}(\zeta_0, \zeta_1).$$

Assume ℓ is nonnegative integer. We choose the $4\ell + 2$ vectors of the space $H_\ell^{\mathbb{C}}$ as follows.

- If ℓ is even, then the elements are

$$\frac{1}{\sqrt{2}}(f_\ell^m(\zeta_0, \zeta_1) - f_\ell^{-m}(\zeta_0, \zeta_1)), \qquad m = \ell, \ell - 2, \ldots, 2, -2, \ldots, -\ell,$$

$$\frac{i}{\sqrt{2}}(f_\ell^m(\zeta_0, \zeta_1) + f_\ell^{-m}(\zeta_0, \zeta_1)), \qquad m = \ell, \ell - 2, \ldots, 2, -2, \ldots, -\ell,$$

$$f_\ell^0(\zeta_0, \zeta_1), \quad if_\ell^0(\zeta_0, \zeta_1),$$

$$\frac{1}{\sqrt{2}}(f_\ell^m(\zeta_0, \zeta_1) - f_\ell^{-m}(\zeta_0, \zeta_1)), \qquad m = \ell - 1, \ell - 3, \ldots, -\ell + 1,$$

$$\frac{i}{\sqrt{2}}(f_\ell^m(\zeta_0, \zeta_1) + f_\ell^{-m}(\zeta_0, \zeta_1)), \qquad m = \ell - 1, \ell - 3, \ldots, -\ell + 1.$$

- If ℓ is odd, then the elements are

$$\frac{1}{\sqrt{2}}(f_\ell^m(\zeta_0, \zeta_1) - f_\ell^{-m}(\zeta_0, \zeta_1)), \quad m = \ell, \ell - 2, \ldots, -\ell,$$

$$\frac{i}{\sqrt{2}}(f_\ell^m(\zeta_0, \zeta_1) + f_\ell^{-m}(\zeta_0, \zeta_1)), \quad m = \ell, \ell - 2, \ldots, -\ell,$$

$$f_\ell^0(\zeta_0, \zeta_1), \quad if_\ell^0(\zeta_0, \zeta_1),$$

$$\frac{1}{\sqrt{2}}(f_\ell^m(\zeta_0, \zeta_1) - f_\ell^{-m}(\zeta_0, \zeta_1)), \quad m = \ell - 1, \ell - 3, \ldots, 2, -2, \ldots, -\ell + 1,$$

$$\frac{i}{\sqrt{2}}(f_\ell^m(\zeta_0, \zeta_1) + f_\ell^{-m}(\zeta_0, \zeta_1)), \quad m = \ell - 1, \ell - 3, \ldots, 2, -2, \ldots, -\ell + 1.$$

All the above chosen vectors are eigenvectors of j for the eigenvalue 1. We can divide them into pairs connected by multiplication by $\pm i$ and choose one vector from each pair in $2^{2\ell+1}$ different ways. For each way, the chosen vectors are linearly independent over $\mathbb{R}$. Our choice is as follows.

$$g_\ell^m(\zeta_0, \zeta_1) = \begin{cases} \frac{i}{\sqrt{2}}(f_\ell^m(\zeta_0, \zeta_1) + (-1)^{1-m} f_\ell^{-m}(\zeta_0, \zeta_1)), & \text{if } -\ell \leq m \leq -1, \\ f_\ell^0(\zeta_0, \zeta_1), & \text{if } m = 0, \\ \frac{1}{\sqrt{2}}(f_\ell^m(\zeta_0, \zeta_1) + (-1)^m f_\ell^{-m}(\zeta_0, \zeta_1)), & \text{otherwise.} \end{cases}$$

$$(3.18)$$

Denote by $R_\ell = H_\ell^{\mathbb{R}}$ the *real* linear span of the vectors $g_\ell^m(\zeta_0, \zeta_1)$. In the so chosen space, the matrices of the real representations U_ℓ in Euler angles take the form TDT^{-1}, where D is the matrix whose entries are the Wigner D-functions, and the matrix T has the same form, as in Lee (2022):

$$T = \frac{1}{\sqrt{2}} \begin{pmatrix} i & 0 & \dots & 0 & \dots & 0 & -i(-1)^\ell \\ 0 & i & \dots & 0 & \dots & -i(-1)^{\ell-1} & 0 \\ \dots & \dots & \dots & \dots & & \dots & \dots \\ 0 & 0 & \dots & \sqrt{2} & \dots & 0 & 0 \\ \dots & \dots & \dots & \dots & & \dots & \dots \\ 0 & 1 & \dots & 0 & \dots & (-1)^{\ell-1} & 0 \\ 1 & 0 & \dots & 0 & \dots & 0 & (-1)^\ell \end{pmatrix}.$$

In the case if either $m \geq 0$ and $m' \geq 0$ or $m < 0$, and $m' < 0$, the matrix entries of the real representation U_ℓ are given by

$$U^\ell_{mm'}(\varphi, \theta, \psi) = -\sin(m\varphi)\sin(m'\psi)u^\ell_{-mm'}(\theta) + \cos(m\varphi)\cos(m'\psi)u^\ell_{mm'}(\theta),$$
$$(3.19a)$$

otherwise

$$U^\ell_{mm'}(\varphi, \theta, \psi) = -\sin(m\varphi)\cos(m'\psi)u^\ell_{-mm'}(\theta) + \cos(m\varphi)\sin(m'\psi)u^\ell_{mm'}(\theta).$$
$$(3.19b)$$

where in the case of $mm' \neq 0$ we have

$$u^\ell_{mm'}(\theta) = (-1)^{m-m'}d^\ell_{|m|\,|m'|}(\theta) + (-1)^m \operatorname{sgn}(m)d^\ell_{|m|\,-|m'|}(\theta), \qquad (3.20a)$$

in the case of $m = m' = 0$

$$u^\ell_0(\theta) = d^\ell_{00}(\theta), \qquad (3.20b)$$

otherwise

$$u^\ell_{mm'}(\theta) = (-1)^{m-m'}\sqrt{2}d^\ell_{|m|\,|m'|}(\theta). \qquad (3.20c)$$

Equations (3.11), (3.19), and (3.20) show the following. The real irreducible representation U_0 (resp., rV_m) of the groups $H = U(1)$ and $H = SO(2)$ acts in the linear span of the vector $g^0_\ell(\zeta_0, \zeta_1)$ (resp., the vectors $g^{-m}_\ell(\zeta_0, \zeta_1)$ and $g^m_\ell(\zeta_0, \zeta_1)$).

According to general theory of Section 1.8, the real linear space $iH^{\mathbb{R}}_\ell$ carries another copy of the real irreducible representation U_ℓ.

Let ℓ be a positive half-integer. Following Leeuwen (2011), make $H^{\mathbb{C}}_\ell$ into a right quaternionic linear space $H^{\mathbb{H}}_\ell$ by defining

$$P(\alpha + \beta i + \gamma j + \delta k) = P(\alpha + \beta i) + j(P(\gamma + \delta i)), \qquad P \in H^{\mathbb{C}}_\ell,$$

where j is the quaternionic structure (3.17). Again, put $\zeta = 1$. The scalar-vector multiplication in the right quaternionic linear space $R = H^{\mathbb{H}}_\ell$ takes the form

$$(P(\alpha + \beta i + \gamma j + \delta k))(\zeta_0, \zeta_1) = (\alpha + \beta i)P(\zeta_0, \zeta_1) + (\gamma - \delta i)\overline{P}(\zeta_1, -\zeta_0), \qquad P \in H^{\mathbb{C}}_\ell.$$

We can choose an orthonormal basis in the space $H^{\mathbb{H}}_\ell$ in $2^{\ell+1/2}$ different way by choosing one vector from each of the $\ell + 1/2$ pairs $(f^m_\ell(\zeta_0, \zeta_1), f^{-m}_\ell(\zeta_0, \zeta_1))$. Our choice is as follows:

$$h^m_\ell(\zeta_0, \zeta_1) = f^m_\ell(\zeta_0, \zeta_1), \qquad m = \frac{1}{2}, \frac{3}{2}, \dots, \ell. \qquad (3.21)$$

Then we have $f_\ell^{-m}(\zeta_0, \zeta_1) = h_\ell^m(\zeta_0, \zeta_1)(-1)^{\ell - m}\mathrm{j}$.

Let $W_\ell^{mm'}(\varphi, \theta, \psi)$ be the matrix entries of the irreducible quaternionic representation W_ℓ in this basis. Then, on the one hand,

$$g \cdot h_\ell^m = \sum_{m'=1/2}^{\ell} W_\ell^{mm'}(\varphi, \theta, \psi) h_\ell^{m'}.$$

On the other hand,

$$g \cdot h_\ell^m = \sum_{m'=-\ell}^{\ell} D_{mm'}^\ell(\varphi, \theta, \psi) f_\ell^{m'}$$

$$= \sum_{m'=1/2}^{\ell} [D_{mm'}^\ell(\varphi, \theta, \psi) h_\ell^{m'} + D_{m-m'}^\ell(\varphi, \theta, \psi) h_\ell^{m'}(-1)^{\ell - m'}\mathrm{j}]$$

$$= \sum_{m'=1/2}^{\ell} [D_{mm'}^\ell(\varphi, \theta, \psi) + (-1)^{\ell - m'}\mathrm{j} \overline{D_{m-m'}^\ell(\varphi, \theta, \psi)}] h_\ell^{m'}$$

That is,

$$W_\ell^{mm'}(\varphi, \theta, \psi) = D_{mm'}^\ell(\varphi, \theta, \psi) + (-1)^{\ell - m'}\mathrm{j} \overline{D_{m-m'}^\ell(\varphi, \theta, \psi)}.$$

This equation shows that upon restriction to $H = \mathrm{U}(1)$, the quaternionic irreducible representation qV_m acts in the linear span of the vector $h_\ell^m(\zeta_0, \zeta_1)$.

To obtain the matrix entries of the real irreducible representation $rc'W_\ell$, we replace each matrix entry

$$W_\ell^{mm'}(\varphi, \theta, \psi) = \operatorname{Re} D_{mm'}^\ell(\varphi, \theta, \psi) + \operatorname{Im} D_{mm'}^\ell(\varphi, \theta, \psi)\mathrm{i}$$

$$+ (-1)^{\ell - m'} \operatorname{Re} D_{m-m'}^\ell(\varphi, \theta, \psi)\mathrm{j} + (-1)^{\ell - m'+1} \operatorname{Im} D_{m-m'}^\ell(\varphi, \theta, \psi)\mathrm{k}$$

with the 4×4 matrix

$$\begin{pmatrix} \operatorname{Re} D_{mm'}^\ell & (-1)^{\ell - m'} \operatorname{Im} D_{m-m'}^\ell \\ (-1)^{\ell - m'+1} \operatorname{Im} D_{m-m'}^\ell & \operatorname{Re} D_{mm'}^\ell \\ (-1)^{\ell - m'} \operatorname{Re} D_{m-m'}^\ell & -\operatorname{Im} D_{mm'}^\ell \\ \operatorname{Im} D_{mm'}^\ell & (-1)^{\ell - m'} \operatorname{Re} D_{m-m'}^\ell \\ (-1)^{\ell - m'+1} \operatorname{Re} D_{m-m'}^\ell & -\operatorname{Im} D_{mm'}^\ell \\ \operatorname{Im} D_{mm'}^\ell & (-1)^{\ell - m'+1} \operatorname{Re} D_{m-m'}^\ell \\ \operatorname{Re} D_{mm'}^\ell & (-1)^{\ell - m'+1} \operatorname{Im} D_{m-m'}^\ell \\ (-1)^{\ell - m'} \operatorname{Im} D_{m-m'}^\ell & \operatorname{Re} D_{mm'}^\ell \end{pmatrix}, \tag{3.22}$$

which corresponds to the following choice of an orthonormal basis in the real linear space $H_\ell^{\mathbb{R}} = rc'H_\ell^{\mathbb{H}}$:

$$g_\ell^{4m-1}(\zeta_0, \zeta_1) = h_\ell^m(\zeta_0, \zeta_1), \qquad g_\ell^{4m}(\zeta_0, \zeta_1) = -h_\ell^m(\zeta_0, \zeta_1)\mathrm{k},$$
$$g_\ell^{4m+1}(\zeta_0, \zeta_1) = -h_\ell^m(\zeta_0, \zeta_1)\mathrm{j}, \quad g_\ell^{4m+2}(\zeta_0, \zeta_1) = -h_\ell^m(\zeta_0, \zeta_1)\mathrm{i}, \tag{3.23}$$

where $m = \frac{1}{2}, \frac{3}{2}, \ldots, \ell$.

We fix our choice as follows. The first (resp., the second) copy of the real irreducible representation rV_m of the group $\mathrm{U}(1)$ acts in the linear span of the vectors $g_\ell^{4m-1}(\zeta_0, \zeta_1)$ and $g_\ell^{4m}(\zeta_0, \zeta_1)$ (resp., $g_\ell^{4m+1}(\zeta_0, \zeta_1)$ and $g_\ell^{4m+1}(\zeta_0, \zeta_1)$).

Finally, for nonnegative integer ℓ we choose the following basis in the space $H_\ell^{\mathbb{H}} = qH_\ell^{\mathbb{C}}$:

$$h_\ell^m(\zeta_0, \zeta_1) = f_\ell^m(\zeta_0, \zeta_1) \otimes_{\mathbb{C}} 1. \tag{3.24}$$

The representation qcU_0 of the groups $\mathrm{U}(1)$ and $\mathrm{SO}(2)$ acts in the linear span of the vector $h_\ell^0(\zeta_0, \zeta_1)$. Let the first (resp., the second) copy of the quaternionic irreducible representation qV_m of the above groups acts in the linear span of the vector $h_\ell^{-m}(\zeta_0, \zeta_1)$ (resp., $h_\ell^m(\zeta_0, \zeta_1)$).

The representations cU_0^g and cU_0^u (resp., U_0^g and U_0^u, resp., qcU_0^g and qcU_0^u) act in the linear span of the vector $f_\ell^0(\zeta_0, \zeta_1)$ (resp., $g_\ell^0(\zeta_0, \zeta_1)$, resp., $h_\ell^0(\zeta_0, \zeta_1)$). The representations cU_m(resp., U_m, resp., qcU_m) act in the linear span of the vectors $f_\ell^{-m}(\zeta_0, \zeta_1)$ and $f_\ell^m(\zeta_0, \zeta_1)$ (resp., $g_\ell^{-m}(\zeta_0, \zeta_1)$ and $g_\ell^m(\zeta_0, \zeta_1)$, resp., $h_\ell^{-m}(\zeta_0, \zeta_1)$ and $h_\ell^m(\zeta_0, \zeta_1)$).

Next, for each irreducible representation L_s of the group H over a skew field $\mathbb{K}$, we write down all irreducible representations R_ℓ of the group G, for which $\dim_{\mathbb{K}'} \mathrm{Hom}_{\mathbb{K}H}(R_\ell \downarrow H, L_s) > 0$. The results are given in Table 3.4.

The figures in the last column of Table 3.4 have been calculated with the help of Equation (3.6).

Definition 117. A *spherical harmonic* is an element of the space R_ℓ.

The construction of a special basis have been described in Section 3.1. The elements of this basis are harmonics. We illustrate their construction in the following examples.

Example 71. In this example, we construct special orthonormal bases in the Hilbert spaces $cU_0 \uparrow G$, $V_s \uparrow G$, and $tV_s \uparrow G$, $s = 1/2, 1, \ldots$, where $G = \mathrm{SU}(2)$, and where cU_0, V_s, and tV_s are complex irreducible representations of the group $H = \mathrm{U}(1)$ shown in the second column of Table 3.1.

Put $s = 0$. Table 3.3 shows that $\ell \in \hat{G}_{\mathbb{C}}(cU_0) = \{0, 1, \ldots\}$. The matrix of a linear operator $f^\sharp \in \mathrm{Hom}_{\mathbb{C}H}(cU_\ell \downarrow H, cU_0)$ has $\dim cU_0 = 1$ row and $\dim cU_\ell = 2\ell + 1$ columns and depends on

$$M(0, \ell)n_0 \dim_{\mathbb{C}} \mathrm{Hom}_{\mathbb{C}G}(cU_\ell, cU_\ell) = 1$$

$\mathbb{C}$-valued parameter, say α. In the chosen basis, it takes the form $f^\sharp = (0 \ldots 0\ \alpha\ 0 \ldots 0)$ with α located in the 0th position. The Frobenius reciprocity gives $\dim_{\mathbb{C}} \mathrm{Hom}_{\mathbb{C}G}(cU_\ell, cU_0 \uparrow G) = 1$.

Denote the vectors of the special basis expressed in the local spin gauge $\psi = 0$ of Example 61 by $\mathbf{Y}_{\ell m}(\theta, \varphi)$. We calculate them in two ways and equate the results.

On the one hand, under the Frobenius isomorphism, an operator $f^\sharp$ becomes the operator $f_\flat \in \mathrm{Hom}_{\mathbb{C}G}(cU_\ell, cU_0 \uparrow G)$ given by Equation (3.7). It maps the basis vector

$$\mathbf{r}_m = (0, \ldots, 0, 1, 0, \ldots, 0)^\top \in R_\ell = H_\ell^{\mathbb{C}}$$

TABLE 3.4
The Structure of Induced Representations

L_s	$R_\ell \in \hat{G}_{\mathbb{K}}(L_s)$	$M = M(s, \ell)$
\multicolumn{3}{c}{$G = \mathrm{SU}(2), H = \mathrm{U}(1)$}		
cU_0	$cU_\ell : \ell \geq 0$	1
$V_s, tV_s : s \geq 1$	$cU_\ell : \ell \geq s$	1
$V_s, tV_s : s \geq \frac{1}{2}$	$c'W_\ell : \ell \geq s$	1
U_0	$U_\ell : \ell \geq 0$	1
$rV_s : s \geq 1$	$U_\ell : \ell \geq s$	2
$rV_s : s \geq \frac{1}{2}$	$rc'W_\ell : \ell \geq s$	1
qcU_0	$qcU_\ell : \ell \geq 0$	1
$qV_s : s \geq 1$	$qcU_\ell : \ell \geq s$	1
$qV_s : s \geq \frac{1}{2}$	$W_\ell : \ell \geq s$	2
\multicolumn{3}{c}{$G = \mathrm{SO}(3), H = \mathrm{SO}(2)$}		
cU_0	$cU_\ell : \ell \geq 0$	1
$V_s, tV_s : s \geq 1$	$cU_\ell : \ell \geq s$	1
U_0	$U_\ell : \ell \geq 0$	1
$rV_s : s \geq 1$	$U_\ell : \ell \geq s$	2
qcU_0	$qcU_\ell : \ell \geq 0$	1
$qV_s : s \geq 1$	$qcU_\ell : \ell \geq s$	1
\multicolumn{3}{c}{$G = \mathrm{O}(3), H = \mathrm{O}(2)$}		
cU_0^+	$cU_{2\ell}^g, cU_{2\ell+1}^u : \ell \geq 0$	1
cU_0^-	$cU_{2\ell+1}^g, cU_{2\ell}^u : \ell \geq 0$	1
$cU_s : s \geq 1$	$cU_\ell^g, cU_\ell^u : \ell \geq s$	1
U_0^+	$U_{2\ell}^g, U_{2\ell+1}^u : \ell \geq 0$	1
U_0^-	$U_{2\ell+1}^g, U_{2\ell}^u : \ell \geq 0$	1
$U_s : s \geq 1$	$U_\ell^g, U_\ell^u : \ell \geq s$	1
qcU_0^+	$qcU_{2\ell}^g, qcU_{2\ell+1}^u : \ell \geq 0$	1
qcU_0^-	$qcU_{2\ell+1}^g, qcU_{2\ell}^u : \ell \geq 0$	1
$qcU_s : s \geq 1$	$qcU_\ell^g, qcU_\ell^u : \ell \geq s$	1

given by Equation (3.14) to the vector

$$f_\flat(\mathbf{r}_m)(g) = (0, \ldots, 0, \alpha, 0, \ldots, 0)(D_{-\ell m}^\ell(g^{-1}), \ldots, D_{\ell m}^\ell(g^{-1}))^\top$$
$$= \alpha \overline{D_{m0}^\ell(\varphi, \theta, 0)} \in cU_0 \uparrow G.$$

On the other hand, the matrix of $f_\flat$ has $2\ell + 1$ rows and columns and is equal to αI. We choose the following one-to-one correspondence:

$$f^\sharp = (0 \ \ldots \ 0 \ \alpha \ 0 \ \ldots \ 0) \leftrightarrow f_\flat = \alpha I. \tag{3.25}$$

With that correspondence, $f_\flat$ maps the vector $\mathbf{r}_m \in R_\ell$ to the vector $\alpha \mathbf{Y}_{\ell m}(\theta, \varphi) \in cU_0 \uparrow G$. It follows that $\mathbf{Y}_{\ell m}(\theta, \varphi)$ is proportional to $\overline{D_{m0}^\ell(\varphi, \theta, 0)}$. To obtain a vector of unit length, we multiply the above vector by $\sqrt{\frac{2\ell+1}{4\pi}}$ and observe that the cross-sections of the chosen basis are *complex-valued spherical harmonics*

$$\mathbf{Y}_{\ell m}(\theta,\varphi) = Y_{\ell,m}(\theta,\varphi) = \sqrt{\frac{2\ell+1}{4\pi}}\overline{D^\ell_{m0}(\varphi,\theta,0)}.\tag{3.26}$$

The notation is taken from Olver et al. (2010), current *de facto* standard in the area of special functions. Observe that

$$Y_{\ell,m}(\theta,\varphi) = \sqrt{\frac{2\ell+1}{4\pi}}\exp(im\varphi)d^\ell_{m0}(\theta).$$

Recall that for $\mu,\nu\in\mathbb{R}$ and $x\in(-1,1)$, the *Ferrers function of the first kind* is given by Olver et al. (2010, (14.3.1)):

$$\mathsf{P}^\mu_\nu(x) = \left(\frac{1+x}{1-x}\right)^{\mu/2}\mathbf{F}(\nu+1,-\nu;1-\mu;\tfrac{1}{2}-\tfrac{1}{2}x),$$

where *Olver's hypergeometric function* $\mathbf{F}(a,b;c;x)$ is given by Olver et al. (2010, (14.3.3)):

$$\mathbf{F}(a,b;c;x) = \frac{1}{\Gamma(c)}F(a,b;c;x),$$

and the hypergeometric function $F(a,b;c;z)$ is given by the *Gauss series*

$$F(a,b;c;\zeta) = \frac{\Gamma(c)}{\Gamma(a)\Gamma(b)}\sum_{s=0}^\infty\frac{\Gamma(a+s)\Gamma(b+s)}{\Gamma(c+s)s!}\zeta^s$$

on the disk $|\zeta|<1$, see Olver et al. (2010, (15.2.1)). In terms of this function, we have

$$Y_{\ell,m}(\theta,\varphi) = \left(\frac{(\ell-m)!(2\ell+1)}{4\pi(\ell+m)!}\right)^{1/2}\exp(im\varphi)\mathsf{P}^m_\ell(\cos\theta),$$

see Olver et al. (2010, (14.30.1)).

Let s be a positive integer. Table 3.3 shows that $\ell\in\hat{G}_\mathbb{C}(cU_0)=\{s,s+1,\dots\}$. The matrix of a linear operator $f^\sharp\in\mathrm{Hom}_{\mathbb{C}H}(cU_\ell\downarrow H,V_s)$ or $f^\sharp\in\mathrm{Hom}_{\mathbb{C}H}(cU_\ell\downarrow H,tV_s)$ has $\dim V_s=\dim tV_s=1$ row and $\dim cU_\ell=2\ell+1$ columns and depends on

$$M(s,\ell)n_s\dim_\mathbb{C}\mathrm{Hom}_{\mathbb{C}G}(V_\ell,V_s)=1$$

$\mathbb{C}$-valued parameter, say α. In the chosen basis, it takes the form $f^\sharp=(0\ \dots\ 0\ \alpha\ 0\ \dots\ 0)$. When $L_s=V_s$, then α is located in the sth position, while for the case of $L_s=tV_s$ it is located in the $-s$th position. The Frobenius reciprocity gives

$$\dim_\mathbb{C}\mathrm{Hom}_{\mathbb{C}G}(cU_\ell,V_s\uparrow G)=\dim_\mathbb{C}\mathrm{Hom}_{\mathbb{C}G}(cU_\ell,tV_s\uparrow G)=1.$$

Denote the vectors of the special basis expressed in the local spin gauge $\psi = 0$ of Example 61 by $_s\mathbf{Y}_{\ell m}(\theta, \varphi)$ when $L_s = tV_s$ and by $_{-s}\mathbf{Y}_{\ell m}(\theta, \varphi)$ when $L_s = V_s$, $s \geq 1$.

On the one hand, under the Frobenius isomorphism, an operator $f^\sharp$ becomes the operator $f_\flat \in \mathrm{Hom}_{\mathbb{C}G}(cU_\ell, V_s \uparrow G)$ (resp., $f_\flat \in \mathrm{Hom}_{\mathbb{C}G}(cU_\ell, tV_s \uparrow G)$) given by Equation (3.7). It maps the basis vector $\mathbf{r}_m$ given by Equation (3.14) to the vector

$$f_\flat(\mathbf{r})(g) = \alpha D^\ell_{sm}(g^{-1}) = \alpha \overline{D^\ell_{m\,-s}(\varphi, \theta, 0)}$$

(resp., to the vector $\alpha \overline{D^\ell_{ms}(\varphi, \theta, 0)}$).

On the other hand, the matrix of $f_\flat$ has $2\ell + 1$ rows and columns and is equal to αI. We choose the one-to-one correspondence (3.25). It follows that $_s\mathbf{Y}_{\ell m}(\theta, \varphi)$ is proportional to $\overline{D^\ell_{m\,-s}(\varphi, \theta, 0)}$. To obtain a vector of unit length, we multiply the above vector by $\sqrt{\frac{2\ell+1}{4\pi}}$ and observe that the cross-sections of the chosen basis are *spin-weighted spherical harmonics*, given by

$$_s Y_{\ell,m}(\theta, \varphi) = \sqrt{\frac{2\ell + 1}{4\pi}} \overline{D^\ell_{m\,-s}(\varphi, \theta, 0)}$$

in spherical coordinates. Observe that this time, in contrast to the case of $s = 0$, the harmonics $_s Y_{\ell,m}(\theta, \varphi)$ are not functions on the sphere, but the cross-sections of a non-trivial complex line bundle.

For positive half-integers s and $\ell \geq s$, we can repeat the last four paragraphs literally, but replace cU_ℓ with $c'W_\ell$.

For future references, we collect here the following symmetry relations for the complex-valued spin-weighted spherical harmonics:

$$_{-m}Y_{\ell,-s}(\theta, \varphi) = (-1)^{m+s}\,_s Y_{\ell,m}(\theta, \varphi), \qquad _s Y_{\ell,-m}(\theta, \varphi) = \mathrm{i}^{2\ell}\,_s Y_{\ell,m}(\theta, \varphi),$$

$$_{-s}Y_{\ell,m}(\theta, \varphi) = \mathrm{i}^{-2\ell}\,_s Y_{\ell,m}(\theta, \varphi), \qquad _{-s}Y_{\ell,-m}(\theta, \varphi) = (-1)^{m+s}\overline{_s Y_{\ell,m}(\theta, \varphi)},$$

$$(3.27)$$

see Penrose and Rindler (1987, Equations (4.15.100)–(4.15.102), (4.15.104)).

Example 72. In this example, we construct special orthonormal bases in the Hilbert spaces $U_0 \uparrow G$ and $rV_s \uparrow G$, $s = 1/2, 1, \ldots$, where $G = \mathrm{SU}(2)$, and where U_0 and rV_s are real irreducible representations of the group $H = \mathrm{U}(1)$ shown in the second column of Table 3.1.

Put $s = 0$. Table 3.3 shows that $\ell \in \hat{G}_\mathbb{R}(U_0) = \{0, 1, \ldots\}$. The matrix of a linear operator $f^\sharp \in \mathrm{Hom}_{\mathbb{R}H}(U_\ell \downarrow H, U_0)$ has $\dim U_0 = 1$ row and $\dim U_\ell = 2\ell + 1$ columns and depends on

$$M(0, \ell)n_0 \dim_\mathbb{R} \mathrm{Hom}_{\mathbb{R}G}(U_\ell, U_\ell) = 1$$

$\mathbb{R}$-valued parameter, say α. In the chosen basis, it takes the form $f^\sharp = (0\ \ldots\ 0\ \alpha\ 0\ \ldots\ 0)$ with α located in the 0th position. The Frobenius reciprocity gives $\dim_\mathbb{R} \mathrm{Hom}_{\mathbb{R}G}(U_\ell, U_0 \uparrow G) = 1$. Again, we can denote the vectors of the special basis by $\mathbf{Y}_{\ell m}(\theta, \varphi)$, as above.

On the one hand, under the Frobenius isomorphism, an operator $f^\sharp$ becomes the operator $f_\flat \in \mathrm{Hom}_{\mathbb{R}G}(U_\ell, U_0 \uparrow G)$ given by Equation (3.7). It maps the basis vector

$$\mathbf{r}_m = (0, \dots, 0, 1, 0, \dots, 0)^\top \in R_\ell = H_\ell^{\mathbb{R}}$$

given by Equation (3.18) to the vector

$$f_\flat(\mathbf{r})(g) = \alpha U_{0m}^\ell(g^{-1}) = \alpha U_{m0}^\ell(\varphi, \theta, 0).$$

On the other hand, the matrix of $f_\flat$ has $2\ell + 1$ rows and columns and is equal to αI. We choose the one-to-one correspondence (3.25), but this time with $\alpha \in \mathbb{R}$. With that correspondence, $f_\flat$ maps the vector $\mathbf{r}_m \in R_\ell$ to the vector $\alpha \mathbf{Y}_{\ell m}(\theta, \varphi) \in U_0 \uparrow G$. It follows that $\mathbf{Y}_{\ell m}(\theta, \varphi)$ is proportional to $U_{m0}^\ell(\varphi, \theta, 0)$. To obtain a vector of unit length, we multiply the above vector by $\sqrt{\frac{2\ell+1}{4\pi}}$ and observe that the cross-sections of the chosen basis are *real-valued spherical harmonics*

$$Y_\ell^m(\theta, \varphi) = \begin{cases} (-1)^m \sqrt{2}\, \mathrm{Im}\, Y_{\ell,m}(\theta, \varphi) & \text{if } m < 0, \\ Y_{\ell,0}(\theta, \varphi), & \text{if } m = 0, \\ (-1)^m \sqrt{2}\, \mathrm{Re}\, Y_{\ell,m}(\theta, \varphi), & \text{otherwise.} \end{cases} \tag{3.28}$$

This follows from Equations (3.19) and (3.20) after simple algebraic manipulations.

Equation (3.28) differs from the definition of the *surface harmonics of the first kind* in Olver et al. (2010, (14.30.2)) by a constant multiple. The functions $Y_\ell^0(\theta, \varphi)$ are called the *zonal spherical harmonics* because the parallels where $Y_\ell^0(\theta, \varphi) = 0$ divide the centred unit sphere into $\ell + 1$ zones. The functions $Y_\ell^{-\ell}(\theta, \varphi)$ and $Y_\ell^\ell(\theta, \varphi)$ are called the *sectorial spherical harmonics* because the meridians where $Y_\ell^{pm\ell}(\theta, \varphi) = 0$ divide the above sphere into 2ℓ sectors. The functions $Y_\ell^m(\theta, \varphi)$ with $m \notin \{-\ell, 0, \ell\}$ are called the *tesseral spherical harmonics* because the parallels and meridians where these functions vanish divide the above sphere into $2m(\ell-m+1)$ cells, and the cells which correspond to positive and negative signs of any function, are arranged in checker order.

Figure 3.1 shows the tesseral spherical harmonic $Y_3^2(\theta, \varphi)$.[1]

Let s be a positive integer. Table 3.3 shows that $\ell \in \hat{G}_{\mathbb{R}}(rV_s) = \{s, s+1, \dots\}$. The matrix of a linear operator $f^\sharp \in \mathrm{Hom}_{\mathbb{R}H}(U_\ell \downarrow H, rV_s)$ with positive integers s and $l \geq s$ has $\dim rV_s = 2$ rows and $\dim U_\ell = 2\ell + 1$ columns and depends on

$$M(s, \ell) n_s \dim_{\mathbb{R}} \mathrm{Hom}_{\mathbb{R}G}(U_\ell, U_\ell) = 2$$

$\mathbb{R}$-valued parameters, say α and β. In the chosen basis, it takes the form

$$f^\sharp = \begin{pmatrix} 0 & \dots & 0 & \alpha & 0 & \dots & 0 & -\beta & 0 & \dots & 0 \\ 0 & \dots & 0 & \beta & 0 & \dots & 0 & \alpha & 0 & \dots & 0 \end{pmatrix}$$

---

1. Source: https://tex.stackexchange.com/questions/268830/drawing-spherical-harmonic-density-plots-on-the-surface-of-a-sphere-in-tikz-pgfp.

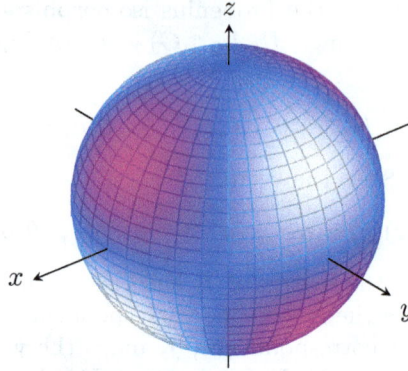

FIGURE 3.1
The density plot of the tesseral spherical harmonic $Y_3^2(\theta, \varphi)$ on the surface of a sphere.

with α, $\beta \in \mathbb{R}$ located in the $-s$th and sth columns. The Frobenius reciprocity gives $\dim_{\mathbb{R}} \mathrm{Hom}_{\mathbb{R}G}(U_\ell, rV_s \uparrow G) = 2$.

Denote the vectors of the special basis expressed in the local spin gauge $\psi = 0$ of Example 61 by ${}_s\mathbf{Y}_{\ell,1}^m(\theta, \varphi)$ and ${}_s\mathbf{Y}_{\ell,2}^m(\theta, \varphi)$.

On the one hand, under the Frobenius isomorphism, an operator $f^\sharp$ becomes the operator $f_\flat \in \mathrm{Hom}_{\mathbb{R}G}(U_\ell, rV_s \uparrow G)$ given by (3.7). It maps the basis vector (3.18) to the vector

$$f_\flat(\mathbf{r})(g) = \begin{pmatrix} \alpha U_{-sm}^\ell(g^{-1}) - \beta U_{sm}^\ell(g^{-1}) \\ \beta U_{-sm}^\ell(g^{-1}) + \alpha U_{sm}^\ell(g^{-1}) \end{pmatrix} = \begin{pmatrix} \alpha U_{m-s}^\ell(g) - \beta U_{ms}^\ell(g) \\ \beta U_{m-s}^\ell(g) + \alpha U_{ms}^\ell(g) \end{pmatrix}.$$

On the other hand, the matrix of $f_\flat$ has $2 \dim U_\ell = 4\ell + 2$ rows and $\dim U_\ell = 2\ell + 1$ columns and is equal to $\left(\begin{smallmatrix} \alpha I \\ \beta I \end{smallmatrix} \right)$. We choose the one-to-one correspondence

$$\begin{pmatrix} 0 \ \dots \ 0 \ \alpha \ 0 \ \dots \ 0 \ -\beta \ 0 \ \dots \ 0 \\ 0 \ \dots \ 0 \ \beta \ 0 \ \dots \ 0 \ \alpha \ 0 \ \dots \ 0 \end{pmatrix} \leftrightarrow \left(\begin{smallmatrix} \alpha I \\ \beta I \end{smallmatrix} \right).$$

With that correspondence, $f_\flat$ maps the vector $\mathbf{r}_m \in R_\ell$ to the vector

$$\alpha_s \mathbf{Y}_{\ell,1}^m(\theta, \varphi) + \beta_s \mathbf{Y}_{\ell,2}^m(\theta, \varphi) \in rV_s \uparrow G.$$

It follows that ${}_s\mathbf{Y}_{\ell,1}^m(\theta, \varphi)$ is proportional to $\begin{pmatrix} U_{m-s}^\ell(\varphi, \theta, 0) \\ U_{ms}^\ell(\varphi, \theta, 0) \end{pmatrix}$, while ${}_s\mathbf{Y}_{\ell,2}^m(\theta, \varphi)$ is proportional to $\begin{pmatrix} -U_{ms}^\ell(\varphi, \theta, 0) \\ U_{m-s}^\ell(\varphi, \theta, 0) \end{pmatrix}$. To obtain a vector of unit length, we multiply the above vectors by $\sqrt{\frac{2\ell+1}{8\pi}}$ and observe that the cross-sections of the chosen basis are *real-valued spin-weighted spherical harmonics*, given by

$$_s\mathbf{Y}^m_{\ell,1}(\theta,\varphi) = \sqrt{\frac{2\ell+1}{8\pi}} \begin{pmatrix} U^\ell_{m-s}(\varphi,\theta,0) \\ U^\ell_{ms}(\varphi,\theta,0) \end{pmatrix},$$

$$_s\mathbf{Y}^m_{\ell,2}(\theta,\varphi) = \sqrt{\frac{2\ell+1}{8\pi}} \begin{pmatrix} -U^\ell_{ms}(\varphi,\theta,0) \\ U^\ell_{m-s}(\varphi,\theta,0) \end{pmatrix}$$

in spherical coordinates.

Let s be a positive half-integer. Table 3.3 shows that $\ell \in \hat{G}_\mathbb{R}(rV_s) = \{s, s+1, \dots\}$. The matrix of a linear operator $f^\sharp \in \mathrm{Hom}_{\mathbb{R}H}(rc'W_\ell \downarrow H, rV_s)$ with positive half-integers s and $l \geq s$ has $\dim rV_s = 2$ rows and $\dim rc'W_\ell = 4\ell+2$ columns and depends on

$$M(s,\ell)n_s \dim_\mathbb{R} \mathrm{Hom}_{\mathbb{R}G}(rc'W_\ell, rc'W_\ell) = 4$$

$\mathbb{R}$-valued parameters, say α, β, γ, and δ. In the chosen basis, it takes the form

$$f^\sharp = \begin{pmatrix} 0 & \dots & 0 & \alpha & -\beta & \gamma & -\delta & 0 & \dots & 0 \\ 0 & \dots & 0 & \beta & \alpha & \delta & \gamma & 0 & \dots & 0 \end{pmatrix},$$

where α, $\beta \in \mathbb{R}$ are located in the $(4s-1)$st and $4s$th columns and γ, $\delta \in \mathbb{R}$ are located in the $4s+1$th and $(4s+2)$nd columns. The Frobenius reciprocity gives $\dim_\mathbb{R} \mathrm{Hom}_{\mathbb{R}G}(rc'W_\ell, rV_s \uparrow G) = 4$.

Denote the vectors of the special basis expressed in the local spin gauge $\psi = 0$ of Example 61 by $_s\mathbf{Y}^m_\ell(\theta,\varphi)$, $1 \leq m \leq 4\ell+2$.

On the one hand, under the Frobenius isomorphism, an operator $f^\sharp$ becomes the operator $f_\flat \in \mathrm{Hom}_{\mathbb{R}G}(rc'W_\ell, rV_s \uparrow G)$ given by (3.7). It maps the basis vector (3.23) to the vector

$$f_\flat(\mathbf{r}_m)(g) = \begin{pmatrix} \alpha U^\ell_{4s-1\,m}(g^{-1}) - \beta U^\ell_{4s\,m}(g^{-1}) + \gamma U^\ell_{4s+1\,m}(g^{-1}) - \delta U^\ell_{4s+2\,m}(g^{-1}) \\ \beta U^\ell_{4s-1\,m}(g^{-1}) + \alpha U^\ell_{4s\,m}(g^{-1}) + \delta U^\ell_{4s+1\,m}(g^{-1}) + \gamma U^\ell_{4s+2\,m}(g^{-1}) \end{pmatrix}$$

$$= \begin{pmatrix} \alpha U^\ell_{m\,4s-1}(g) - \beta U^\ell_{m\,4s}(g) + \gamma U^\ell_{m\,4s+1}(g) - \delta U^\ell_{m\,4s+2}(g) \\ \beta U^\ell_{m\,4s-1}(g) + \alpha U^\ell_{m\,4s}(g) + \delta U^\ell_{m\,4s+1}(g) + \gamma U^\ell_{m\,4s+2}(g) \end{pmatrix}.$$

On the other hand, the matrix of $f_\flat$ has $\dim rc'W_\ell = 4\ell+2$ rows and columns and is equal to

$$f_\flat = \begin{pmatrix}
\alpha & -\delta & -\gamma & -\beta & 0 & 0 & 0 & 0 & \dots & 0 & 0 & 0 & 0 \\
\delta & \alpha & \beta & -\gamma & 0 & 0 & 0 & 0 & \dots & 0 & 0 & 0 & 0 \\
\gamma & -\beta & \alpha & \delta & 0 & 0 & 0 & 0 & \dots & 0 & 0 & 0 & 0 \\
\beta & \gamma & -\delta & \alpha & 0 & 0 & 0 & 0 & \dots & 0 & 0 & 0 & 0 \\
0 & 0 & 0 & 0 & \alpha & -\delta & -\gamma & -\beta & \dots & 0 & 0 & 0 & 0 \\
0 & 0 & 0 & 0 & \delta & \alpha & \beta & -\gamma & \dots & 0 & 0 & 0 & 0 \\
0 & 0 & 0 & 0 & \gamma & -\beta & \alpha & \delta & \dots & 0 & 0 & 0 & 0 \\
0 & 0 & 0 & 0 & \beta & \gamma & -\delta & \alpha & \dots & 0 & 0 & 0 & 0 \\
\dots & \dots & \dots & \dots & \dots & \dots & \dots & \dots & \dots & \dots & \dots & \dots & \dots \\
0 & 0 & 0 & 0 & 0 & 0 & 0 & 0 & \dots & \alpha & -\delta & -\gamma & -\beta \\
0 & 0 & 0 & 0 & 0 & 0 & 0 & 0 & \dots & \delta & \alpha & \beta & -\gamma \\
0 & 0 & 0 & 0 & 0 & 0 & 0 & 0 & \dots & \gamma & -\beta & \alpha & \delta \\
0 & 0 & 0 & 0 & 0 & 0 & 0 & 0 & \dots & \beta & \gamma & -\delta & \alpha
\end{pmatrix}.$$

We choose the one-to-one correspondence

$$
\begin{pmatrix} 0 \; \cdots \; 0 \; \alpha \; -\beta \; \gamma \; -\delta \; 0 \; \cdots \; 0 \\ 0 \; \cdots \; 0 \; \beta \; \alpha \; \delta \; \gamma \; 0 \; \cdots \; 0 \end{pmatrix} \leftrightarrow
\begin{pmatrix}
\alpha & -\delta & -\gamma & -\beta & 0 & 0 & 0 & 0 & \cdots & 0 & 0 & 0 & 0 \\
\delta & \alpha & \beta & -\gamma & 0 & 0 & 0 & 0 & \cdots & 0 & 0 & 0 & 0 \\
\gamma & -\beta & \alpha & \delta & 0 & 0 & 0 & 0 & \cdots & 0 & 0 & 0 & 0 \\
\beta & \gamma & -\delta & \alpha & 0 & 0 & 0 & 0 & \cdots & 0 & 0 & 0 & 0 \\
0 & 0 & 0 & 0 & \alpha & -\delta & -\gamma & -\beta & \cdots & 0 & 0 & 0 & 0 \\
0 & 0 & 0 & 0 & \delta & \alpha & \beta & -\gamma & \cdots & 0 & 0 & 0 & 0 \\
0 & 0 & 0 & 0 & \gamma & -\beta & \alpha & \delta & \cdots & 0 & 0 & 0 & 0 \\
0 & 0 & 0 & 0 & \beta & \gamma & -\delta & \alpha & \cdots & 0 & 0 & 0 & 0 \\
\cdots & \cdots & \cdots & \cdots & \cdots & \cdots & \cdots & \cdots & & \cdots & \cdots & \cdots & \cdots \\
0 & 0 & 0 & 0 & 0 & 0 & 0 & 0 & \cdots & \alpha & -\delta & -\gamma & -\beta \\
0 & 0 & 0 & 0 & 0 & 0 & 0 & 0 & \cdots & \delta & \alpha & \beta & -\gamma \\
0 & 0 & 0 & 0 & 0 & 0 & 0 & 0 & \cdots & \gamma & -\beta & \alpha & \delta \\
0 & 0 & 0 & 0 & 0 & 0 & 0 & 0 & \cdots & \beta & \gamma & -\delta & \alpha
\end{pmatrix}.
$$

In order to be sure that f_b has the above form, we choose a basis of *real-valued spin-weighted spherical harmonics* with half-integer s in such a way that

$$
f_b(\mathbf{r}_m)(g) = \alpha_s \mathbf{Y}_\ell^{4k-1}(\theta, \varphi) + \delta_s \mathbf{Y}_\ell^{4k}(\theta, \varphi) \\
+ \gamma_s \mathbf{Y}_\ell^{4k+1}(\theta, \varphi) + \beta_s \mathbf{Y}_\ell^{4k+2}(\theta, \varphi).
$$

It follows that

$$
{}_s\mathbf{Y}_\ell^{4k-1}(\theta, \varphi) = \sqrt{\frac{2\ell+1}{4\pi}} \begin{pmatrix} U_{4k-1\;4s-1}^\ell(\theta,\varphi,0) \\ U_{4k-1\;4s}^\ell(\theta,\varphi,0) \end{pmatrix},
$$

$$
{}_s\mathbf{Y}_\ell^{4k}(\theta, \varphi) = \sqrt{\frac{2\ell+1}{4\pi}} \begin{pmatrix} -U_{4k\;4s+2}^\ell(\theta,\varphi,0) \\ U_{4k\;4s+1}^\ell(\theta,\varphi,0) \end{pmatrix},
$$

$$
{}_s\mathbf{Y}_\ell^{4k+1}(\theta, \varphi) = \sqrt{\frac{2\ell+1}{4\pi}} \begin{pmatrix} U_{4k+1\;4s+1}^\ell(\theta,\varphi,0) \\ U_{4k+1\;4s+2}^\ell(\theta,\varphi,0) \end{pmatrix},
$$

$$
{}_s\mathbf{Y}_\ell^{4k+2}(\theta, \varphi) = \sqrt{\frac{2\ell+1}{4\pi}} \begin{pmatrix} -U_{4k+2\;4s}^\ell(\theta,\varphi,0) \\ U_{4k+2\;4s-1}^\ell(\theta,\varphi,0) \end{pmatrix},
$$

where $k = \frac{1}{2}, \ldots, \ell$. Using Equation (3.22), we may write these equations in the following form

$$
{}_s\mathbf{Y}_\ell^{4k-1}(\theta, \varphi) = \sqrt{\frac{2\ell+1}{4\pi}} \begin{pmatrix} \mathrm{Re}\, D_{k\;s}^\ell(\theta,\varphi,0) \\ (-1)^{\ell-k}\, \mathrm{Im}\, D_{k\;-s}^\ell(\theta,\varphi,0) \end{pmatrix},
$$

$$
{}_s\mathbf{Y}_\ell^{4k}(\theta, \varphi) = \sqrt{\frac{2\ell+1}{4\pi}} \begin{pmatrix} (-1)^{\ell-k}\, \mathrm{Re}\, D_{k\;-s}^\ell(\theta,\varphi,0) \\ \mathrm{Im}\, D_{k\;s}^\ell(\theta,\varphi,0) \end{pmatrix},
$$

$$
{}_s\mathbf{Y}_\ell^{4k+1}(\theta, \varphi) = \sqrt{\frac{2\ell+1}{4\pi}} \begin{pmatrix} \mathrm{Re}\, D_{k\;s}^\ell(\theta,\varphi,0) \\ (-1)^{\ell-k+1}\, \mathrm{Im}\, D_{k\;-s}^\ell(\theta,\varphi,0) \end{pmatrix},
$$

$$
{}_s\mathbf{Y}_\ell^{4k+2}(\theta, \varphi) = \sqrt{\frac{2\ell+1}{4\pi}} \begin{pmatrix} (-1)^{\ell-k+1}\, \mathrm{Re}\, D_{k\;-s}^\ell(\theta,\varphi,0) \\ \mathrm{Im}\, D_{k\;s}^\ell(\theta,\varphi,0) \end{pmatrix}.
$$

Observe that the above introduced real-valued spin spherical harmonics differ from those by (Bishop 2005; Bishop and Rezzolla 2016; Zlochower et al. 2003). The latter harmonics have the form

$$
{}_sZ_{\ell m}(\theta,\varphi) = \begin{cases} \frac{1}{\sqrt{2}}\left({}_sY_{\ell m}(\theta,\varphi) + (-1)^m \, {}_sY_{\ell-m}(\theta,\varphi)\right), & \text{if } m > 0, \\ {}_sY_{\ell 0}(\theta,\varphi), & \text{if } m = 0, \\ \frac{i}{\sqrt{2}}\left((-1)^m \, {}_sY_{\ell m}(\theta,\varphi) - {}_sY_{\ell-m}(\theta,\varphi)\right), & \text{otherwise.} \end{cases} \quad (3.29)
$$

Example 73. In this example, we construct special orthonormal bases in the Hilbert spaces $qcU_0 \uparrow G$ and $qV_s \uparrow G$, $s = 1/2, 1, \dots$, where $G = \mathrm{SU}(2)$, and where qcU_0 and qV_s are quaternionic irreducible representations of the group $H = \mathrm{U}(1)$ shown in the second column of Table 3.1.

Put $s = 0$. Table 3.3 shows that $\ell \in \hat{G}_{\mathbb{H}}(qcU_0) = \{0, 1, \dots\}$. The matrix of a linear operator $f^\sharp \in \mathrm{Hom}_{\mathbb{H}H}(qcU_\ell \downarrow H, qcU_0)$ has $\dim qcU_0 = 1$ row and $\dim qcU_\ell = 2\ell + 1$ columns and depends on

$$
M(0,\ell)n_0 \dim_{\mathbb{R}} \mathrm{Hom}_{\mathbb{H}G}(qcU_\ell, qcU_\ell) = 4
$$

$\mathbb{R}$-valued parameters, say α, β, γ, and δ. In the chosen basis, it takes the form $f^\sharp = (0 \ \dots \ 0 \ q \ 0 \ \dots \ 0)$ with $q = \alpha + \beta i + \gamma j + \delta k$ located in the 0th position. The Frobenius reciprocity gives $\dim_{\mathbb{R}} \mathrm{Hom}_{\mathbb{H}G}(qcU_\ell, qcU_0 \uparrow G) = 4$. Again, we can denote the vectors of the special basis by $\mathbf{Y}_{\ell m}(\theta, \varphi)$, as above.

On the one hand, under the Frobenius isomorphism, an operator $f^\sharp$ becomes the operator $f_\flat \in \mathrm{Hom}_{\mathbb{H}G}(qcU_\ell, qcU_0 \uparrow G)$ given by Equation (3.7). It maps the basis vector

$$
\mathbf{r}_m = (0, \dots, 0, 1, 0, \dots, 0)^\top \in R_\ell = H_\ell^{\mathbb{H}}
$$

given by Equation (3.24) to the vector

$$
f_\flat(\mathbf{r}_m)(g) = qU_{0m}^\ell(g^{-1}) = qU_{m0}^\ell(\varphi, \theta, 0).
$$

On the other hand, the matrix of $f_\flat$ has $2\ell + 1$ rows and columns and is equal to qI. We choose the one-to-one correspondence (3.25), but this time with $q \in \mathbb{H}$ in place of $\alpha \in \mathbb{C}$. With that correspondence, $f_\flat$ maps the vector $\mathbf{r}_m \in R_\ell$ to the vector $\mathbf{Y}_{\ell m}(\theta, \varphi)q \in qcU_0 \uparrow G$. It follows that $\mathbf{Y}_{\ell m}(\theta, \varphi)$ is proportional to $U_{m0}^\ell(\varphi, \theta, 0)$. To obtain a vector of unit length, we multiply the above vector by $\sqrt{\frac{2\ell+1}{4\pi}}$ and observe that the cross-sections of the chosen basis are quaternion-valued spherical harmonics which coincide with the real-valued ones.

For positive integers s and $\ell \geq s$, we can repeat the last three paragraphs literally, but replace qcU_0 with qV_s.

Let s be a positive half-integer. Table 3.3 shows that $\ell \in \hat{G}_{\mathbb{H}}(qV_s) = \{s, s+1, \dots\}$. The matrix of a linear operator $f^\sharp \in \mathrm{Hom}_{\mathbb{H}H}(W_\ell \downarrow H, qV_s)$ with positive half-integers s and $l \geq s$ has $\dim qV_s = 1$ row and $\dim W_\ell = \ell + 1/2$ columns and depends on

$$
M(s,\ell)n_s \dim_{\mathbb{R}} \mathrm{Hom}_{\mathbb{H}G}(W_\ell, W_\ell) = 2
$$

$\mathbb{R}$-valued parameters, say α and β. In the chosen basis, it takes the form

$$f^\sharp = (0 \ \ldots \ 0 \ \alpha + \beta\mathrm{i} \ 0 \ \ldots \ 0),$$

where $\alpha + \beta\mathrm{i}$ is located in the $(s+1/2)$th column. The Frobenius reciprocity gives $\dim_{\mathbb{R}} \operatorname{Hom}_{\mathbb{H}G}(W_\ell, qV_s \uparrow G) = 2$.

Denote the vectors of the special basis expressed in the local spin gauge $\psi = 0$ of Example 61 by ${}_sY_{\ell,1}^m(\theta, \varphi)$ and ${}_sY_{\ell,2}^m(\theta, \varphi)$, $1/2 \le m \le \ell + 1/2$.

On the one hand, under the Frobenius isomorphism, an operator $f^\sharp$ becomes the operator $f_\flat \in \operatorname{Hom}_{\mathbb{H}G}(W_\ell, qV_s \uparrow G)$ given by (3.7). It maps the basis vector $\mathbf{r}_m$ given by Equation (3.21) to the vector

$$f_\flat(\mathbf{r}_m)(g) = (\alpha + \beta\mathrm{i})\overline{W_{ms}^\ell(\varphi, \theta, 0)} \in qV_s \uparrow G.$$

On the other hand, the matrix of $f_\flat$ has $M(s, \ell) \dim W_\ell = 2\ell + 1$ rows and $\dim W_\ell = \ell + 1/2$ columns and is equal to $\left(\begin{smallmatrix} \alpha I \\ \beta I \end{smallmatrix}\right)$. We choose the one-to-one correspondence

$$(0 \ \ldots \ 0 \ \alpha + \beta\mathrm{i} \ 0 \ \ldots \ 0) \leftrightarrow \left(\begin{smallmatrix} \alpha I \\ \beta I \end{smallmatrix}\right).$$

With that correspondence, $f_\flat$ maps the vector $\mathbf{r}_m \in R_\ell$ to the vector

$$\alpha_s Y_{\ell,1}^m(\theta, \varphi) + \beta_s Y_{\ell,2}^m(\theta, \varphi) \in qV_s \uparrow G,$$

and we obtain

$$_sY_{\ell,1}^m(\theta, \varphi) = \sqrt{\frac{\ell + 1/2}{4\pi}} \overline{W_{ms}^\ell(\varphi, \theta, 0)},$$

$$_sY_{\ell,2}^m(\theta, \varphi) = \sqrt{\frac{\ell + 1/2}{4\pi}} W_{ms}^\ell(\varphi, \theta, 0)\mathrm{i}.$$

Observe that the special orthonormal bases in the Hilbert spaces $L \uparrow \mathrm{SO}(3)$, where L is an irreducible representation of the group $H = \mathrm{SO}(2)$ over a (skew) field $\mathbb{K}$, coincide with those described above that correspond to the integer values of ℓ and s.

Example 74. In this example, we construct the special orthonormal bases in the Hilbert spaces $L \uparrow \mathrm{O}(3)$, where L is an irreducible representation of the group $H = \mathrm{O}(2)$ over a (skew) field $\mathbb{K}$.

Consider the irreducible representation cU_0^+ of the group H. Table 3.3 shows that $\hat{G}_{\mathbb{H}}(cU_0^+) = \{cU_0^g, cU_1^u, cU_2^g, \ldots\}$. The matrix of a linear operator $f^\sharp \in \operatorname{Hom}_{\mathbb{C}H}(cU_\ell^g \downarrow H, cU_0^+)$ or $f^\sharp \in \operatorname{Hom}_{\mathbb{C}H}(cU_\ell^u \downarrow H, cU_0^+)$ has $\dim cU_0^+ = 1$ row and $\dim qcU_\ell^g = \dim qcU_\ell^u = 2\ell + 1$ columns and depends on

$$M(0, \ell)n_0 \dim_{\mathbb{C}} \operatorname{Hom}_{\mathbb{C}G}(cU_\ell^g, cU_\ell^g) = M(0, \ell)n_0 \dim_{\mathbb{C}} \operatorname{Hom}_{\mathbb{C}G}(cU_\ell^u, cU_\ell^u) = 1$$

$\mathbb{R}$-valued parameters, say α. In the chosen basis, it takes the form $f^\sharp = (0 \ \ldots \ 0 \ \alpha \ 0 \ \ldots \ 0)$ with α located in the 0th position. The Frobenius reciprocity gives

$$\dim_{\mathbb{R}} \operatorname{Hom}_{\mathbb{C}G}(cU_\ell^g, cU_0^+ \uparrow G) = \dim_{\mathbb{R}} \operatorname{Hom}_{\mathbb{C}G}(cU_\ell^u, cU_0^+ \uparrow G) = 1.$$

Denote the vectors of the special basis by $Y_{\ell m}(\theta, \varphi)$ as above.

Under the Frobenius isomorphism, an operator $f^\sharp$ becomes the operator $f_\flat \in \mathrm{Hom}_{\mathbb{C}G}(cU_\ell, cU_0 \uparrow G)$ given by Equation (3.7). This time, for each non-negative integer m, we can freely choose one of the two local gauges (3.13) and map the basis vector

$$\mathbf{r}_m = (0, \ldots, 0, 1, 0, \ldots, 0)^\top \in R_\ell = H_\ell^{\mathbb{C}}$$

given by Equation (3.14) to one of the two vectors

$$f_{\flat,\pm}(\mathbf{r}_m)(g) = \pm(0, \ldots, 0, \alpha, 0, \ldots, 0)(D_{-\ell m}^\ell(g^{-1}), \ldots, D_{\ell m}^\ell(g^{-1}))^\top$$
$$= \pm \alpha \overline{D_{m0}^\ell(\varphi, \theta, 0)} \in cU_0^+ \uparrow G.$$

We fix our choice as follows. For the representations $cU_0^\pm$, $U_0^\pm$, and $qcU_0^\pm$, we choose the local gauge $\psi_\pm$ for all values of m. In that case, the complex-valued spherical harmonics $Y_{\ell m}^+(\theta, \varphi)$ that correspond to the representation cU_0^+ are given by Equation (3.26), while the harmonics $Y_{\ell m}^-(\theta, \varphi)$ that correspond to the representation cU_0^- are given by $Y_{\ell m}^-(\theta, \varphi) = -Y_{\ell m}^+(\theta, \varphi)$. Likewise, the real-valued spherical harmonics $Y_\ell^{m+}(\theta, \varphi)$ that correspond to the representation U_0^+, are given by Equation (3.28), while the harmonics $Y_\ell^{m-}(\theta, \varphi)$ that correspond to the representation U_0^-, are given by $Y_\ell^{m-}(\theta, \varphi) = -Y_\ell^{m+}(\theta, \varphi)$. The quaternionic case coincides with the real one.

Let s be a positive integer. Table 3.3 shows that $\hat{G}_{\mathbb{C}}(cU_s) = \{cU_\ell^g, cU_\ell^u : \ell \geq s\}$. The matrix of a linear operator $f^\sharp \in \mathrm{Hom}_{\mathbb{C}H}(cU_\ell^g \downarrow H, cU_s)$ (resp., $f^\sharp \in \mathrm{Hom}_{\mathbb{C}H}(cU_\ell^u \downarrow H, cU_s)$) with positive integers s and $l \geq s$ has $\dim cU_s = 2$ rows and $\dim cU_\ell^g = \dim cU_\ell^u = 2\ell + 1$ columns and depends on

$$M(s, \ell)n_s \dim_{\mathbb{C}} \mathrm{Hom}_{\mathbb{C}G}(cU_\ell^g, cU_\ell^g) = M(s, \ell)n_s \dim_{\mathbb{C}} \mathrm{Hom}_{\mathbb{C}G}(cU_\ell^u, cU_\ell^u) = 1$$

$\mathbb{C}$-valued parameters, say α. In the chosen basis, it takes the form

$$f^\sharp = \begin{pmatrix} 0 & \cdots & \alpha & \cdots & 0 & \cdots & 0 \\ 0 & \cdots & 0 & \cdots & \alpha & \cdots & 0 \end{pmatrix}$$

with $\alpha \in \mathbb{R}$ located in the intersection of the first line and the $-s$th column as well as in the intersection of the second line and the sth column. The Frobenius reciprocity gives

$$\dim_{\mathbb{C}} \mathrm{Hom}_{\mathbb{C}G}(cU_\ell^g, cU_s \uparrow G) = \dim_{\mathbb{C}} \mathrm{Hom}_{\mathbb{C}G}(cU_\ell^u, cU_s \uparrow G) = 1.$$

Denote the vectors of the special basis expressed in the local spin gauge $\psi = 0$ of Example 61 by $_s\mathbf{Y}_{\ell,1}^m(\theta, \varphi)$ and $_s\mathbf{Y}_{\ell,2}^m(\theta, \varphi)$, $-s \leq m \leq s$.

On the one hand, under the Frobenius isomorphism, an operator $f^\sharp$ becomes the operator $f_\flat \in \mathrm{Hom}_{\mathbb{C}G}(cU_\ell^g, cU_s \uparrow G)$ or $f_\flat \in \mathrm{Hom}_{\mathbb{C}G}(cU_\ell^u, cU_s \uparrow G)$ given by (3.7). It maps the basis vector $\mathbf{r}_m$ given by Equation (3.14) to the vector

$$f_\flat(\mathbf{r}_m)(g) = \alpha \begin{pmatrix} \overline{D_{m\,-s}^\ell(\varphi, \theta, 0)} \\ -\overline{D_{ms}^\ell(\varphi, \theta, 0)} \end{pmatrix} \in cU_s \uparrow G.$$

On the other hand, the matrix of $f_\flat$ has

$$M(s,\ell)\dim cU_\ell^g + M(s,\ell)\dim cU_\ell^u = 4\ell + 2$$

rows and $\dim cU_\ell^g = \dim cU_\ell^u = 2\ell + 1$ columns and is equal to $\left(\begin{smallmatrix}\alpha I\\\alpha I\end{smallmatrix}\right)$. We choose the one-to-one correspondence

$$\left(\begin{smallmatrix}0 & \cdots & \alpha & \cdots & 0 & \cdots & 0\\0 & \cdots & 0 & \cdots & \alpha & \cdots & 0\end{smallmatrix}\right) \leftrightarrow \left(\begin{smallmatrix}\alpha I\\\alpha I\end{smallmatrix}\right).$$

With that correspondence, $f_\flat$ maps the vector $\mathbf{r}_m \in R_\ell$ to the vector

$$\alpha_s \mathbf{Y}_{\ell,1}^m(\theta,\varphi) + \alpha_s \mathbf{Y}_{\ell,2}^m(\theta,\varphi) \in cU_s \uparrow G,$$

and we obtain

$$_s\mathbf{Y}_{\ell,1}^m(\theta,\varphi) = \sqrt{\frac{2\ell+1}{8\pi}}\left(\begin{smallmatrix}\overline{D_{m-s}^\ell(\varphi,\theta,0)}\\0\end{smallmatrix}\right),$$

$$_s\mathbf{Y}_{\ell,2}^m(\theta,\varphi) = \sqrt{\frac{2\ell+1}{8\pi}}\left(\begin{smallmatrix}0\\-D_{ms}^\ell(\varphi,\theta,0)\end{smallmatrix}\right).$$

Table 3.3 shows that $\hat{G}_\mathbb{R}(U_s) = \{U_\ell^g, U_\ell^u : \ell \geq s\}$. The matrix of a linear operator $f^\sharp \in \operatorname{Hom}_{\mathbb{R}H}(U_\ell^g \downarrow H, U_s^g)$ (resp., $f^\sharp \in \operatorname{Hom}_{\mathbb{R}H}(U_\ell^u \downarrow H, U_s^u)$) with positive integers s and $l \geq s$ has $\dim U_s = 2$ rows and $\dim cU_\ell^g = \dim cU_\ell^u = 2\ell + 1$ columns and depends on

$$M(s,\ell)n_s \dim_\mathbb{R} \operatorname{Hom}_{\mathbb{R}G}(U_\ell^g, U_\ell^g) = M(s,\ell)n_s \dim_\mathbb{R} \operatorname{Hom}_{\mathbb{R}G}(U_\ell^u, U_\ell^u) = 1$$

$\mathbb{C}$-valued parameters, say α. In the chosen basis, it takes the form

$$f^\sharp = \left(\begin{smallmatrix}0 & \cdots & \alpha & \cdots & 0 & \cdots & 0\\0 & \cdots & 0 & \cdots & \alpha & \cdots & 0\end{smallmatrix}\right)$$

with $\alpha \in \mathbb{R}$ located in the intersection of the first line and the $-s$th column as well as in the intersection of the second line and the sth column. The Frobenius reciprocity gives

$$\dim_\mathbb{R}\operatorname{Hom}_{\mathbb{R}G}(U_\ell^g, U_s \uparrow G) = \dim_\mathbb{R}\operatorname{Hom}_{\mathbb{R}G}(U_\ell^u, U_s \uparrow G) = 1.$$

On the one hand, under the Frobenius isomorphism, an operator $f^\sharp$ becomes the operator $f_\flat \in \operatorname{Hom}_{\mathbb{R}G}(U_\ell^g, U_s \uparrow G)$ or $f_\flat \in \operatorname{Hom}_{\mathbb{R}G}(U_\ell^u, U_s \uparrow G)$ given by (3.7). It maps the basis vector $\mathbf{r}_m$ given by Equation (3.14) to the vector

$$f_\flat(\mathbf{r})(g) = \alpha\left(\begin{smallmatrix}U_{-sm}^\ell(g^{-1})\\U_{sm}^\ell(g^{-1})\end{smallmatrix}\right) = \alpha\left(\begin{smallmatrix}U_{m-s}^\ell(g)\\U_{ms}^\ell(g)\end{smallmatrix}\right).$$

On the other hand, the matrix of $f_\flat$ has $M(s,\ell)\dim U_\ell^g = 2\ell + 1$ rows and columns and is equal to αI. We choose the one-to-one correspondence

$$\left(\begin{smallmatrix}0 & \cdots & \alpha & \cdots & 0 & \cdots & 0\\0 & \cdots & 0 & \cdots & \alpha & \cdots & 0\end{smallmatrix}\right) \leftrightarrow \alpha I.$$

With that correspondence, $f_\flat$ maps the vector $\mathbf{r}_m \in R_\ell$ to the vector

$$\alpha_s \mathbf{Y}_\ell^m(\theta,\varphi) \in U_s \uparrow G,$$

and we obtain

$$_s\mathbf{Y}_\ell^m(\theta,\varphi) = \sqrt{\frac{2\ell+1}{8\pi}}\left(\begin{smallmatrix}U_{m-s}^\ell(g)\\U_{ms}^\ell(g)\end{smallmatrix}\right).$$

The quaternionic case coincides with the complex one.

3.3 Physical Application of the Irreducible Case

Example 75. Consider the basis (3.26). Theorem 24 gives that the spectral expansion of a $\mathbb{C}^1$-valued isotropic mean-square continuous random field on the sphere S^2 in the local chart given by the standard spherical coordinates, has the form

$$X(\theta, \varphi) = \sum_{\ell=0}^{\infty} \sum_{m=-\ell}^{\ell} a_{\ell m} Y_{\ell,m}(\theta, \varphi),$$

where $a_{\ell m}$ are complex-valued uncorrelated random variable satisfying the following conditions: $\mathsf{E}[a_{\ell m}] = 0$ unless $\ell = 0$, there are nonnegative real numbers C_ℓ such that

$$\mathsf{E}[a_{\ell m}\overline{a_{\ell' m'}}] = \delta_{\ell\ell'}\delta_{mm'}C_\ell$$

and

$$\sum_{\ell=0}^{\infty}(2\ell + 1)C_\ell < \infty.$$

Denote by $T(\theta, \varphi)$ the absolute temperature of the Cosmic Microwave Background at the point (θ, φ) on the sky sphere. Its spectral expansion has the form

$$T(\theta, \varphi) = \sum_{\ell=0}^{\infty} \sum_{m=-\ell}^{\ell} a_{\ell m} Y_{\ell,m}(\theta, \varphi). \tag{3.30}$$

To be sure that this quantity is *real-valued*, we use the relation $Y_{\ell,-m}(\theta, \varphi) = (-1)^m \overline{Y_{\ell,m}(\theta, \varphi)}$ and conclude that the random field (3.30) is real-valued if and only if the *multipole coefficients* $a_{\ell m}$ satisfy the relation

$$a_{\ell-m} = (-1)^m \overline{a_{\ell m}},$$

called the *reality condition*. In particular, the coefficient $a_{\ell 0}$ is a real-valued random variable.

The remaining physical terms in the above expansion are coined as follows. The multipole coefficient a_{00} is called the *monopole*, the coefficients a_{1m} are *dipoles*, the number m inside the ℓ-pole $\{\, a_{\ell m}\colon\ -\ell \leq m \leq \ell \,\}$ is called the *asimuthal number*, the numbers C_ℓ constitute the *CMB temperature power spectrum*.

Sometimes, it is convenient to write the expansion (3.30) in terms of a unit vector $\mathbf{n} \in S^2 \subset \mathbf{R}^3$ as follows:

$$T(\mathbf{n}) = \sum_{\ell=0}^{\infty} \sum_{m=-\ell}^{\ell} a_{\ell m} Y_{\ell,m}(\mathbf{n}).$$

With this notation, it is easy to calculate the two-point correlation tensor of the CMB:

$$
\begin{aligned}
\mathsf{E}[T(\mathbf{n}_1)T(\mathbf{n}_2)] &= \sum_{\ell,\ell'=0}^{\infty}\sum_{m=-\ell}^{\ell}\sum_{m'=-\ell'}^{\ell'}\mathsf{E}[a_{\ell m}\overline{a_{\ell' m'}}]Y_{\ell,m}(\mathbf{n}_1)\overline{Y_{\ell',m'}(\mathbf{n}_2)} \\
&= \sum_{\ell=0}^{\infty}C_{\ell}\sum_{m=-\ell}^{\ell}Y_{\ell,m}(\mathbf{n}_1)\overline{Y_{\ell',m'}(\mathbf{n}_2)} \\
&= \sum_{\ell=0}^{\infty}C_{\ell}\frac{2\ell+1}{4\pi}P_{\ell}(\mathbf{n}_1\cdot\mathbf{n}_2) \\
&= \frac{1}{4\pi}\sum_{\ell=0}^{\infty}(2\ell+1)C_{\ell}P_{\ell}(\mathbf{n}_1)\cdot\mathbf{n}_2),
\end{aligned}
$$

where we use the *addition theorem for spherical harmonics*

$$
\sum_{m-\ell}^{\ell}Y_{\ell,m}(\mathbf{n}_1)\overline{Y_{\ell',m'}(\mathbf{n}_2)} = \frac{2\ell+1}{4\pi}P_{\ell}(\mathbf{n}_1\cdot\mathbf{n}_2),
$$

and where the P_{ℓ}s are the *Legendre polynomials.* In this case, the two-point correlation tensor is a $\mathbb{R}^1$-valued function called the *two-point correlation function* of the CMB. It depends only on the angle between the vectors $\mathbf{n}_1$ and $\mathbf{n}_2$. This is the reason to call the random field (3.30) isotropic.

The expected value $\mathsf{E}[a_{00}]$ is the mean temperature of the CMB. According to Workman (2022), it has the value of $T = 2.7255(6)$ K at the confidence level 68 %.

The dipole is dominated by the motion of the Solar System with respect to the comoving observer. The velocity of that motion is $v = 369.82(11)$ km s^{-1}.

It is customary to plot the quantity $\frac{\ell(\ell+1)C_{\ell}}{2\pi}$ expressed in µK^2 against the multipole ℓ, see, for example, Workman (2022, Fig. 29.2). The description of physics underlying the CMB temperature power spectrum may be found in Durrer (2020), Workman (2022, Section 29.5), and many other papers and books.

Figure 3.2[2] shows in blue (resp., in red) the zones where the absolute temperature of the CMB is less (resp., greater) than T.

Theoretical models predict that the multipole coefficients are normal random variables. It follows that they are independent and the random field (3.30) is Gaussian, that is, its finite-dimensional distributions are normal random vectors. According to Acrami (2020), the statistical hypothesis 'The CMB is a single realisation of a Gaussian random field' cannot be rejected. In all examples below, the random fields will be Gaussian.

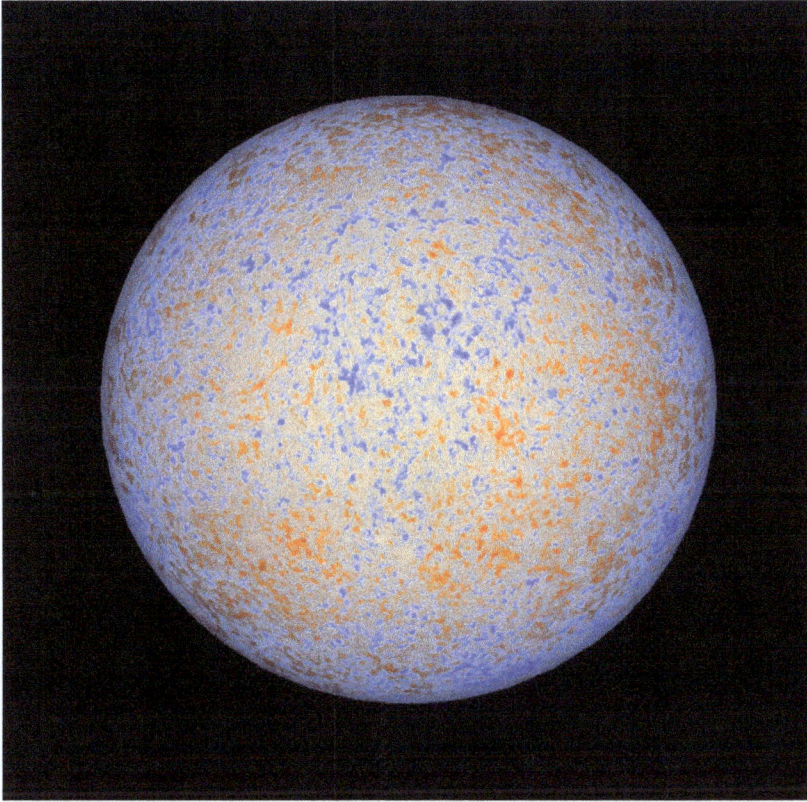

FIGURE 3.2
The CMB sphere mapped by ESA and the Planck Collaboration in 2013 showing hot (red) and cold (blue) zones.

Equation (3.30) may also describe the intensity I of all three cosmic backgrounds: microwave, gravitational, and neutrino ones. The second one is reported to be discovered in Agazie (2023). The theoretical power spectrum of the massless Cosmic Neutrino Background was computed by Hu et al. (1995). For the massive one, see (Hannestad and Brandbyge 2010; Michney and Caldwell 2007) and the discussion in Lesgourgues et al. (2013, Subsection 7.2).

Example 76. Consider the basis (3.28). Theorem 24 gives that the spectral expansion of a $\mathbb{R}^1$-valued isotropic mean-square continuous Gaussian random field on the sphere S^2 in the local chart given by the standard spherical coordinates has the form

$$X(\theta, \varphi) = \sum_{\ell=0}^{\infty} \sum_{m=-\ell}^{\ell} a_{\ell m} Y_\ell^m(\theta, \varphi),$$

where $a_{\ell m}$ are independent normal random variables satisfying the conditions described in Example 75. This equation can also describe the absolute temperature of the CMB and the intensity of all three Cosmic Backgrounds. Without reference to Theorem 24, its proof may use the Funck–Hecke theorem, as in Yadrenko (1959), or the real version of Schur's Lemma, which is not well-known to most physicists. In the author's opinion, by this reason it does not use so often as its complex counterpart (3.30).

According to Example 73, the quaternionic spherical harmonics coincide with real-valued ones, no new cases appears here.

Remark 13. The circular polarisation of both the Microwave and Gravitational Cosmic Backgrounds can be expanded exactly as their intensity. Indeed, according to Example 74, the basis in the space of the square-integrable functions on the sphere that correspond to the representations U_0^-, cU_0^-, and qcU_0^- of the group O(2), can be obtained from the usual basis of spherical harmonics by multiplying them by -1. The random multipole coefficients are centred normal and therefore symmetric, hence the finite-dimensional distributions of the circular polarisation random field does not change under multiplication by -1.

Example 77. Following Durrer (2020), modify Example 44. Using the angular spherical coordinates on S^2, define the basis vectors of the tangent plane T_pS^2 as follows:

$$\frac{\partial}{\partial\theta} = (\cos\theta\cos\varphi, \cos\theta\sin\varphi, -\sin\theta)^\top,$$

$$\frac{1}{\sin\theta}\frac{\partial}{\partial\varphi} = (-\sin\varphi, \cos\varphi, 0)^\top.$$

The inner products of the vectors $\mathbf{e}_\theta = \frac{\partial}{\partial\theta}$ and $\mathbf{e}_\varphi = \frac{1}{\sin\theta}\frac{\partial}{\partial\varphi}$ form the 2×2 identity matrix. The orthonormal basis of the tangent plane formed by the above vectors is called *canonical*. The vectors

$$\mathbf{e}^\pm = \frac{1}{\sqrt{2}}(\mathbf{e}_\theta \mp i\mathbf{e}_\varphi) \tag{3.31}$$

form an orthonormal basis of the *complexified* tangent plane called the *helicity basis*.

In the helicity basis, the normalised Stokes parameters that describe the linear polarisation of the Cosmic Microwave Background can be represented by the following spectral expansion:

$$(s_2 \pm is_3)(\theta,\varphi) = \sum_{\ell=2}^{\infty}\sum_{m=-\ell}^{\ell} a_{\ell m}^{(\pm 2)}{}_{\pm 2}Y_{\ell m}(\theta,\varphi). \tag{3.32}$$

Compare this result against Theorem 24. We see that the *complex polarisation* $(s_2 + is_3)(\theta,\varphi)$ (resp., $(s_2 - is_3)(\theta,\varphi)$) is an isotropic random cross-section of the spin bundle $\xi[V_2]$ (resp., $\xi[tV_2]$). Moreover,

$$\mathsf{E}[a_{\ell m}^{(2)}] = 0, \qquad \mathsf{E}[a_{\ell m}^{(2)}\overline{a_{\ell'm'}^{(2)}}] = \delta_{\ell\ell'}\delta_{mm'}C_{2\ell}$$

with

$$\sum_{\ell=2}^{\infty}(2\ell+1)C_{2\ell}<\infty,$$

and similarly for $a_{\ell m}^{(-2)}$.

Similarly, according to Conneely, Jaffe, and Mingarelli (2019), the complex polarisation of the Cosmic Gravitational Background can be represented by the following spectral expansion:

$$(s_2\pm is_3)(\theta,\varphi)=\sum_{\ell=2}^{\infty}\sum_{m=-\ell}^{\ell}a_{\ell m}^{(\pm4)}{}_{\pm4}Y_{\ell m}(\theta,\varphi).$$

That is, this time the random field $(s_2+is_3)(\theta,\varphi)$ (resp., $(s_2-is_3)(\theta,\varphi)$) is an isotropic random cross-section of the spin bundle $\xi[V_4]$ (resp., $\xi[tV_4]$).

Since $s_2(\theta,\varphi)$ and $s_3(\theta,\varphi)$ are real-valued, the multipole coefficients $a_{\ell m}^{(\pm2)}$ must satisfy the *reality condition*. It follows from Equation (3.27) that this condition has the form

$$a_{\ell-m}^{(\pm2)}=(-1)^m\overline{a_{\ell m}^{(\mp2)}}.$$

The two-point correlation tensor of the random field (3.32) has the form

$$\mathsf{E}[(s_2+is_3)(\theta_1,\varphi_1)(s_2-is_3)(\theta_2,\varphi_2)]=\frac{1}{4\pi}\sum_{\ell=2}^{\infty}C_{2\ell}\sqrt{2\ell+1}\,{}_2Y_{\ell2}(\beta,\alpha)\mathrm{e}^{-2\mathrm{i}\gamma},$$

where (α,β,γ) are the Euler angles of the rotation $g_2^{-1}g_1\in\mathrm{SO}(3)$, g_1 (resp., g_2) is the rotation with Euler's angles $(\varphi_1,\theta_1,0)$ (resp., $(\varphi_2,\theta_2,0)$). Here we use the addition theorem for the spin-weighted spherical harmonics:

$$\sqrt{\frac{4\pi}{2\ell+1}}\sum_{m'=-\ell}^{\ell}{}_sY_{\ell m'}(\theta_1,\varphi_1)\overline{{}_{-m}Y_{\ell m'}(\theta_2,\varphi_2)}={}_sY_{\ell m}(\beta,\alpha)\mathrm{e}^{-\mathrm{i}s\gamma},$$

see Durrer (2020, Equation (A.4.101)). A proof of the equation for the two-point correlation tensor of the linear polarisation of the Cosmic Gravitational Background is similar and can be left to the reader.

The above introduced random cross-sections are not rotationally invariant, but rather depend on the space orientation of the device that measure them. We overcome this difficulty, following Zaldarriaga and Seljak (1997). In what follows, it will be convenient to work in terms of a unit vector **n**. It is well-known that

$$\eth_{-1}\eth_{-2}{}_{-2}Y_{\ell m}(\mathbf{n})=\overline{\eth}_1\overline{\eth}_2{}_2Y_{\ell m}(\mathbf{n})=\sqrt{\frac{(\ell+2)!}{(\ell-2)!}}Y_{\ell m}(\mathbf{n}),$$

see Durrer (2020, Equations (A.4.108) and (A.4.109)). Assume for a moment that

$$\sum_{\ell=2}^{\infty}\frac{(2\ell+1)(\ell+2)!}{(l-2)!}C_{2\ell}<\infty. \tag{3.33}$$

Then, it is possible to act with $\eth_{-2}$ and then with $\eth_{-1}$ on both sides of (3.32) and to interchange differentiation and summation:

$$\eth_{-1}\eth_{-2}(s_2 - is_3)(\mathbf{n}) = \eth_{-1}\eth_{-2}\sum_{\ell=2}^{\infty}\sum_{m=-\ell}^{\ell} a_{\ell m}^{(-2)}{}_{-2}Y_{\ell m}(\mathbf{n})$$

$$= \sum_{\ell=2}^{\infty}\sum_{m=-\ell}^{\ell} a_{\ell m}^{(-2)}\eth_{-1}\eth_{-2}{}_{-2}Y_{\ell m}(\mathbf{n})$$

$$= \sum_{\ell=2}^{\infty}\sum_{m=-\ell}^{\ell}\sqrt{\frac{(\ell+2)!}{(\ell-2)!}}a_{\ell m}^{(-2)}Y_{\ell m}(\mathbf{n}).$$

Similarly,

$$\overline{\eth_1\eth_2}(s_2 + is_3)(\mathbf{n}) = \sum_{\ell=2}^{\infty}\sum_{m=-\ell}^{\ell}\sqrt{\frac{(\ell+2)!}{(\ell-2)!}}a_{\ell m}^{(2)}Y_{\ell m}(\mathbf{n}).$$

We see that the random fields

$$E(\mathbf{n}) = \frac{1}{2}(\eth_{-1}\eth_{-2}(s_2 - is_3)(\mathbf{n}) + \overline{\eth_1\eth_2}(s_2 + is_3)(\mathbf{n}))$$

$$= \sum_{\ell=2}^{\infty}\sum_{m=-\ell}^{\ell} e_{\ell m}\sqrt{\frac{(\ell+2)!}{(\ell-2)!}}Y_{\ell m}(\mathbf{n}),$$

$$B(\mathbf{n}) = -\frac{i}{2}(\eth_{-1}\eth_{-2}(s_2 - is_3)(\mathbf{n}) - \overline{\eth_1\eth_2}(s_2 + is_3)(\mathbf{n})) \qquad (3.34)$$

$$= \sum_{\ell=2}^{\infty}\sum_{m=-\ell}^{\ell}\sqrt{\frac{(\ell+2)!}{(\ell-2)!}}Y_{\ell m}(\mathbf{n}),$$

where

$$e_{\ell m} = \frac{1}{2}(a_{\ell m}^{(2)} + a_{\ell m}^{(-2)}), \qquad b_{\ell m} = -\frac{i}{2}(a_{\ell m}^{(2)} - a_{\ell m}^{(-2)})$$

are real-valued isotropic random fields on the sphere S^2.

A *parity transformation* is the map $S^2 \to S^2$, $\mathbf{n} \mapsto -\mathbf{n}$. Under this transformation, the vectors of the helicity basis interchange, $\mathbf{e}^{\pm} \mapsto \mathbf{e}^{\mp}$. Then, the random multipole coefficients of the complex polarisation change as follows: $a_{\ell m}^{(2)} \mapsto (-1)^{\ell}a_{\ell-m}^{-2}$ and $a_{\ell m}^{(-2)} \mapsto (-1)^{\ell}a_{\ell-m}^{2}$. The random multipole coefficients of the random fields $E(\mathbf{n})$ and $B(\mathbf{n})$ change as $e_{\ell m} \mapsto (-1)^{\ell}e_{\ell-m}$ and $b_{\ell m} \mapsto (-1)^{\ell+1}b_{\ell-m}$. In physical terms, $E(\mathbf{n})$ measures the gradient-type, or electric type polarisation, while $B(\mathbf{n})$ measures the curl-type, or magnetic type polarisation, see explanation in Durrer (2020, Section 5.1).

Example 78. This example is adapted from Malyarenko and Ostoja-Starzewski (2023). It is proved there that the real-valued spin spherical harmonics are connected to their complex-valued counterparts as follows:

$$_s\mathbf{Y}_{\ell,1}^m(\mathbf{n}) = C\begin{pmatrix} {}_sY_{\ell m}(\mathbf{n}) + (-1)^m {}_{-s}Y_{\ell-m}(\mathbf{n}) \\ -\mathrm{i}_s Y_{\ell m}(\mathbf{n}) + \mathrm{i}(-1)^m {}_{-s}Y_{\ell-m}(\mathbf{n}) \end{pmatrix},$$

$$_s\mathbf{Y}_{\ell,2}^m(\mathbf{n}) = C\begin{pmatrix} \mathrm{i}_s Y_{\ell m}(\mathbf{n}) - \mathrm{i}(-1)^m {}_{-s}Y_{\ell-m}(\mathbf{n}) \\ {}_sY_{\ell m}(\mathbf{n}) + (-1)^m {}_{-s}Y_{\ell-m}(\mathbf{n}) \end{pmatrix},$$

where C is a constant. We obtain, that the Stokes parameters $s_2(\mathbf{n})$ and $s_3(\mathbf{n})$ of the Cosmic Microwave Background can be expanded in the basis of real spherical harmonics as

$$\begin{pmatrix} s_2 & s_3 \end{pmatrix}(\mathbf{n}) = \sum_{\ell=2}^{\infty}\sum_{m=-\ell}^{\ell}\left(_2a_{\ell,12}^m\mathbf{Y}_{\ell,1}^m(\mathbf{n}) + _2a_{\ell,22}^m\mathbf{Y}_{\ell,2}^m(\mathbf{n})\right)$$

where the multipole coefficients are independent centred normal random variables with cross-correlations given by

$$\mathsf{E}[(_2a_{\ell,1}^m, _2a_{\ell,2}^m)(_2a_{\ell',1}^{m'}, _2a_{\ell',2}^{m'})^\top] = \delta_{\ell\ell'}\delta_{mm'}C_\ell,$$

and the symmetric 2×2 nonnegative-definite matrices C_ℓ satisfy the condition

$$\sum_{\ell=2}^{\infty}(2\ell+1)\operatorname{tr}C_\ell < \infty.$$

3.4   Examples of Special Bases: The Reducible Case

In the reducible case, it is possible to give a lot of examples of special bases. We include only those connected with the spectral expansions of cosmic backgrounds and start from an explanation of necessary physical foundations.

3.4.1   The Polarisation Matrix

It is possible to write down spectral expansions of isotropic tensor-valued random cross-sections in a coordinate-free form, see (Challinor 2000a, 2000b). We prefer the more traditional coordinate approach.

Let $\mathbf{n} \in S^2$ be a point on the sky sphere. The electric field of the Cosmic Microwave Background, $\mathbf{E}(\mathbf{n})$, is a random cross-section of the tangent bundle. The time average of the tensor product $\mathbf{E}(\mathbf{n})\otimes\overline{\mathbf{E}(\mathbf{n})}$ over the historical accidents that produced a particular pattern of fluctuations is the polarisation tensor. Following Durrer (2020), we denote it by the symbol P:

$$P(\mathbf{n}) = \langle\mathbf{E}(\mathbf{n})\otimes\overline{\mathbf{E}(\mathbf{n})}\rangle,$$

where the angular brackets denote the time average.

We will work in spherical coordinates (θ, φ). At each point in the domain of this chart, we introduce a basis $\varepsilon^{(1)}$ and $\varepsilon^{(2)}$ in the tangent space and call it the *polarisation basis*. The *standard* polarisation basis is

$$\varepsilon^{(1)} = \mathbf{e}_\varphi, \qquad \varepsilon^{(2)} = \mathbf{e}_\theta,$$

the *helicity* polarisation basis is

$$\varepsilon^{(+)} = \mathbf{e}^+, \qquad \varepsilon^{(-)} = \mathbf{e}^-.$$

We denote by E_1 and E_2 (resp., by E_+ and E_-) the coordinates of the electric field in the standard (resp., helicity) polarisation basis.

By Durrer (2020, Equation (5.2)), the polarisation tensor of the CMB in the standard polarisation basis becomes the 2×2 Hermitian *polarisation matrix*

$$P = \frac{1}{2}(I\sigma_0 + U\sigma_1 + V\sigma_2 + Q\sigma_3) = \frac{1}{2}\begin{pmatrix} I+Q & U-iV \\ U+iV & I-Q \end{pmatrix}, \qquad (3.35)$$

that is, up to a constant, the Stokes parameters are the coefficients of the expansion of an arbitrary 2×2 Hermitian matrix with respect to the basis given by the Pauli matrices.

It is easy to calculate the polarisation matrix in the helicity basis:

$$P = \frac{1}{2}(I\sigma_0 + Q\sigma_1 + U\sigma_2 + V\sigma_3) = \frac{1}{2}\begin{pmatrix} I+V & Q-iU \\ Q+iU & I-V \end{pmatrix}.$$

On the one hand, by Jackson (1975, Equation (7.27)), in term of the electric field, the Stokes parameters are the following time averages:

$$I = \langle E_1^1 + E_2^2 \rangle, \qquad\qquad Q = \langle E_1^2 - E_2^2 \rangle,$$
$$U = \langle 2\operatorname{Re}(\overline{E_1}E_2) \rangle, \qquad\qquad V = \langle 2\operatorname{Im}(\overline{E_1}E_2) \rangle.$$

On the other hand, by Conneely, Jaffe, and Mingarelli (2019, Equation (1)), the Stokes parameters for the Cosmic Gravitational Background are the averages of the amplitudes h_+ and $h_\times$ of the plus- and cross-polarisation as follows:

$$I = \langle |h_+|^2 + |h_\times|^2 \rangle, \qquad\qquad Q = \langle |h_+|^2 - |h_\times|^2 \rangle,$$
$$U = \langle 2\operatorname{Re}(h_+\overline{h_\times}) \rangle, \qquad\qquad V = -\langle 2\operatorname{Im}(h_+\overline{h_\times}) \rangle.$$

By this reason, the polarisation matrix of the Cosmic Gravitational Background has the same form as for the CMB.

3.4.2   Vector and Tensor Spherical Harmonics

In order to study the random multipole expansions of the polarisation tensor of the Cosmic Microwave and Gravitational Backgrounds, we must induce certain *reducible* representation of the group H in the linear space of the 2×2 Hermitian matrices to the group G. In that case, the special bases in the Hilbert space $L \uparrow G$ are constructed with the help of special numbers called the *Clebsch–Gordan coefficients*.

In fact, this work has already been done by several authors *without* mentioning the induced representations. Thorne (1980) contains an excellent review of special bases in the Hilbert spaces $L \uparrow \mathrm{SU}(2)$, where L is a finite-dimensional complex representation of the group $H = \mathrm{U}(1)$. We describe them in our terms as follows.

Put $L = cU_0$. The Hilbert space $cU_0 \uparrow \mathrm{SU}(2)$ is the Hilbert direct sum of the spaces cU_ℓ, $\ell \geq 0$. The special orthonormal basis of cU_ℓ consists of complex-valued spherical harmonics $Y_{\ell,m}$ constructed in Example 71. In Thorne (1980), they are called *scalar spherical harmonics*. Observe that $Y_{\ell,m}(-\mathbf{n}) = (-1)^\ell Y_{\ell,m}(\mathbf{n})$, we say that the scalar spherical harmonics have 'electric-type' *parity* or *even parity*.

Consider the representation cU_1 of the group $G = \mathrm{SU}(2)$. Its restriction to H is the direct sum $L = tV_1 \oplus cU_0 \oplus V_1$. The elements of L are vectors. Table 3.4 shows that the Hilbert space $L \uparrow \mathrm{SU}(2)$ is the Hilbert direct sum of the subspaces R_ℓ, where R_0 carries a single copy of the irreducible representation cU_0, while each of the spaces R_ℓ, $\ell \geq 1$, carries three copies of the irreducible representation cU_ℓ.

The *pure-orbital vector spherical harmonics* have been constructed by (Edmonds 1957; Mathews 1962; Rose 1955) with the help of the Clebsch–Gordan coefficients. They are given by Thorne (1980, Equation (2.16)). The pure-orbital vector spherical harmonic $\mathbf{Y}^{1,00}(\theta, \varphi)$ spans the space R_0, while the $6\ell + 3$ harmonics $\mathbf{Y}^{\ell',\ell m}$ with $\ell \geq 1$, $\ell - 1 \leq \ell' \leq \ell + 1$, and $-\ell \leq m \leq \ell$ span the space R_ℓ.

Thorne (1980) calls a vector spherical harmonic *pure radial* (resp., *pure transverse*) if it belongs to the subspace of R_ℓ where the representation of the group G induced by the representation cU_0 (resp., $tV_1 \oplus V_1$) of the group H acts. The pure-orbital vector spherical harmonics $\mathbf{Y}^{\ell\pm1,\ell m}$ are neither pure radial nor pure transverse. The *pure-spin vector spherical harmonics* are constructed by Thorne (1980) as follows:

$$
\begin{aligned}
\mathbf{Y}^{E,\ell m} &= \sqrt{\frac{\ell+1}{2\ell+1}}\,\mathbf{Y}^{\ell-1,\ell m} + \sqrt{\frac{\ell}{2\ell+1}}\,\mathbf{Y}^{\ell+1,\ell m}, \\
\mathbf{Y}^{B,\ell m} &= \mathrm{i}\mathbf{Y}^{\ell,\ell m}, \\
\mathbf{Y}^{R,\ell m} &= \sqrt{\frac{\ell}{2\ell+1}}\,\mathbf{Y}^{\ell-1,\ell m} - \sqrt{\frac{\ell+1}{2\ell+1}}\,\mathbf{Y}^{\ell+1,\ell m}.
\end{aligned}
\tag{3.36}
$$

The harmonics $\mathbf{Y}^{E,\ell m}$ have electric-type parity, $\mathbf{Y}^{B,\ell m}$ have 'magnetic-type' *parity* or *odd parity*, that is, $\mathbf{Y}^{B,\ell m}(-\mathbf{n}) = (-1)^{\ell+1}\mathbf{Y}^{B,\ell m}(-\mathbf{n})$. Both are purely transverse, while $\mathbf{Y}^{R,\ell m}$ are radial. Moreover, there is a connection between the pure-spin vector and spin-weighted spherical harmonics given by

$$\mathbf{Y}^{E,\ell m} = \frac{1}{\sqrt{2}}(_{-1}Y_{\ell,m}\mathbf{e}^- - {}_1Y_{\ell,m}\mathbf{e}^+),$$

$$\mathbf{Y}^{B,\ell m} = -\frac{i}{\sqrt{2}}(_{-1}Y_{\ell,m}\mathbf{e}^- + {}_1Y_{\ell,m}\mathbf{e}^+),$$

$$\mathbf{Y}^{R,\ell m} = Y_{\ell,m}\mathbf{n},$$

see Thorne (1980, Equation (2.22)), where $\mathbf{n} = \frac{\partial}{\partial r}$. Note that Thorne denotes the vectors of the helicity basis by $\mathbf{m}$ and $\mathbf{m}^*$.

Consider the representation $cU_1 \otimes cU_1$ of the group $G = \mathrm{SU}(2)$. Its restriction to H is the direct sum

$$L = tV_2 \oplus 2tV_1 \oplus 2cU_0 \oplus 2V_1 \oplus V_2.$$

The elements of L are rank 2 tensors. Table 3.4 shows that the Hilbert space $L \uparrow \mathrm{SU}(2)$ is the Hilbert direct sum of the subspaces R_ℓ, where R_0 carries 3 copies of the irreducible representation cU_0, R_1 carries 7 copies of cU_1, while each of the spaces R_ℓ, $\ell \geq 2$, carries 9 copies of the irreducible representation cU_ℓ.

The *pure-orbital tensor spherical harmonics* have been constructed by (Mathews 1962; Zerilli 1970b) again with the help of the Clebsch–Gordan coefficients. They are given by Thorne (1980, Equation (2.28)). The 3 pure-orbital tensor spherical harmonics $\mathsf{T}^{00,00}(\theta,\varphi)$, $\mathsf{T}^{11,00}(\theta,\varphi)$, and $\mathsf{T}^{22,00}(\theta,\varphi)$ span the space R_0, the 21 harmonics $\mathsf{T}^{01,1m}(\theta,\varphi)$, $\mathsf{T}^{10,1m}(\theta,\varphi)$, $\mathsf{T}^{11,1m}(\theta,\varphi)$, $\mathsf{T}^{12,1m}(\theta,\varphi)$, $\mathsf{T}^{20,1m}(\theta,\varphi)$, $\mathsf{T}^{21,1m}(\theta,\varphi)$, and $\mathsf{T}^{21,1m}(\theta,\varphi)$ with $-1 \leq m \leq 1$ span the space R_1, while the $18\ell + 9$ harmonics $\mathbf{Y}^{0\ell,\ell m}$, $\mathbf{Y}^{1\ell',\ell m}$ with $\ell - 1 \leq \ell' \leq \ell + 1$, $\mathbf{Y}^{2\ell',\ell m}$ with $\ell \geq 2$, $\ell - 2 \leq \ell' \leq \ell + 1$, and $-\ell \leq m \leq \ell$ span the space R_ℓ.

The *pure-spin tensor spherical harmonics* are constructed by Zerilli (1970b) and given in Thorne (1980) using the unitary transformation similar to Equation (3.36). Of these, we are mostly interested in the harmonics $\mathsf{T}^{E2,\ell m}$ and $\mathsf{T}^{B2,\ell m}$, because they are pure transverse, traceless, and are connected with the spin-weighted spherical harmonics by the relations

$$\mathsf{T}^{E2,\ell m} = \frac{1}{\sqrt{2}}(_{-2}Y_{\ell,m}\mathbf{e}^- \otimes \mathbf{e}^- + {}_2Y_{\ell,m}\mathbf{e}^+ \otimes \mathbf{e}^+),$$

$$\mathsf{T}^{B2,\ell m} = -\frac{i}{\sqrt{2}}(_{-2}Y_{\ell,m}\mathbf{e}^- \otimes \mathbf{e}^- - {}_2Y_{\ell,m}\mathbf{e}^+ \otimes \mathbf{e}^+),$$

$$(3.37)$$

see Thorne (1980, Equation (2.38)). Naturally, $\mathsf{T}^{E2,\ell m}$ (resp., $\mathsf{T}^{B2,\ell m}$) have electric (resp., magnetic) type parity.

The reason for our interest is as follows. Denote by $\tilde{\mathsf{T}}^{E2,\ell m}$ and $\tilde{\mathsf{T}}^{B2,\ell m}$ the projection of the harmonics $\mathsf{T}^{E2,\ell m}$ and $\mathsf{T}^{B2,\ell m}$ to the tensor square of the two-dimensional linear subspace of cU_1 generated by the vectors of the helicity basis. The tensor spherical harmonics $Y^G_{(\ell m)}$ and $Y^C_{(\ell m)}$ used by Kamionkowski, Kosowsky, and Stebbins (1997) for the spectral expansion of the linear polarisation matrix of the CMB are identical to the above projections:

$$Y^G_{(\ell m)} = \tilde{\mathsf{T}}^{E2,\ell m}, \qquad Y^C_{(\ell m)} = \tilde{\mathsf{T}}^{B2,\ell m}. \tag{3.38}$$

Indeed, on the one hand, the fifth equation in Thorne (1980, Equation (2.31)) has the form $\mathsf{f}_{\ell m} = \mathsf{T}^{E2,\ell m}$. On the other hand, we compare Kamionkowski, Kosowsky, and Stebbins (1997, Equation (2.20)) and Zerilli (1970a, Equation (A2j)) and obtain $Y^G_{\ell m} = \tilde{\mathsf{f}}_{\ell m}$. Similarly, the fourth equation in Thorne (1980, Equation (2.31)) has the form $\mathsf{d}_{\ell m} = -\mathrm{i}\mathsf{T}^{E2,\ell m}$, while the comparison of Kamionkowski, Kosowsky, and Stebbins (1997, Equation (2.21)) and Zerilli (1970a, Equation (A2h)) shows that $\tilde{\mathsf{d}}_{\ell m} = -\mathrm{i}Y^C_{\ell m}$.

In the above notation, the letter G (resp., C) stands for 'gradient' (resp., 'curl').

3.4.3 The Bases

In this subsection, the symbol S denotes the two-dimensional complex linear space of spinors, s a positive integer or half-integer number. The symbols $\mathbf{e}_\theta$ and $\mathbf{e}_\varphi$ denote the orthonormal basis of a tangent plane to the sphere S^2 introduced in Example 77.

Example 79. Consider the Clifford algebra $\mathrm{Cl}_2 = \mathbb{C}(2)$. Its Dirac spinor representation acts on S by Equation (1.23): $\rho(A)\varphi = A\varphi$, $A \in \mathrm{Cl}_2$, $\varphi \in S$. By Example 11, its restriction to the Clifford algebra $\mathrm{Cl}_2^+ = \mathbb{C} \oplus \mathbb{C}$ splits into two irreducible components. The left-handed Weyl spinor representation ρ_+ (resp., the right-handed Weyl spinor representation ρ_-) acts in the $+1$-eigenspace $S_+ = \mathbb{C}^1$ (resp., -1-eigenspace $S_- = \mathbb{C}^1$) of the physical chirality operator Γ_3 given by Equation (1.13): $\Gamma_3 = \left(\begin{smallmatrix} 1 & 0 \\ 0 & -1 \end{smallmatrix}\right)$. The restriction of ρ to the spinor group $H = \mathrm{Spin}(2) = \mathrm{U}(1)$ is the direct sum $V_{1/2} \oplus tV_{1/2}$.

Consider the representation $V_s \oplus tV_s$ instead. The space $S^* \otimes S$ is the space of 2×2 matrices with complex entries. It carries the representation

$$(tV_s \oplus V_s) \otimes (V_s \oplus tV_s) = 2cU_0 \oplus tV_{2s} \oplus V_{2s} \tag{3.39}$$

of the group $U(1)$. In coordinates, the irreducible component tV_{2s} acts in the complex linear space spanned by the Pauli matrix σ_2, V_{2s} in the space spanned by σ_1, $2cU_0$ in the space spanned by σ_0 and σ_3.

In the case of positive integer s, it is possible to eliminate the multiplicity in Equation (3.39). Indeed, the space S carries the representation cU_s of the group $O(2)$. The space $S^* \otimes S$ carries the representation

$$cU_s \otimes cU_s = cU_0^+ \oplus cU_0^- \oplus cU_{2s}$$

of the group $O(2)$. In coordinates, the irreducible component cU_0^+ acts in the complex linear space spanned by the Pauli matrix σ_0, cU_0^- in the space spanned by σ_3, cU_{2s} in the space spanned by σ_1 and σ_2.

A special basis of the space $S^* \otimes S \uparrow \mathrm{SU}(2)$ is given by the tensor-valued cross-sections

$$_{2s}\mathsf{T}^0_{\ell,m}(\theta,\varphi) = \frac{1}{\sqrt{2}}\sigma_0 Y_{\ell,m}(\theta,\varphi),$$

$$_{2s}\mathsf{T}^1_{\ell,m}(\theta,\varphi) = \frac{1}{2}(\sigma_1 + i\sigma_2)_{-2s}Y_{\ell,m}(\theta,\varphi),$$

$$_{2s}\mathsf{T}^2_{\ell,m}(\theta,\varphi) = \frac{1}{2}(\sigma_1 - i\sigma_2)_{2s}Y_{\ell,m}(\theta,\varphi),$$

$$_{2s}\mathsf{T}^3_{\ell,m}(\theta,\varphi) = -\frac{1}{\sqrt{2}}\sigma_3 Y_{\ell,m}(\theta,\varphi).$$

Example 80. The map $j\colon S \to S$, $j(\zeta_1,\zeta_2)^\top = (\overline{\zeta_2},\overline{\zeta_1})^\top$, clearly commutes with the representation $V_s \oplus tV_s$. The eigenspace S_+ of j that corresponds to the eigenvalue 1 carries the representation rV_s: $e^{i\psi/2} \mapsto \begin{pmatrix} \cos(s\psi) & -\sin(s\psi) \\ \sin(s\psi) & \cos(s\psi) \end{pmatrix}$ of the group H. The character of the representation rV_s is equal to $2\cos(s\psi)$. The space $S_+^* \otimes S_+$ is the space of 2×2 matrices with real entries. The character of the representation $rV_s \otimes rV_s$ that acts by

$$e^{i\psi/2} \cdot A = \begin{pmatrix} \cos(s\psi) & -\sin(s\psi) \\ \sin(s\psi) & \cos(s\psi) \end{pmatrix} A \begin{pmatrix} \cos(s\psi) & \sin(s\psi) \\ -\sin(s\psi) & \cos(s\psi) \end{pmatrix}, \qquad A \in S_+^* \otimes S_+$$

is equal to $4\cos^2(s\psi) = 2 + 2\cos(2s\psi)$. It follows that $rV_s \otimes rV_s = 2U_0 \oplus rV_{2s}$. The component $2U_0$ acts in the real linear space of diagonal 2×2 matrices with real entries, while the component rV_{2s} in the space of 2×2 matrices with real entries and zero main diagonal. A special basis of the space $S_+^* \otimes S_+ \uparrow \mathrm{SU}(2)$ is given by the matrix-valued cross-sections

$$_s\mathsf{T}^{\mathrm{R},1}_{\ell,m}(\theta,\varphi) = \mathbf{e}_- \otimes \mathbf{e}_- Y_\ell^m(\theta,\varphi),$$

$$_s\mathsf{T}^{\mathrm{R},2}_{\ell,m}(\theta,\varphi) = \mathbf{e}_- \otimes \mathbf{e}_+{}_{-s}Z_{\ell m}(\theta,\varphi),$$

$$_s\mathsf{T}^{\mathrm{R},3}_{\ell,m}(\theta,\varphi) = \mathbf{e}_+ \otimes \mathbf{e}_-{}_sZ_{\ell m}(\theta,\varphi),$$

$$_s\mathsf{T}^{\mathrm{R},4}_{\ell,m}(\theta,\varphi) = -\mathbf{e}_+ \otimes \mathbf{e}_+ Y_\ell^m(\theta,\varphi),$$

where $_{\pm s}Z_{\ell m}(\theta,\varphi)$ are the real-valued spin-weighted spherical harmonics (3.29).

The physical gamma matrix $\Gamma_2 = \begin{pmatrix} 0 & -1 \\ 1 & 0 \end{pmatrix}$ defines a quaternionic structure in the complex linear space S. The resulting quaternionic linear space W carries a representation whose character is equal to $2\cos(s\psi)$. Table 3.1 gives $W = qV_s$. The tensor product $W \otimes W$ is a real representation considered in Example 80.

The case of $G = \mathrm{SO}(3)$, $H = \mathrm{SO}(2)$ does not give new bases.

3.5 Physical Application of the Reducible Case

In this section, s is a positive integer, the group $O(2)$ acts in the complex linear space $cU = S^* \otimes S$ by the representation $cU_0^+ \oplus cU_0^- \oplus cU_{2s}$ and in its real subspace U by the representation $U_0^+ \oplus U_0^- \oplus U_{2s}$. The orthonormal basis in both spaces cU and U is given by the Pauli matrices multiplied by $\frac{1}{\sqrt{2}}$. Let $(E_{\mathbb{C}}, \pi_{cU}, S^2)$ and $(E_{\mathbb{R}}, \pi_U, S^2)$ be the corresponding fibre bundles. We will study isotropic random sections of the above bundles and their sub-bundles. The particular case of $s = 1$ (resp., $s = 2$) describes the Stokes parameters of the Cosmic Microwave (resp., Gravitational) Background.

Example 81. This is a version of Malyarenko and Ostoja-Starzewski (2023, Example 7.1). In the helicity basis and angular spherical coordinates, the polarisation matrix of the CMB takes the form

$$P(\theta, \varphi) = \frac{1}{2} \sum_{i=0}^{3} \sum_{\ell=0}^{\infty} \sum_{m=-\ell}^{\ell} {}_2a_{\ell m}^{(i)} {}_2\mathsf{T}_{\ell,m}^{i}(\theta, \varphi).$$

The random vectors ${}_2\mathbf{a}_{\ell m}$ with components ${}_2a_{\ell m}^{(i)}$, $0 \le i \le 3$, are complex-valued normal. Moreover, we have

$${}_2a_{0m}^{(1)} = {}_2a_{1m}^{(1)} = {}_2a_{0m}^{(2)} = {}_2a_{1m}^{(2)} = 0$$

by the properties of spin-valued spherical harmonics. Other dependencies are reality conditions:

$${}_2a_{\ell-m}^{(0)} = (-1)^m \overline{{}_2a_{\ell m}^{(0)}}, \qquad {}_2a_{\ell-m}^{(3)} = (-1)^m \overline{{}_2a_{\ell m}^{(3)}},$$

$${}_2a_{\ell-m}^{(1)} = (-1)^m \overline{{}_2a_{\ell m}^{(2)}}, \qquad {}_2a_{\ell-m}^{(2)} = (-1)^m \overline{{}_2a_{\ell m}^{(1)}},$$

which follow from Equation (3.27) and the fact that $P(\theta, \varphi)$ is a Hermitian matrix. Finally, the covariance matrix C_ℓ of the random vector ${}_2\mathbf{a}_{\ell m}$ does not depend on m and satisfies the condition

$$\sum_{\ell=0}^{\infty} (2\ell + 1) \operatorname{tr} C_\ell < \infty. \tag{3.40}$$

The polarisation matrix of the Cosmic Gravitational Background takes the form

$$P(\theta, \varphi) = \frac{1}{2} \sum_{i=0}^{3} \sum_{\ell=0}^{\infty} \sum_{m=-\ell}^{\ell} {}_4a_{\ell m}^{(i)} {}_4\mathsf{T}_{\ell,m}^{i}(\theta, \varphi).$$

This time, the multipole coefficients ${}_4a_{\ell m}^{(1)}$ and ${}_4a_{\ell m}^{(2)}$ are equal to 0 for $0 \le \ell \le 3$.

Example 82. This is a version of Malyarenko and Ostoja-Starzewski (2023, Example 7.3). Let $E(\theta, \varphi)$ and $B(\theta, \varphi)$ be the random fields (3.34). The function $(I, E, B, V)^\top(\theta, \varphi)$ is a random section of the Cartesian product $S^2 \times \mathbb{R}^4$. Its spectral expansion takes the form

$$(I, E, B, V)^\top(\theta, \varphi) = \sum_{i=0}^{3} \sum_{\ell=0}^{\infty} a_{\ell m}^{(i)} Y_{\ell m}(\theta, \varphi) \mathbf{e}_i, \qquad (3.41)$$

where $\{\mathbf{e}_i : 0 \leq i \leq 3\}$ is the standard basis of the space $\mathbb{R}^4$. The random vectors $\mathbf{a}_{\ell m}$ with components $a_{\ell m}^{(i)}$, $0 \leq i \leq 3$, are complex-valued normal. Moreover, we have

$$a_{0m}^{(1)} = a_{1m}^{(1)} = a_{0m}^{(2)} = a_{1m}^{(2)} = 0,$$

and the reality conditions

$$a_{\ell -m}^{(i)} = (-1)^m \overline{a_{\ell m}^{(i)}}, \qquad 0 \leq i \leq 3.$$

The 4×4 covariance matrix C_ℓ of the vector $\mathbf{a}_{\ell m}$ does not depend on m and satisfies condition (3.40).

The diagonal (resp., off-diagonal) elements of the matrix C_ℓ are four *auto-power spectra* (resp., *cross-power spectra*). It is customary to replace the auto-power spectrum C_ℓ^{II} with C_ℓ^{TT}, where T is the absolute temperature of the CMB. The two quantities are connected: the intensity I is proportional to the fourth power of the absolute temperature T by the *Stefan–Boltzmann law*.

The random fields $T(\theta, \varphi)$ and $E(\theta, \varphi)$ are scalars, while $B(\theta, \varphi)$ and $V(\theta, \varphi)$ are pseudo-scalars. That is, the former (resp., the latter) belongs to the one-dimensional space where the representation U_0^+ (resp., U_0^-) acts. By parity conservation, we expect that the cross-power spectra C_ℓ^{TB}, C_ℓ^{TV}, C_ℓ^{EB}, and C_ℓ^{EV} vanish.

It is possible to repeat the calculations of Example 77 for the case of $s = 2$. We obtain the following result. If

$$\sum_{\ell=4}^{\infty} \frac{(2\ell+1)(\ell+4)!}{(\ell-4)!} C_{4\ell} < \infty,$$

then the random fields

$$E(\mathbf{n}) = \sum_{\ell=4}^{\infty} \sum_{m=-\ell}^{\ell} e_{\ell m} \sqrt{\frac{(\ell+4)!}{(\ell-4)!}} Y_{\ell m}(\mathbf{n}),$$

$$B(\mathbf{n}) = \sum_{\ell=4}^{\infty} \sum_{m=-\ell}^{\ell} \sqrt{\frac{(\ell+4)!}{(\ell-4)!}} Y_{\ell m}(\mathbf{n}),$$

where

$$e_{\ell m} = \frac{1}{2}(a_{\ell m}^{(4)} + a_{\ell m}^{(-4)}), \qquad b_{\ell m} = -\frac{i}{2}(a_{\ell m}^{(4)} - a_{\ell m}^{(-4)})$$

are real-valued isotropic random fields on the sphere S^2. The spectral expansion of the Cosmic Gravitational Background is given by Equation (3.41), but this time

$$a_{0m}^{(1)} = a_{1m}^{(1)} = a_{2m}^{(1)} = a_{3m}^{(1)} = 0,$$

and similarly for $a_{\ell m}^{(2)}$, $0 \le \ell \le 3$.

Example 83. Consider the linear polarisation part of Equation (3.35). It takes the form

$$P(\theta, \varphi) = \frac{1}{2} \begin{pmatrix} Q(\theta, \varphi) & U(\theta, \varphi) \\ U(\theta, \varphi) & -Q(\theta, \varphi) \end{pmatrix}.$$

This is a random section of the fibre bundle $\xi[cU_2]$ over the sky sphere. A special basis in the Hilbert space $cU_2 \uparrow O(3)$ of the square-integrable cross-sections of the above bundle is given by the pure transverse and traceless tensor spherical harmonics $\tilde{T}^{E2,\ell m}$ and $\tilde{T}^{B2,\ell m}$, or, equivalently, by the gradient and curl tensor spherical harmonics $Y^G_{(\ell m)}$ and $Y^C_{(\ell m)}$ through the equalities (3.38). Specifically, the multipole expansion of the so defined polarisation matrix has either the form

$$P(\theta, \varphi) = \sum_{\ell=2}^{\infty} \sum_{m=-\ell}^{\ell} [a^G_{(\ell m)} Y^G_{(\ell m)}(\theta, \varphi) + a^C_{(\ell m)} Y^C_{(\ell m)}(\theta, \varphi)], \tag{3.42}$$

where we absorbed the coefficient $\frac{1}{2}$ into the sum, following Kamionkowski, Kosowsky, and Stebbins (1997), or the equivalent form

$$P(\theta, \varphi) = \sum_{\ell=2}^{\infty} \sum_{m=-\ell}^{\ell} [a^G_{(\ell m)} \tilde{T}^{E2,\ell m}(\theta, \varphi) + a^C_{(\ell m)} \tilde{T}^{B2,\ell m}(\theta, \varphi)], \tag{3.43}$$

which follows from Equation (3.38).

The multipoles $a^G_{(\ell m)}$ and $a^C_{(\ell m)}$ satisfy the reality condition

$$\overline{a^G_{(\ell m)}} = (-1)^m a^G_{(\ell-m)}, \qquad \overline{a^C_{(\ell m)}} = (-1)^m a^C_{(\ell-m)},$$

while the multipoles $a^{E2,\ell m}$ and $a^{B2,\ell m}$ satisfy

$$\overline{a^{E2,\ell m}} = (-1)^m a^{E2,\ell-m}, \qquad \overline{a^{B2,\ell m}} = (-1)^m a^{B2,\ell-m}.$$

Theorem 26. *Equations (3.32) and (3.42) represent two equivalent expansions of the linear polarisation of the CMB.*

Proof. Equation (3.37) gives

$$\begin{aligned} P(\theta, \varphi) = \frac{1}{\sqrt{2}} \sum_{\ell=2}^{\infty} \sum_{m=-\ell}^{\ell} & [(a^G_{(\ell m)} - ia^C_{(\ell m)})_{-2} Y_{\ell,m}(\theta, \varphi) \mathbf{e}^- \otimes \mathbf{e}^- \\ & + (a^G_{(\ell m)} + ia^C_{(\ell m)})_2 Y_{\ell,m}(\theta, \varphi) \mathbf{e}^+ \otimes \mathbf{e}^+)]. \end{aligned}$$

Using the definition of the helicity basis (3.31), we obtain, that the tensors $\mathbf{e}^- \otimes \mathbf{e}^-$ and $\mathbf{e}^+ \otimes \mathbf{e}^+$ in the canonical basis have the form

$$\mathbf{e}^- \otimes \mathbf{e}^- = \frac{1}{2}\begin{pmatrix} 1 & i \\ i & -1 \end{pmatrix}, \qquad \mathbf{e}^+ \otimes \mathbf{e}^+ = \frac{1}{2}\begin{pmatrix} 1 & -i \\ -i & -1 \end{pmatrix}.$$

It follows that

$$Q(\theta, \varphi) = \frac{1}{2\sqrt{2}}\sum_{\ell=2}^{\infty}\sum_{m=-\ell}^{\ell}(a^G_{(\ell m)} - ia^C_{(\ell m)})_{-2}Y_{\ell,m}(\theta,\varphi)$$
$$+ (a^G_{(\ell m)} + ia^C_{(\ell m)})_2 Y_{\ell,m}(\theta,\varphi),$$

$$iU(\theta, \varphi) = \frac{1}{2\sqrt{2}}\sum_{\ell=2}^{\infty}\sum_{m=-\ell}^{\ell}(-a^G_{(\ell m)} + ia^C_{(\ell m)})_{-2}Y_{\ell,m}(\theta,\varphi)$$
$$+ (a^G_{(\ell m)} + ia^C_{(\ell m)})_2 Y_{\ell,m}(\theta,\varphi),$$

and finally

$$(Q \pm iU)(\theta, \varphi) = \frac{1}{\sqrt{2}}\sum_{\ell=2}^{\infty}\sum_{m=-\ell}^{\ell}(a^G_{(\ell m)} \pm ia^C_{(\ell m)})_{\pm 2}Y_{\ell,m}(\theta,\varphi),$$

which coincides with Equation (3.32), where

$$a^{(\pm 2)}_{\ell m} = \frac{1}{\sqrt{2}}(a^G_{(\ell m)} \pm ia^C_{(\ell m)}).$$

$\square$

It is possible to write down a similar random multipole expansion for the linear polarisation of the Cosmic Gravitational Background. Indeed, *define* the gradient and curl tensor spherical harmonics by

$$Y^{G,2s}_{(\ell m)}(\theta, \varphi) = \frac{1}{\sqrt{2}}(_{-2s}Y_{\ell,m}\mathbf{e}^- \otimes \mathbf{e}^- + {}_{2s}Y_{\ell,m}\mathbf{e}^+ \otimes \mathbf{e}^+),$$

$$Y^{C,2s}_{(\ell m)}(\theta, \varphi) = -\frac{i}{\sqrt{2}}(_{-2s}Y_{\ell,m}\mathbf{e}^- \otimes \mathbf{e}^- - {}_{2s}Y_{\ell,m}\mathbf{e}^+ \otimes \mathbf{e}^+),$$

For the case of $s = 2$, the expansion

$$P(\theta, \varphi) = \sum_{\ell=4}^{\infty}\sum_{m=-\ell}^{\ell}[a^{G,4}_{(\ell m)}Y^{G,4}_{(\ell m)}(\theta, \varphi) + a^{C,4}_{(\ell m)}Y^{C,4}_{(\ell m)}(\theta, \varphi)]$$

describes the above polarisation.

Example 84. The polarisation matrix of the CMB can be expanded with respect to the basis of Example 80 as follows:

$$P(\theta, \varphi) = \frac{1}{2}\sum_{i=0}^{3}\sum_{\ell=0}^{\infty}\sum_{m=-\ell}^{\ell}{}_2a^{\mathbb{R},(i)}_{\ell m}\,{}_2T^{\mathbb{R},i}_{\ell,m}(\theta, \varphi).$$

The random vectors $_2\mathbf{a}_{\ell m}^{\mathbb{R}}$ with components $_2a_{\ell m}^{\mathbb{R},(i)}$, $0 \le i \le 3$, are normal. Moreover, we have

$$_2a_{0m}^{\mathbb{R},(1)} = {_2a_{1m}^{\mathbb{R},(1)}} = {_2a_{0m}^{\mathbb{R},(2)}} = {_2a_{1m}^{\mathbb{R},(2)}} = 0$$

by the properties of spin-valued spherical harmonics. The covariance matrix C_ℓ of the random vector $_2\mathbf{a}_{\ell m}^{\mathbb{R}}$ does not depend on m and satisfies condition (3.40).

The polarisation matrix of the Cosmic Gravitational Background takes the form

$$P(\theta, \varphi) = \frac{1}{2} \sum_{i=0}^{3} \sum_{\ell=0}^{\infty} \sum_{m=-\ell}^{\ell} {_4a_{\ell m}^{\mathbb{R},(i)}} {_4T_{\ell,m}^i}(\theta, \varphi).$$

This time, the multipole coefficients $_4a_{\ell m}^{\mathbb{R},(1)}$ and $_4a_{\ell m}^{\mathbb{R},(2)}$ are equal to 0 for $0 \le \ell \le 3$.

Example 85. The function $(I, E, B, V)^\top(\theta, \varphi)$ of Example 82 can be expanded as follows:

$$(I, E, B, V)^\top(\theta, \varphi) = \sum_{i=0}^{3} \sum_{\ell=0}^{\infty} a_{\ell m}^{\mathbb{R},(i)} Y_\ell^m(\theta, \varphi) \mathbf{e}_i.$$

The random vectors $\mathbf{a}_{\ell m}^{\mathbb{R}}$ with components $a_{\ell m}^{\mathbb{R},(i)}$, $0 \le i \le 3$, are normal. Moreover, we have

$$a_{0m}^{\mathbb{R},(1)} = a_{1m}^{\mathbb{R},(1)} = a_{0m}^{\mathbb{R},(2)} = a_{1m}^{\mathbb{R},(2)} = 0.$$

The 4×4 covariance matrix C_ℓ of the vector $\mathbf{a}_{\ell m}^{\mathbb{R}}$ does not depend on m and satisfies condition (3.40).

The corresponding expansion for the Cosmic Gravitational Background has the form, but this time

$$a_{0m}^{\mathbb{R},(1)} = a_{1m}^{\mathbb{R},(1)} = a_{2m}^{\mathbb{R},(1)} = a_{3m}^{\mathbb{R},(1)} = 0,$$

and similarly for $a_{\ell m}^{\mathbb{R},(2)}$, $0 \le \ell \le 3$.

3.6 Concluding Remarks: Where to Go Next?

Here, we describe a couple of directions in current research where the knowledge obtained from this book, may be useful. The following information is not exhaustive, many other interesting directions exist.

An interesting direction is connected with the investigations in the theory of geometry and topology of random sections of several bundles over the sphere. For a survey in this area, see the book Marinucci and Peccati (2011), the survey papers (Carrón Duque and Marinucci 2023; Marinucci 2023), and the research paper Lerario et al. (2022).

Both scalar and spin random fields on the sphere can be analysed with the help of wavelets and their relatives, see (Durastanti 2017; Geller and Marinucci 2010, 2011; Iglewska-Nowak 2019, 2022; Lê Gia et al. 2017; McEwen, Durastanti, and Wiaux 2018)

The stochastic nature of the cosmic backgrounds is caused by two groups of effects: primary and secondary. An effect is *primary* if it influenced the background *before* its decoupling from the rest of the Universe, and *secondary* otherwise. All of them carry important cosmological information. See the surveys (Bucher 2015; Cabella and Marinucci 2009; Puget 2021; Straumann 2006).

The problems of extracting cosmological information from primary effects are considered in (Chandra and Pal 2018; Clarke, Copeland, and Moss 2020; Giovannini 2018; Guan and Kosowsky 2022; Inomata and Kamionkowski 2019; Ota 2022; Sachs and Wolfe 2007; Switzer and Watts 2016) among others.

The problems of extracting cosmological information from secondary effects are considered in (Beck, Fabbian, and Errard 2018; Cai et al. 2023; Calabrese et al. 2015; Cayuso et al. 2023; Prince et al. 2018; Wibig and Wolfendale 2016) among others.

According to the ΛCDM model, the Cosmic Microwave Background is a single realisation of a Gaussian isotropic random section of a certain homogeneous bundle over the sky sphere. Possible deviations from Gaussianity and isotropy can point to new physics beyond the model. For investigations in this direction, see (Hamann et al. 2021; Komatsu 2010; Marinucci 2004; Ravenni et al. 2017) among others.

Currently, the most intensive application area of stochastic models of the CMB is the solution of two 'opposite' problems. The first one is calculating the theoretical power spectrum of the CMB given the parameters of the ΛCDM model. The so-called cosmological Boltzmann codes are actively used for that purposes. For introduction to that area of research, see (Blas, Lesgourgues, and Tram 2011; Das and Phan 2020; Doran 2005; Seljak and Zaldarriaga 1996; Zaldarriaga, Seljak, and Bertschinger 1998) among others.

The second problem, 'opposite' to the first one, is to calculate the best-fit cosmological parameters from the CMB observation data. See (Das and Souradeep 2014; Lewis 2013; Doran and Müller 2004; Lewis and Bridle 2002; Paykari, Starck, and Fadili 2013).

3.7   Bibliographical Remarks

The treatment of isotropic random cross-sections of homogeneous fibre bundles in Section 3.1 is a slightly modified version of that in Malyarenko and Ostoja-Starzewski (2023). See also Leonenko and Sakhno (2012).

The Wigner D-functions and spherical harmonics are subject of Special Functions. For this theory, see (Torres del Castillo 2003; Vilenkin and Klimyk 1991, 1992, 1993, 1995)

The Wigner D-functions were introduced by Wigner (1927). Being the matrix entries of irreducible unitary representations of the group SU(2), they depend on the choice of a basis in the linear space, where the representation acts. We follow the choice made by (Brink and Satchler 1993; de-Shalit and Talmi 1963; Edmonds 1957; Messiah 1959; Newton 1966; Rose 1957; Tinkham 1964; Wawrzyńczyk 1984) among others, and the description by Man (2022). Different choices have been made by (Bohr and Mottelson 1998; Davydov 1976; Fano and Racah 1959; Gel'fand and Šapiro 1952; Wigner 1959; Yutsis, Levinsonas, and Vanagas 1962) among others. Wigner (1959) coins the name of Hermann Weyl to the method we used.

The name *spherical harmonics* appeared for the first time in Thomson and Tait (1867). The first author is also known as Lord Kelvin. See also (Atkinson and Han 2012; Byerly 1959; Efthimiou and Frye 2014; Ferrers 1877; Hobson 1955; Müller 1966; MacRobert 1947; Sternberg and Smith 1944) and a very comprehensive review by Thorne (1980).

The adjective 'spin-weighted' is coined to Newman and Penrose (1966), but the spin-weighted spherical harmonics were known at least to Gel'fand and Šapiro (1952). Curtis and Lerner (1978) identified them with cross-sections of a certain fibre bundle. See also (Boyle 2016; Dray 1985, 1986; Goldberg et al. 1967; Gómez et al. 1997; Michel, Plattner, and Seibert 2022; Penrose and Rindler 1987, 1988).

For the clear description of the irreducible real representations of SU(2), see Itzkowitz, Rothman, and Strassberg (1991). For matrix entries of the representations with nonnegative integer ℓ, see also (Blanco, Flórez, and Bermejo 1997; Choi et al. 1999; Gimbutas and Greengard 2009; Ivanic and Ruedenberg 1996, 1998; Pinchon and Hoggan 2007). The $\mathbb{R}^2$-valued spin-weighted spherical harmonics with nonzero spin appeared in Malyarenko and Ostoja-Starzewski (2023).

The theoretical models that describe the polarisation of the Cosmic Microwave Background, appeared in (Kamionkowski, Kosowsky, and Stebbins 1997; Zaldarriaga and Seljak 1997). The first (resp., the second) paper expands the polarisation tensor (resp., the complex polarisation) of the CMB into a multipole expansion with respect to the tensor-valued (resp., spin-weighted) spherical harmonics.

More mathematical description can be found in (Baldi and Rossi 2014; Geller and Marinucci 2010; Malyarenko 2011, 2013). The first cited paper uses the algebraic definition (1.59) of an induced representation, the second one uses methods of Complex Analysis in spirit of Example 64, while the last two use the geometric definition of an induced representation described in Section 1.10. See also a different treatment in Stecconi (2022).

The polarisation maps of the CMB are usually given in terms of the variances of the random multipole coefficients $e_{\ell m}$ and $b_{\ell m}$.

Vector spherical harmonics were also introduced by Regge and Wheeler (1957). Thorne (1980) denotes them by $\boldsymbol{\Psi}^{\ell m}$ and $\boldsymbol{\Phi}^{\ell m}$. They are not orthonormal, but the former one is proportional to $\mathbf{Y}^{E,\ell m}$ and the latter to $\mathbf{Y}^{B,\ell m}$, see Thorne (1980, Equation (2.21)) for exact coefficients. Martel and Poisson (2005) denote the harmonics $\boldsymbol{\Psi}^{\ell m}$ (resp., $\boldsymbol{\Phi}^{\ell m}$) by $Y_A^{\ell m}$ (resp., $X_A^{\ell m}$). Wardell and Warburton (2015) denote the harmonics $\boldsymbol{\Psi}^{\ell m}$ (resp., $\boldsymbol{\Phi}^{\ell m}$) by $Z_A^{\ell m}$ (resp., $X_A^{\ell m}$). Thompson, Wardell, and Whiting (2019) denote them by $Y_a^{E,\ell m}$ and $Y_a^{B,\ell m}$.

In alternative theories of gravitation, gravitational waves may have up to six modes of polarisation, see (Eardley, Lee, and Lightman 1973; Abbott 2018) and a very comprehensive review Clifton et al. (2012). In that case, all six types of tensor spherical harmonics described in Thorne (1980) should be used in random multipole expansions.

Tensor spherical harmonics also come in different notation. Martel and Poisson (2005) denote by $Y_{AB}^{\ell m}$ (resp., $X_{AB}^{\ell m}$) the harmonics proportional to $\mathsf{T}^{E2,\ell m}$ (resp., $\mathsf{T}^{B2,\ell m}$), see their Appendix A for exact coefficients. Thompson, Wardell, and Whiting (2019) denote them by $T_{ab}^{E2,\ell m}$ and $T_{ab}^{B2,\ell m}$, see also Pound and Wardell (2020).

Observe that the terms 'electric-type' parity and 'magnetic-type' parity are used in different senses in some papers, see Zerilli (1970a, Appendix B).

For higher rank spherical harmonics, see (Freeden, Michel, and Simons 2018; Hill 1954)

Currently, the auto-power spectra C_ℓ^{TT} (resp., C_ℓ^{EE} as well as the cross-power spectrum C_ℓ^{TE}) of the CMB are measured with good accuracy up to the multipole $\ell = 2508$ (resp., $\ell = 1996$), see Aghanim (2020a). The other possible auto-power spectra are not yet discovered. That is, the statistical hypothesis $C_\ell^{BB} = C_\ell^{VV} = 0$ cannot be rejected.

The electric-type auto-power spectrum C_ℓ^{EE} of the CMB is induced by the physical process called *Thomson scattering*, see Durrer (2020, Section 5.3). This is a primary effect that does not introduce the circular polarisation. The (until now) hypothetical magnetic-type auto-power spectrum C_ℓ^{BB} can be induced by inflationary gravitational waves. Its discovery, if it happens, would become a direct support of the inflation theory, see Kamionkowski and Kovetz (2016).

In contrast to the case of the CMB, the multipole coefficients of the Cosmic Gravitational Background may depend on the frequency f of the gravitational waves. In that case, its multipole expansion takes the form

$$(I, E, B, V)^\top(\theta, \varphi; f) = \sum_{i=0}^{3} \sum_{\ell=0}^{\infty} a_{\ell m}^{(i)}(f) Y_{\ell m}(\theta, \varphi) \mathbf{e}_i.$$

Again, we expect that the cross-power spectra C_ℓ^{TB}, C_ℓ^{TV}, C_ℓ^{EB}, and C_ℓ^{EV} vanish.

A

Additional Topics

Several mathematical concepts described in the main text can be formulated using different 'mathematical languages'. One of these languages, the commutative diagrams, is described in Section A.1. Non-formally, a commutative diagram is a collection of maps between sets in which all map compositions starting from the same set, say A, and ending with the same set, say B, give the same result.

Another 'language' has been established in order to give the exact sense to the expressions like 'there are no natural isomorphism between a finite-dimensional linear space and its dual, but the natural isomorphism to its second dual exists'. This leads to a general theory of mathematical structures where the main notions are *categories* and *functors*. We give a short introduction to the above theory is Section A.2.

The constructions of many mathematical objects are complicated, especially for physicists and other applied specialists. Introducing the concept of an *universal property* makes many such constructions easier to understand. We consider these questions in Section A.3.

Finally, a rigourous description of spin manifolds using some concepts of algebraic topology is introduced in Section A.4.

A.1 Commutative Diagrams

A *formal definition* of a commutative diagram may be formulated using categories and functors, see (Riehl 2016) and Section A.2.

Non-formally, a *diagram* is a directed graph. It is *commutative* if any two paths of arrows with common source and target have the same composite.

Example 86. Definition 15 of equivalent representations of algebras can be drawn as

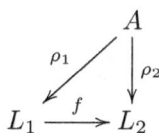

$$
\begin{array}{ccc}
 & & A \\
 & \swarrow^{\rho_1} & \downarrow^{\rho_2} \\
L_1 & \xrightarrow{f} & L_2
\end{array}
$$

In this diagram, there are two paths of arrows with common source A and common target L_2. They have the same composite, which means that $f \circ \rho_1 = \rho_2$.

Example 87. Similarly, Definition 34 of isomorphic coverings can be drawn as

$$
\begin{array}{ccc}
E_1 & \xrightarrow[u^{-1}]{\quad u \quad} & E_2 \ , \\
{\scriptstyle \pi_1} \downarrow & \swarrow {\scriptstyle \pi_2} & \\
B & &
\end{array}
$$

while Definition 55 of equivalent principal bundles becomes

$$
\begin{array}{ccc}
P_1 & \xrightarrow[\Pi^{-1}]{\quad \Pi \quad} & P_2 \ . \\
{\scriptstyle \pi_1} \downarrow & \swarrow {\scriptstyle \pi_2} & \\
M & &
\end{array}
$$

Example 88. Definition 87 of a spin manifold can be drawn as

$$
\begin{array}{ccc}
\tilde{P} & \xrightarrow{\ \Pi\ } & P \\
{\scriptstyle \tilde{\pi}} \downarrow & \swarrow {\scriptstyle \pi} & \\
B & &
\end{array}
$$

Example 89. The relation between the operators ∂ and $\eth$ of Example 64 can be drawn as follows.

$$
\begin{array}{ccc}
\mathcal{S}(-2s) & \xrightarrow{\ \partial\ } & \mathcal{S}(-2s-2) \\
{\scriptstyle \times|\zeta|^{2s}} \uparrow & & \downarrow {\scriptstyle \times|\zeta|^{-2s+2}} \\
\mathcal{B}(s) & \xrightarrow{\ \eth\ } & \mathcal{B}(s+1)
\end{array}
$$

According to (Eastwood and Tod 1982),

'...$\eth$ is just ∂ in disguise'.

Similarly, the relation between the operators $\bar{\partial}$ and $\bar{\eth}$ is as follows.

$$
\begin{array}{ccc}
\mathcal{B}(s-1) & \xleftarrow{\ \bar{\eth}\ } & \mathcal{B}(s) \\
{\scriptstyle \times|\zeta|^{2s+2}} \uparrow & & \downarrow {\scriptstyle \times|\zeta|^{2s}} \\
\mathcal{S}(-2s+2) & \xleftarrow{\ \bar{\partial}\ } & \mathcal{S}(-2s)
\end{array}
$$

A.2   Categories

Let U be a finite-dimensional real linear space and $U^* = \mathrm{Hom}_{\mathbb{R}}(L, \mathbb{R}^1)$ be its *dual space*. The spaces U and U^* have the same dimension. It is clear that they are isomorphic. However, an isomorphism between U and U^* cannot be constructed without using additional structures on U. In other words, any such construction is not 'natural'.

On the other hand, a 'natural' isomorphism between U and the second conjugate space U^{**} is constructed in Subsection 1.2.2.

In order to give the exact sense to the above and similar statements, a special language has been established.

Definition 118. A *category* C consists of data and conditions. The data are:

- a collection Ob C whose elements are called the *objects* of the category C;

- a collection Mor C whose elements are called the *morphisms* of the category C;

- for every ordered pair $\langle X, Y \rangle$ of objects, a set $\mathrm{Hom}_{\mathsf{C}}(X, Y)$ whose elements are called morphisms from X to Y and are denoted by $f \colon X \to Y$ or $X \xrightarrow{f} Y$;

- for every ordered triple $\langle X, Y, Z \rangle$ of objects, a map

$$\mathrm{Hom}_{\mathsf{C}}(X, Y) \times \mathrm{Hom}_{\mathsf{C}}(Y, Z) \to \mathrm{Hom}_{\mathsf{C}}(X, Z), \qquad (f, g) \mapsto fg,$$

which maps the pair (f, g) to the composition fg.

The conditions are:

- foe every morphism f of C there exist unique objects X, Y of C such that $f \in \mathrm{Hom}_{\mathsf{C}}(X, Y)$;

- the composition of morphisms is associative;

- for any object X of C there exists an identity morphism 1_X of C such that for all objects Y, Z of C, every morphism $f \in \mathrm{Hom}_{\mathsf{C}}(X, Y)$, $g \in \mathrm{Hom}_{\mathsf{C}}(Z, X)$ we have $1_X f = f$ and $g 1_X = g$.

The difference between *collections* and *sets* is discussed in axiomatic set theory, see, e.g., Cenzer et al. (2020).

Definition 119. A morphism $f \in \mathrm{Hom}_{\mathsf{C}}(X, Y)$ is called an *isomorphism* if there exists a morphism $g \in \mathrm{Hom}_{\mathsf{C}}(Y, X)$ such that $fg = 1_X$ and $gf = 1_Y$.

Example 90. The objects of the category Set are sets, the elements of the set $\mathrm{Hom}_{\mathsf{Set}}(X, Y)$ are maps $f \colon X \to Y$.

Example 91. The objects of the category Gr are groups, the elements of the set $\mathrm{Hom}_{\mathsf{Gr}}(G, H)$ are maps $f \colon G \to H$ that respect the group multiplication: $f(g_1 g_2) = f(g_1) f(g_2)$.

Example 92. The objects of the category $\mathsf{Lin}_{\mathbb{K}}$ are right finite-dimensional linear spaces over a skew field $\mathbb{K}$. The elements of the set $\mathrm{Hom}_{\mathbb{K}}(L_1, L_2)$ are morphisms from L_1 to L_2.

Let C and D be two categories.

Definition 120. An *object map* is a map T which assigns an object TX of D to each object X of C.

Example 93. Let $\mathsf{C} = \mathsf{D} = \mathsf{Lin}_{\mathbb{R}}$. Put $TU = U^*$.

We observe that the object map $T = *$ is natural. This means that the *morphism map* can be constructed. It is customary to denote both functions by the same symbol. A morphism map assigns a morphism Tf of D to each morphism f of C. For any morphism f of $\mathsf{Lin}_{\mathbb{R}}$ put $Tf = f^*$, the conjugate operator. We see that $1_X^* = 1_{X^*}$ and whenever the composition fg is defined in $\mathsf{Lin}_{\mathbb{R}}$, we have $(fg)^* = g^* f^*$.

Definition 121. A pair, that consists of an object map T and a morphism map T is called a *contravariant functor* if

$$T1_X = 1_{TX}, \qquad T(fg) = T(g)T(f)$$

for all objects X of C and whenever the composition fg is defined in C.

Example 94. The map $*$ is a contravariant functor from the category $\mathsf{Lin}_{\mathbb{R}}$ to itself called the *dual functor*.

For every real linear space U, put $TU = \mathrm{Hom}_{\mathbb{R}}(U, U_0)$, where U_0 is a fixed linear space. For every morphism $f \colon U_1 \to U_2$, let Tf be the morphism from $\mathrm{Hom}_{\mathbb{R}}(U_1, U_0)$ to $\mathrm{Hom}_{\mathbb{R}}(U_2, U_0)$ acting by $Tf(g) = gf$. We obtain a contravariant functor.

It is customary to present functors with commutative diagrams of arrows, see Section A.1. For example:

$$
\begin{array}{ccc}
U_1 & \xrightarrow{\ f\ } & U_2 \\
\downarrow{\scriptstyle *} & & \downarrow{\scriptstyle *} \\
U_1^* & \xleftarrow{\ f^*\ } & U_2^*
\end{array}
$$

Definition 122. A pair, that consists of an object function T and a morphism function T is called a *covariant functor* if

$$T1_X = 1_{TX}, \qquad T(fg) = T(f)T(g)$$

for all objects X of C and whenever the composition fg is defined in C.

Example 95. The *identity functor* id_C that maps every object and every morphism of a category C to itself, is a covariant functor.

The *constant functor*, that maps every linear space $U \in \text{Ob Lin}_\mathbb{R}$ to a fixed space $U_0 \in \text{Ob Lin}_\mathbb{R}$ and every morphism f of $\text{Lin}_\mathbb{R}$ to 1_{U_0} is a covariant functor.

For every real linear space U, put $TU = \text{Hom}_\mathbb{R}(U_0, U)$. For every morphism $f \colon U_1 \to U_2$, let Tf be the morphism from $\text{Hom}_\mathbb{R}(U_0, U_1)$ to $\text{Hom}_\mathbb{R}(U_0, U_2)$ acting by $Tf(g) = fg$. We obtain a covariant functor.

For every real linear space U, put $TU = U_0 \otimes U$. For every morphism $f \colon U_1 \to U_2$, let Tf be the morphism from $U_0 \otimes U_1$ to $U_0 \otimes U_2$ acting on the tensor product of vectors by $Tf(\mathbf{x}_0 \otimes \mathbf{x}_1) = \mathbf{x}_0 \otimes f(\mathbf{x}_1)$. We obtain a covariant functor.

Similarly, for every real linear space U, put $TU = U \otimes U_0$. For every morphism $f \colon U_1 \to U_2$, let Tf be the morphism from $U_1 \otimes U_0$ to $U_2 \otimes U_0$ acting on the tensor product of vectors by $Tf(\mathbf{x}_1 \otimes \mathbf{x}_0) = f(\mathbf{x}_1) \otimes \mathbf{x}_0$. We obtain a covariant functor.

The *double dual functor* takes a linear space U to its double dual U^{**} and a morphism $f \in \text{Hom}_\mathbb{R}(U_1, U_2)$ to the morphism $f^{**} \in \text{Hom}_\mathbb{R}(U_1^{**}, U_2^{**})$ given by $f^{**}(g)h = ghf$. It is a covariant functor.

For every real linear space U, put $TU = cU$. This object function can be extended to a covariant functor from $\text{Lin}_\mathbb{R}$ to $\text{Lin}_\mathbb{C}$ by defining for every morphism $f \in \text{Hom}_\mathbb{R}(U_1, U_2)$ the morphism $cf \in \text{Hom}_\mathbb{C}(cU_1, cU_2)$ as the unique $\mathbb{C}$-linear map such that for $\mathbf{x} \in U_1$ and $\zeta \in \mathbb{C}$, we have

$$cf(\mathbf{x} \otimes \zeta) = f(\mathbf{x}) \otimes \zeta.$$

It is well-defined. Indeed, if $\mathbf{x} \in U_1$, $\alpha \in \mathbb{R}$, and $\zeta \in \mathbb{C}$, then

$$cf(\alpha\mathbf{x} \otimes \zeta) = f(\alpha\mathbf{x}) \otimes \zeta = \alpha f(\mathbf{x}) \otimes \zeta = f(\mathbf{x}) \otimes (\alpha\zeta) = cf(\mathbf{x} \otimes (\alpha\zeta)).$$

The functor c (resp., r) is called *extension of scalars* (resp., *restriction of scalars*).

Similarly, for every complex linear space V, put $TV = qV$. This object function can be extended to a covariant functor from $\text{Lin}_\mathbb{C}$ to $\text{Lin}_\mathbb{H}$, the map $W \mapsto c'W$ can be extended to a covariant functor from $\text{Lin}_\mathbb{H}$ to $\text{Lin}_\mathbb{C}$, the map $V \mapsto rV$ can be extended to a covariant functor from $\text{Lin}_\mathbb{C}$ to $\text{Lin}_\mathbb{R}$, and the map $V \mapsto tV$ can be extended to a covariant functor from $\text{Lin}_\mathbb{C}$ to itself.

Let T_1 and T_2 be two covariant functors from C to D. Assume that for any object X of C there is a morphism $\varphi(X)$ of D.

Definition 123. φ is called a *natural transformation* from T_1 to T_2 if for every object X of C the morphism $\varphi(X)$ maps $T_1(X)$ to $T_2(X)$, and for any morphism $f \in \text{Hom}_C(X, Y)$ we have

$$T_2(f)\varphi(X) = \varphi(Y)T_1(f).$$

In other words, the diagram

$$
\begin{array}{ccc}
T_1(X) & \xrightarrow{\varphi(X)} & T_2(X) \\
\downarrow {\scriptstyle T_1(f)} & & \downarrow {\scriptstyle T_2(f)} \\
T_1(Y) & \xrightarrow{\varphi(Y)} & T_2(Y)
\end{array}
$$

must be *commutative*.

Similarly, let T_1 and T_2 be two contravariant functors from C to D. Assume that for any object X of C there is a morphism $\varphi(X)$ of D.

Definition 124. φ is called a *natural transformation* from T_1 to T_2 if for every object X of C, the morphism $\varphi(X)$ maps $T_1(X)$ to $T_2(X)$, and for any morphism $f \in \mathrm{Hom}_{\mathsf{C}}(X, Y)$ we have

$$
\varphi(X)T_1(f) = T_2(f)\varphi(Y).
$$

In other words, the diagram

$$
\begin{array}{ccc}
T_1(X) & \xrightarrow{\varphi(X)} & T_2(X) \\
\uparrow {\scriptstyle T_1(f)} & & \uparrow {\scriptstyle T_2(f)} \\
T_1(Y) & \xrightarrow{\varphi(Y)} & T_2(Y)
\end{array}
$$

must be commutative.

Definition 125. Two functors T_1 and T_2 from C to D which are either both covariant or both contravariant are called *naturally isomorphic* if there exists a natural transformation $\varphi\colon T_1 \to T_2$ such that for every object X of C, the morphism $\varphi(X)$ is an isomorphism.

Example 96. The identity functor is naturally isomorphic to the double dual functor. Indeed, define a map φ as follows. For every real linear space U, let $\varphi(U)\colon U \to U^{**}$ be the map

$$
\varphi(U)\mathbf{x}(\mathbf{x}^*) = \mathbf{x}^*(\mathbf{x}), \qquad \mathbf{x} \in U, \quad \mathbf{x}^* \in U^*.
$$

It is proved in any course of linear algebra that the morphism $\varphi(U)$ is an isomorphism. The reader may check that for any $f \in \mathrm{Hom}_{\mathbb{R}}(U_1, U_2)$ the diagram

$$
\begin{array}{ccc}
U_1 & \xrightarrow{\varphi(U_1)} & U_1^{**} \\
\downarrow {\scriptstyle f} & & \downarrow {\scriptstyle f^{**}} \\
U_2 & \xrightarrow{\varphi(U_2)} & U_2^{**}
\end{array}
$$

is commutative. Usually, this situation is described as 'a finite-dimensional real linear space is naturally isomorphic to its double dual.'

On the other hand, the dual functor is contravariant, while the identity functor is covariant. The notion of a natural transformation between these functors is undefined. It makes no sense even to ask whether these functors are naturally isomorphic. In other words, there exist no natural isomorphisms between U and U^*.

Example 97. Consider the object function that sends a finite-dimensional complex linear space V to the space $V \oplus tV$. The reader can check that this function can be extended to a functor from the category $\mathsf{Lin}_\mathbb{C}$ to itself. Denote this functor by $1 + t$.

On the other hand, consider the functor cr whose object function sends V to crV. For every complex linear space V, define a map $\varphi(V) \colon (1+t)V \to crV$ by

$$\varphi(V)(\mathbf{x}, \mathbf{y}) = \mathbf{x} \otimes 1 - \mathbf{x}\mathrm{i} \otimes \mathrm{i} + \mathbf{y} \otimes 1 + \mathbf{y}\mathrm{i} \otimes \mathrm{i}.$$

The reader may check, that the map $\varphi(V)$ is an isomorphism between $(1+t)V$ and crV. Moreover, this isomorphism is natural: for any $f \in \mathrm{Hom}_\mathbb{C}(V_1, V_2)$, the diagram

$$
\begin{array}{ccc}
(1+t)V_1 & \xrightarrow{\varphi(V_1)} & crV_1 \\
\downarrow{\scriptstyle (1+t)f} & & \downarrow{\scriptstyle crf} \\
(1+t)V_2 & \xrightarrow{\varphi(V_2)} & crV_2
\end{array}
$$

is commutative. That is, the functors $1 + t$ and cr are naturally isomorphic.

Similarly, denote by 2 the functor whose object function sends a finite-dimensional quaternionic linear space W to the space $W \oplus W$. Denote by qc' the functor whose object function sends W to $qc'W$.

For every quaternionic linear space W, define a map $\varphi(W) \colon 2W \to qc'W$ by

$$\varphi(W)(\mathbf{x} \oplus \mathbf{y}) = \mathbf{x} \otimes 1 - \mathbf{x}\mathrm{j} \otimes \mathrm{j} + \mathbf{y}\mathrm{i} \otimes 1 + \mathbf{y}\mathrm{k} \otimes \mathrm{j}.$$

The reader may check, that the map $\varphi(W)$ is an isomorphism between $2W$ and $qc'W$. Moreover, this isomorphism is natural: for any $f \in \mathrm{Hom}_\mathbb{H}(W_1, W_2)$ the diagram

$$
\begin{array}{ccc}
2W_1 & \xrightarrow{\varphi(W_1)} & qc'W_1 \\
\downarrow{\scriptstyle 2f} & & \downarrow{\scriptstyle qc'f} \\
2W_2 & \xrightarrow{\varphi(W_2)} & qc'W_2
\end{array}
$$

is commutative. That is, the functors 2 and qc' are naturally isomorphic.

Definition 126. The categories C and D are called *equivalent* if there exist two functors $T_1 \colon \mathsf{C} \to \mathsf{D}$ and $T_2 \colon \mathsf{D} \to \mathsf{C}$ such that the functor $T_1 T_2$ is naturally equivalent to the identity functor in T_2, while the functor $T_2 T_1$ is naturally equivalent to the identity functor in T_1.

Example 98. Let $\mathsf{Lin}_{\mathbb{C},+}$ be the category, where the objects are the pairs (V, j). Here, V is a finite-dimensional complex linear space, and j is a real structure on V. The morphisms of the category $\mathsf{Lin}_{\mathbb{C},+}$ are linear maps that commute with real structures.

We show that the categories $\mathsf{Lin}_{\mathbb{C},+}$ and $\mathsf{Lin}_{\mathbb{R}}$ are equivalent.

Consider the object map s_+ that maps a pair (V, j) to the $(+1)$ eigenspace of j, say V^+. Let (V_1, j_1) and (V_2, j_2) be objects of the category $\mathsf{Lin}_{\mathbb{C},+}$, and let $f \in \mathrm{Hom}_{\mathbb{C}}(V_1, V_2)$ be a morphism of the above category, that is, for every $\mathbf{x} \in V_1$ we have $f(j_1\mathbf{x}) = j_2 f(\mathbf{x})$. For $\mathbf{x} \in V_1^+$ we have: $j_1\mathbf{x} = \mathbf{x}$ and $f(\mathbf{x}) = j_2 f(\mathbf{x})$, that is, $F(V_1^+) \subset V_2^+$. Denote the restriction of f to V_1^+ by $s_+ f$. It is the extension of the object map s_+ to a functor.

On the other hand, consider the object map c that maps a real finite-dimensional linear space U to the complex linear space $U \otimes_{\mathbb{R}} \mathbb{C}$. Define a real structure j on cU by $j(\mathbf{x} \otimes \zeta) = \mathbf{x} \otimes \bar{\zeta}$. Let $f \in \mathrm{Hom}_{\mathbb{R}}(U_1, U_2)$. Define a linear operator $cf \in \mathrm{Hom}_{\mathbb{C}}(cU_1, cU_2)$ by $cf(\mathbf{x} \otimes \zeta) = f(\mathbf{x}) \otimes \zeta$. For this operator, we have

$$cf(j_1(\mathbf{x} \otimes \zeta)) = cf(\mathbf{x} \otimes \bar{\zeta}) = f(\mathbf{x}) \otimes \bar{\zeta} = j_2(f(\mathbf{x}) \otimes \zeta) = j_2 cf(\mathbf{x} \otimes \zeta),$$

that is, cf is a morphism of the category $\mathsf{Lin}_{\mathbb{C},+}$ and the extension of the object map c to a functor.

Consider the map $\varphi \colon \mathrm{Ob}\, \mathsf{Lin}_{\mathbb{C},+} \to \mathrm{Mor}\, \mathsf{Lin}_{\mathbb{R}}$ which maps a pair (V, j) to the morphism $\varphi(V, j) \colon V^+ \to (V^+ \otimes_{\mathbb{R}} \mathbb{C})_+$, $\varphi(V, j)\mathbf{x} = \mathbf{x} \otimes 1$. The reader may check that the above map is an isomorphism and the diagram

$$
\begin{array}{ccc}
V_1^+ & \xrightarrow{\varphi(V_1^+)} & (V_1^+ \otimes_{\mathbb{R}} \mathbb{C})_+ \\
\downarrow{\scriptstyle f} & & \downarrow{\scriptstyle s_+ cf} \\
V_2^+ & \xrightarrow{\varphi(V_2^+)} & (V_2^+ \otimes_{\mathbb{R}} \mathbb{C})_+
\end{array}
$$

is commutative for any $f \in \mathrm{Hom}_{\mathbb{R}}(V_1^+, V_2^+)$. That is, the map φ is a natural isomorphism between the functor $s_+ c$ and the identity functor of the category $\mathsf{Lin}_{\mathbb{R}}$.

On the other hand, consider the map $\psi \colon \mathrm{Ob}\, \mathsf{Lin}_{\mathbb{R}} \to \mathrm{Mor}\, \mathsf{Lin}_{\mathbb{C},+}$ which maps a space $V^+ \otimes_{\mathbb{R}} \mathbb{C}$ to the morphism $\psi(V^+ \otimes_{\mathbb{R}} \mathbb{C}) \colon V^+ \otimes_{\mathbb{R}} \mathbb{C} \to V$, $\psi(V^+ \otimes_{\mathbb{R}} \mathbb{C})(\mathbf{x} \otimes \zeta) = \mathbf{x}\zeta$. The reader may check that the above map is an isomorphism and the diagram

$$
\begin{array}{ccc}
V_1^+ \otimes_{\mathbb{R}} \mathbb{C} & \xrightarrow{\psi(V_1^+ \otimes_{\mathbb{R}}\mathbb{C})} & V_1 \\
\downarrow{\scriptstyle f} & & \downarrow{\scriptstyle s_+ cf} \\
V_2^+ \otimes_{\mathbb{R}} \mathbb{C} & \xrightarrow{\psi(V_2^+ \otimes_{\mathbb{R}}\mathbb{C})} & V_2
\end{array}
$$

is commutative for any $f \in \mathrm{Hom}_{\mathbb{C}}(V_1^+ \otimes_{\mathbb{R}} \mathbb{C}, V_2^+ \otimes_{\mathbb{R}} \mathbb{C})$. That is, the map φ is a natural isomorphism between the functor $s_+ c$ and the identity functor of the category $\mathsf{Lin}_{\mathbb{C},+}$.

Example 99. Let $\mathsf{Lin}_{\mathbb{C},-}$ be the category, where the objects are the pairs (V, j). Here, V is a finite-dimensional complex linear space and j is a quaternionic structure on V. The morphisms of the category $\mathsf{Lin}_{\mathbb{C},-}$ are linear maps that commute with quaternionic structures.

We show that the categories $\mathsf{Lin}_{\mathbb{C},-}$ and $\mathsf{Lin}_{\mathbb{H}}$ are equivalent.

Consider the object map s_- that maps a pair (V, j) to the right quaternionic linear space V^-, where $V^- = V$ as sets, and the product ζq of $\zeta \in V$ and $q = \alpha + \beta i + \gamma j + \delta k$ is defined by Equation (1.48). Let (V_1, j_1) and (V_2, j_2) be objects of the category $\mathsf{Lin}_{\mathbb{C},-}$, and let $f \in \mathrm{Hom}_{\mathbb{C}}(V_1, V_2)$ be a morphism of the above category, that is, for every $\zeta \in V_1$ we have $f(j_1 \zeta) = j_2 f(\zeta)$. Define the map $s_- f \colon V_1^- \to V_2^-$ by $s_- f(\zeta) = f(\zeta)$. To prove that $s_- f \in \mathrm{Hom}_{\mathbb{H}}(V_1^-, V_2^-)$, we calculate

$$
\begin{aligned}
s_- f(\zeta q) &= f(\alpha \zeta + \beta \zeta i + \gamma j_1(\zeta) + \delta j_1(\zeta i)) \\
&= f(\zeta)\alpha + f(\zeta)\beta i + f(j_1 \zeta)\gamma + f(j_1(\zeta i))\delta \\
&= f(\zeta)\alpha + f(\zeta)\beta i + j_2 f(\zeta)\gamma + j_2 f(\zeta i)\delta \\
&= f(\zeta)\alpha + f(\zeta)\beta i + f(\zeta)\gamma j + f(\zeta)\delta k = s_- f(\zeta)q.
\end{aligned}
$$

It is the extension of the object map s_- to a functor.

On the other hand, consider the object map c' that maps a quaternionic finite-dimensional linear space W to the complex linear space $c'W$, where $c'W = W$ as sets, but the skew field of scalars is restricted from $\mathbb{H}$ to $\mathbb{C}$. Define a quaternionic structure j on $c'W$ by $j(q) = qj$. Let $f \in \mathrm{Hom}_{\mathbb{H}}(W_1, W_2)$. Define a linear operator $c'f \in \mathrm{Hom}_{\mathbb{C}}(c'W_1, c'W_2)$ by $e_- f(q) = f(q)$. For this operator, we have

$$
c'f(j_1(q)) = f(j_1 q) = f(qj) = j_2(f(q)) = j_2 c' f(q),
$$

that is, $c'f$ is a morphism of the category $\mathsf{Lin}_{\mathbb{C},-}$ and the extension of the object map c' to a functor.

Consider the map $\varphi \colon \mathrm{Ob}\,\mathsf{Lin}_{\mathbb{C},-} \to \mathrm{Mor}\,\mathsf{Lin}_{\mathbb{H}}$ which maps a pair (V, j) to the morphism $\varphi(V, j) \colon V^- \to s_- c' V^-$, $\varphi(V, j)\zeta = \zeta$. The reader may check that the above map is an isomorphism and the diagram

$$
\begin{array}{ccc}
V_1^- & \xrightarrow{\;\varphi(V_1, j)\;} & V_1 \\
\downarrow{\scriptstyle f} & & \downarrow{\scriptstyle s_- c' f} \\
V_2^- & \xrightarrow{\;\varphi(V_2, j)\;} & V_2
\end{array}
$$

is commutative for any $f \in \mathrm{Hom}_{\mathbb{H}}(V_1^-, V_2^-)$. That is, the map φ is a natural isomorphism between the functor $s_- c'$ and the identity functor of the category $\mathsf{Lin}_{\mathbb{H}}$.

On the other hand, consider the map $\psi \colon \mathrm{Ob}\,\mathsf{Lin}_{\mathbb{H}} \to \mathrm{Mor}\,\mathsf{Lin}_{\mathbb{C},-}$ which maps a space V to the morphism $\psi(V) \colon V \to c' s_- V$, $\psi(V)(\zeta) = \zeta$. The reader may check that the above map is an isomorphism and the diagram

$$
\begin{array}{ccc}
V_1 & \xrightarrow{\ \psi(V_1)\ } & c's_-V_1 \\
\downarrow{\scriptstyle f} & & \downarrow{\scriptstyle c's_-f} \\
V_2 & \xrightarrow{\ \psi(V_1)\ } & c's_-V_2
\end{array}
$$

is commutative for any $f \in \mathrm{Hom}_{\mathbb{H}}(V_1, V_2)$. That is, the map ψ is a natural isomorphism between the functor $c's_-$ and the identity functor of the category $\mathsf{Lin}_{\mathbb{C},-}$.

A.3   Universality

A.3.1   Initial and Terminal Objects

Example 100. Consider the object $\varnothing$ of Set. For an arbitrary set X, the set $\mathrm{Hom}_{\mathsf{Set}}(\varnothing, X)$ is a singleton, its unique element is the 'empty function'. No other set has such a property. We say that the empty set is the unique *initial object* in the category of sets.

On the other hand, an arbitrary singleton $\{x\}$ has the following property: for an arbitrary set Y, the set $\mathrm{Hom}_{\mathsf{Set}}(Y, \{x\})$ is again a singleton: its unique element is a map $f: Y \to \{x\}$ with $f(y) = x$ for all $y \in Y$. Any singleton has this property, but any two singletons $\{x\}$ and $\{y\}$ are isomorphic as sets: the map $f: \{x\} \to \{y\}$ with $f(x) = y$ is the isomorphism. We say that an arbitrary singleton is a *terminal object* in the category of sets, but it is unique only up to an isomorphism.

The reader may check that the group $\{e\}$ that contains only the identity element has both of the above described properties: for an arbitrary group G both sets $\mathrm{Hom}_{\mathsf{Gr}}(\{e\}, G)$ and $\mathrm{Hom}_{\mathsf{Gr}}(G, \{e\})$ are singletons. The group $\{e\}$ is unique up to an isomorphism initial and the unique terminal object of the category Gr. In other words, it is the *zero object* of the above category.

The linear space $\{0\}$ with $0 \in \mathbb{K}$ is the zero object of the category $\mathsf{Lin}_{\mathbb{K}}$.

Definition 127. An object X of C is called *initial* (resp., *terminal*) if for all objects Y of C, the set $\mathrm{Hom}_{\mathsf{C}}(X, Y)$ (resp., $\mathrm{Hom}_{\mathsf{C}}(Y, X)$) is a singleton. X is called *zero object* if it is both initial and terminal.

Theorem 27. *Two initial (resp., terminal) objects of a category are isomorphic.*

Proof. Let X and Y be two initial objects of a category C. By definition, there exist unique morphisms $f: X \to Y$ and $g: Y \to X$. The composition fg is the morphism from Y to itself. Since Y is initial, $fg = 1_Y$. Similarly, $gf = 1_X$, so f and g are two-sided inverse of each other.

Proof for the case of terminal objects is similar and can be left to the reader. □

Remark 14. Some authors use the adjective 'final' in place of 'terminal'.

A.3.2 Comma Categories

Let F be a functor from a category C to a category D, and let X be an object of the category D. The objects of the *comma category* $(X \to F)$ (resp., $(X \leftarrow F)$) are the pairs (Y, f) (resp., (f, Y)), where Y is an object of C, and f is a morphism from X to $F(Y)$ (resp., from $F(Y)$ to X).

Every morphism of the comma category $(X \to F)$ is just a morphism $\tau \in \operatorname{Hom}_{\mathsf{C}}(Y_1, Y_2)$, for which the diagram

$$
\begin{array}{ccc}
X & \xrightarrow{f_1} & F(Y_1) \\
{\scriptstyle f_2} \downarrow & \swarrow {\scriptstyle F\tau} & \\
F(Y_2) & &
\end{array}
$$

is commutative in D.

Similarly, every morphism of the comma category $(X \leftarrow F)$ is just a morphism $\tau \in \operatorname{Hom}_{\mathsf{C}}(Y_1, Y_2)$, for which the diagram

$$
\begin{array}{ccc}
X & \xleftarrow{f_2} & F(Y_1) \\
{\scriptstyle f_1} \uparrow & \nearrow {\scriptstyle F\tau} & \\
F(Y_2) & &
\end{array}
$$

is commutative in D.

Example 101. Let F be the identity functor on the category Set. For an arbitrary set X, the objects of the comma category $(X \to F)$ are the pairs (Y, f), where Y is a set, and where f is an arbitrary map from X to Y. The morphisms of the category $(X \to F)$ are such maps $\tau \colon Y_1 \to Y_2$, that the diagram

$$
\begin{array}{ccc}
X & \xrightarrow{f_1} & Y_1 \\
{\scriptstyle f_2} \downarrow & \swarrow {\scriptstyle \tau} & \\
Y_2 & &
\end{array}
$$

is commutative. Such a map τ exists if and only if for any point $y_1 \in f_1(X)$, the map f_2 maps all points of the inverse image $f_1^{-1}(y_1)$ to the same point $y_2 \in Y_2$.

Indeed, assume, on the one hand, that the above condition holds true. Then, we define a map τ as follows: for any $y_1 \in f_1(X)$ we have $\tau(y_1) = y_2$, otherwise, $f(y_1)$ is an arbitrary point in Y_2.

On the other hand, if the above condition fails, then there exist two different points x and y in X such that $f_1(x) = f_1(y)$, but $f_2(x) \neq f_2(y)$. Had the map τ existed, we should have $\tau(f_1(x)) = f_2(x) = f_2(y)$, which is impossible.

Example 102. Let F be the underlying set functor from $\mathsf{Lin}_{\mathbb{K}}$ to Set. This functor juts 'forgets' all linear structures of the $\mathbb{K}$-linear space L. For an arbitrary set L, the objects of the comma category $(L \to F)$ are the pairs (L_1, f), where L_1 is a finite-dimensional linear space over $\mathbb{K}$, and where f is an *arbitrary* (not necessarily $\mathbb{K}$-linear) map from L to L_1. The morphisms of the category $(L \to F)$ are such *linear* maps $\tau : L_1 \to L_2$, and the diagram

$$
\begin{array}{ccc}
L & \xrightarrow{\;f_1\;} & L_1 \\
{\scriptstyle f_2}\downarrow & \swarrow{\scriptstyle \tau} & \\
L_2 & &
\end{array}
$$

is commutative.

Example 103. Let L_1 and L_2 be two finite-dimensional linear spaces over a field $\mathbb{F}$. Define the objects of the category $\mathsf{Lin}_{\mathbb{F}}^{\times}$ as the finite-dimensional linear spaces over $\mathbb{F}$ *and* the Cartesian product $L_1 \times L_2$. The morphisms of that category are the linear maps between the objects of $\mathsf{Lin}_{\mathbb{F}}$ *and* the sets

$$\mathrm{Hom}_{\mathsf{Lin}_{\mathbb{F}}^{\times}}(L_1 \times L_2, L), \quad \mathrm{Hom}_{\mathsf{Lin}_{\mathbb{F}}^{\times}}(L, L_1 \times L_2) = \varnothing, \quad \mathrm{Hom}_{\mathsf{Lin}_{\mathbb{F}}^{\times}}(L_1 \times L_2, L_1 \times L_2),$$

where, by definition, the first (resp., the third) mentioned set contains all bilinear maps $L_1 \times L_2 \to L$ (resp., only the identity map).

Let F be the inclusion functor from $\mathsf{Lin}_{\mathbb{F}}$ to $\mathsf{Lin}_{\mathbb{F}}^{\times}$. By definition, it maps all objects and morphisms of $\mathsf{Lin}_{\mathbb{F}}$ to themselves.

For the object $L_1 \times L_2$ of $\mathsf{Lin}_{\mathbb{F}}^{\times}$, the objects of the category $(L_1 \times L_2 \to F)$ are the pairs (L, f), where L is a finite-dimensional linear space over $\mathbb{F}$, and where f is a $\mathbb{F}$-linear map from $L_1 \times L_2 \to L$. The morphisms of the category $(L_1 \times L_2 \to F)$ are such $\mathbb{F}$-linear maps $\tau : M_1 \to M_2$, and the diagram

$$
\begin{array}{ccc}
L_1 \times L_2 & \xrightarrow{\;f_1\;} & M_1 \\
{\scriptstyle f_2}\downarrow & \swarrow{\scriptstyle \tau} & \\
M_2 & &
\end{array}
$$

is commutative.

Example 104. Let $\mathsf{A}_{\mathbb{F}}$ be the category whose objects are the finite-dimensional unital associative algebras over $\mathbb{F}$ and whose morphisms are linear maps that respect the identity elements and the multiplication. Let B be a symmetric bilinear form on a finite-dimensional linear space L over $\mathbb{F}$ given by Equation (1.1) (resp., Equation (1.3)) if $\mathbb{F} = \mathbb{R}$ (resp., $\mathbb{F} = \mathbb{C}$).

Define the objects of the category $\mathsf{A}_{\mathbb{F}}^{(L,B)}$ as the finite-dimensional unital associative algebras over $\mathbb{F}$ *and* the set L. The morphisms of that category are linear maps between the objects of $\mathsf{A}_{\mathbb{F}}$ that respect the identity elements and the multiplication *and* the sets

$$\mathrm{Hom}_{\mathsf{A}_{\mathbb{F}}^{(L,B)}}(L, A), \quad \mathrm{Hom}_{\mathsf{A}_{\mathbb{F}}^{(L,B)}}(A, L) = \varnothing, \quad \mathrm{Hom}_{\mathsf{A}_{\mathbb{F}}^{(L,B)}}(L, L),$$

where, by definition, the first (resp., the third) mentioned set contains all linear maps $j\colon V \to A$ satisfying the condition

$$j(\mathbf{x})^2 = -B(\mathbf{x}, \mathbf{x})1_A, \qquad \mathbf{x} \in L,$$

(resp., only the identity map).

Let F be the inclusion functor from $\mathsf{A}_{\mathbb{F}}$ to $\mathsf{A}_{\mathbb{F}}^{(L,B)}$. For the object L of $\mathsf{A}_{\mathbb{F}}^{(L,B)}$, the objects of the category $(L \to F)$ are the pairs (A, j), where A is a finite-dimensional unital associative algebra over $\mathbb{F}$, and where $j \in \mathrm{Hom}_{\mathsf{A}_{\mathbb{F}}^{(L,B)}}(L, A)$. The morphisms of the category $(L \to F)$ are such morphisms $\tau\colon A_1 \to A_2$ of the category $\mathsf{A}_{\mathbb{F}}$, and the diagram

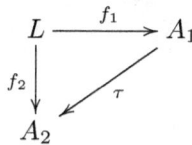

$$
\begin{array}{ccc}
L & \xrightarrow{f_1} & A_1 \\
{\scriptstyle f_2}\downarrow & \swarrow{\scriptstyle \tau} & \\
A_2 & &
\end{array}
$$

is commutative.

A.3.3   Universal Objects and Universal Maps

A comma category $(X \to F)$ may have or not have an initial object. If such an object, say, (Y_0, f_0), exists, then, for *any* object (Y, f) of $(X \to F)$, there is a *unique* morphism $\tau \in \mathrm{Hom}_{(X \to F)}(Y_0, Y)$ for which the diagram

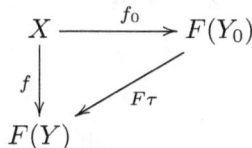

$$
\begin{array}{ccc}
X & \xrightarrow{f_0} & F(Y_0) \\
{\scriptstyle f}\downarrow & \swarrow{\scriptstyle F\tau} & \\
F(Y) & &
\end{array}
$$

is commutative.

In this case, we say that the pair (Y_0, f_0) has the *universal mapping property*, the object Y_0 is called the *universal object*, the map f_0 is a *universal map*, the pair (Y_0, f_0) is a *universal pair* for (X, F).

In other words, a universal pair for (X, F) is an initial object of the comma category $(X \to F)$.

Similarly, a comma category $(X \leftarrow F)$ may have or not have a terminal object. If such an object, say, (f_0, Y_0), exists, then, for *any* object (f, Y) of $(X \leftarrow F)$, there is a *unique* morphism $\tau \in \mathrm{Hom}_{(X \leftarrow F)}(Y, Y_0)$ for which the diagram

$$X \xleftarrow{\quad f_0 \quad} F(Y_0)$$

$$f \Big\uparrow \qquad \nearrow F\tau$$

$$F(Y)$$

is commutative.

In this case, we say that the pair (Y_0, f_0) has the *couniversal mapping property*, the object Y_0 is called the *couniversal object*, the map f_0 is a *couniversal map*, the pair (f_0, Y_0) is a *couniversal pair* for (X, F).

In other words, a couniversal pair for (X, F) is a terminal object of the comma category $(X \leftarrow F)$.

Example 105. A pair (Y_0, f_0) is universal for (X, F), where $(X \to F)$ is the comma category of Example 101 if any map $f \colon X \to Y$ can be uniquely factored through f_0, that is if there exists a unique map $\tau_f \colon Y_0 \to Y$ such that the diagram

$$X \xrightarrow{\quad f_0 \quad} Y_0$$

$$f \Big\downarrow \qquad \swarrow \tau_f$$

$$Y$$

is commutative.

This happens if and only if the map f_0 is one-to-one. Indeed, if the above diagram is commutative, then Example 101 shows that for any point $y_0 \in f_0(X)$, the map f maps all points of the inverse image $f_0^{-1}(y_0)$ to the same point $y \in Y$. This applies to any map f, so f_0 must be injective.

To prove that f_0 is surjective, assume that it is not. Then any map τ_f for which the diagram is commutative may be changed by changing its value on an arbitrary element of the set $Y_0 \setminus f_0(X)$. The uniqueness of τ_f fails.

Conversely, if f_0 is one-to-one, then the diagram is commutative if and only if $\tau_f = f \circ f_0^{-1}$.

In other words, any one-to-one map $f_0 \colon X \to Y_0$ has the universal mapping property, the set Y_0 is the universal object, and the map f_0 is universal.

Example 106. Let $\mathcal{B}$ be a basis of a finite-dimensional linear space L over a skew field $\mathbb{K}$, and let $i \colon \mathcal{B} \to L$ be the inclusion map. The pair (L, i) is universal for $(\mathcal{B}, F)$, where F is the underlying set functor of Example 102.

Indeed, we need to prove that any linear map $f \colon L \to L_1$ is uniquely determined by its values on the elements of $\mathcal{B}$ and these values can be assigned arbitrarily. This is a well-known result of linear algebra.

Example 107. Let L_1 and L_2 be two finite-dimensional linear spaces over a field $\mathbb{F}$. Define the map $f_0 \colon L_1 \times L_2 \to L_1 \otimes L_2$ by $f_0(\mathbf{x}_1, \mathbf{x}_2) = \mathbf{x}_1 \otimes \mathbf{x}_2$, $\mathbf{x}_1 \in L_1$, $\mathbf{x}_2 \in L_2$. The pair $(L_1 \otimes L_2, f_0)$ is universal for $(F, L_1 \times L_2)$, where $(L_1 \times L_2 \to F)$ is the comma category of Example 103.

Indeed, we need to prove that any bilinear map $f \colon L_1 \times L_2 \to L$ can be uniquely factored through f_0, that is, there exists a unique linear map $\tau_f \colon L_1 \otimes L_2 \to L$ such that the diagram

$$
\begin{array}{ccc}
L_1 \times L_2 & \xrightarrow{\;f_0\;} & L_1 \otimes L_2 \\
{\scriptstyle f}\downarrow & \swarrow{\scriptstyle \tau_f} & \\
L & &
\end{array}
$$

is commutative.

The map $\tau_f(\mathbf{x}_1 \otimes \mathbf{x}_2) = f(\mathbf{x}_1, \mathbf{x}_2)$ extends by linearity to the map in question.

Example 108. Let B be a symmetric bilinear form on a finite-dimensional linear space L over $\mathbb{F}$ given by Equation (1.1) (resp., Equation (1.3)) if $\mathbb{F} = \mathbb{R}$ (resp., $\mathbb{F} = \mathbb{C}$). Let $(\mathrm{Cl}(L, B), \gamma)$ be the Clifford algebra for the pair (L, B). It is universal for (F, L), where $(L \to F)$ is the comma category of Example 104.

Indeed, we need to prove that any finite-dimensional unital associative algebra A over $\mathbb{F}$, and for any Clifford map $\varphi \colon L \to A$, there exists a unique linear map $\tau \colon \mathrm{Cl}(L, B) \to A$ that respects multiplication and identity elements such that the diagram

$$
\begin{array}{ccc}
L & \xrightarrow{\;\gamma\;} & \mathrm{Cl}(L, B) \\
{\scriptstyle \varphi}\downarrow & \swarrow{\scriptstyle \tau} & \\
A & &
\end{array}
$$

is commutative.

A.3.4   Adjoint and Coadjoint Functors

We start with an example.

Example 109. Consider the functors $c \colon \mathrm{Lin}_{\mathbb{R}} \to \mathrm{Lin}_{\mathbb{C}}$ and $r \colon \mathrm{Lin}_{\mathbb{C}} \to \mathrm{Lin}_{\mathbb{R}}$ of Example 95. Let U be a real finite-dimensional linear space. The objects of the comma category $(U \to r)$ are the pairs (V, f), where V is a complex finite-dimensional linear space, and $f \colon U \to rV$ is a $\mathbb{R}$-linear map. Every morphism of the category $(U \to r)$ is just a morphism $\tau \in \mathrm{Hom}_{\mathbb{C}}(V_1, V_2)$, for which the diagram

$$
\begin{array}{ccc}
U & \xrightarrow{\;f_1\;} & rV_1 \\
{\scriptstyle f_2}\downarrow & \swarrow{\scriptstyle r\tau} & \\
rV_2 & &
\end{array}
$$

is commutative in $\mathrm{Lin}_{\mathbb{R}}$.

Observe that the category $(U \to r)$ has an initial object (cU, η_U), where $\eta_U(\mathbf{x}) = \mathbf{x} \otimes 1$, $\mathbf{x} \in U$. To prove that, we need to show that any $\mathbb{R}$-linear map $f \colon U \to rV$ can be uniquely factored through η_U, that is, there exists a unique $\mathbb{C}$-linear map $\tau_f \colon cU \to V$ such that the diagram

$$
\begin{array}{ccc}
U & \xrightarrow{\;\eta_U\;} & rcU \\
{\scriptstyle f}\big\downarrow & \swarrow{\scriptstyle r\tau_f} & \\
rV & &
\end{array}
\tag{A.1}
$$

is commutative. Define $\tau_f(\mathbf{x} \otimes \zeta) = \zeta f(\mathbf{x})$, $\mathbf{x} \in U$, $\zeta \in \mathbb{C}$. The pair (cU, η_U) is indeed a universal pair for (U, r).

The maps $\eta_U \colon U \to rcU$ assemble to a natural transformation $\eta \colon \mathrm{id}_{\mathsf{Lin}_{\mathbb{R}}} \to rc$. That is, for all real linear spaces U' and for all linear operators $h \colon U \to U'$, the diagram

$$
\begin{array}{ccc}
U & \xrightarrow{\;\eta_U\;} & rcU \\
{\scriptstyle h}\big\downarrow & & \big\downarrow{\scriptstyle rch} \\
U' & \xrightarrow{\;\eta_{U'}\;} & rcU'
\end{array}
$$

commutes in $\mathsf{Lin}_{\mathbb{R}}$.

Dually, the objects of the comma category $(V \leftarrow c)$ are the pairs (f, U), where U is a real finite-dimensional linear space, and $f \colon cU \to V$ is a $\mathbb{C}$-linear map. Every morphism of the category $(V \leftarrow c)$ is just a morphism $\tau \in \mathrm{Hom}_{\mathbb{R}}(U_1, U_2)$, for which the diagram

$$
\begin{array}{ccc}
V & \xleftarrow{\;f_1\;} & cU_1 \\
{\scriptstyle f_2}\big\uparrow & \swarrow{\scriptstyle c\tau} & \\
cU_2 & &
\end{array}
$$

is commutative in $\mathsf{Lin}_{\mathbb{C}}$.

Observe that the category $(V \leftarrow c)$ has a terminal object (rV, ε_V), where $\varepsilon_V(\mathbf{y} \otimes \zeta) = \zeta \mathbf{y}$, $\mathbf{y} \in V$, $\zeta \in \mathbb{C}$. To prove that, we need to show that any $\mathbb{C}$-linear map $f \colon cU \to V$ can be uniquely factored through ε_V, that is, there exists a unique $\mathbb{R}$-linear map $\tau_f \colon U \to rV$ such that the diagram

$$
\begin{array}{ccc}
V & \xleftarrow{\;f\;} & cU \\
{\scriptstyle \varepsilon_V}\big\uparrow & \swarrow{\scriptstyle c\tau_f} & \\
crV & &
\end{array}
\tag{A.2}
$$

is commutative. Define $\tau_f(\mathbf{x}) = f(\mathbf{x} \otimes 1)$, $\mathbf{x} \in U$. The pair (rV, ε_V) is indeed a couniversal pair for (V, c).

The maps $\varepsilon_V \colon crV \to V$ assemble to a natural transformation $\varepsilon \colon cr \to \mathrm{id}_{\mathsf{Lin}_{\mathbb{C}}}$. That is, for all complex linear spaces V' and for all linear operators $h \colon V \to V'$, the diagram

$$
\begin{array}{ccc}
crV & \xrightarrow{\ \varepsilon_V\ } & V \\
{\scriptstyle crh}\big\downarrow & & \big\downarrow {\scriptstyle h} \\
crV' & \xrightarrow{\ \varepsilon_{V'}\ } & V'
\end{array}
$$

commutes in $\mathsf{Lin}_{\mathbb{C}}$.

Finally, for any real linear space U and complex linear space V, the diagrams

$$
\begin{array}{ccc}
cU & \xrightarrow{\ \eta_U^{\sharp}\ } & crcU \\
{\scriptstyle \mathrm{id}_{cU}}\big\downarrow & \swarrow{\scriptstyle \varepsilon_{cU}} & \\
cU & &
\end{array}
\quad\text{and}\quad
\begin{array}{ccc}
rV & \xrightarrow{\ \eta_{rV}\ } & rcrV \\
{\scriptstyle \mathrm{id}_{rV}}\big\downarrow & \swarrow{\scriptstyle (\varepsilon_V)_{\flat}} & \\
rV & &
\end{array}
\tag{A.3}
$$

commute.

Moreover, we can formulate and prove a toy variant of the Frobenius reciprocity. Specifically, consider the set $\mathrm{Hom}_{\mathbb{C}}(cU, V)$ of all $\mathbb{C}$-linear maps $f^{\sharp} \colon cU \to V$. The commutative diagram (A.1) states that the isomorphism

$$
\flat \colon f^{\sharp} \to f_{\flat} = rf^{\sharp} \circ \eta_U \in \mathrm{Hom}_{\mathbb{R}}(U, rV)
$$

is one-to-one. In particular, we have $\eta_U = (\mathrm{id}_{cU})_{\flat}$ and $\eta_U^{\sharp} = \mathrm{id}_{cU}$.

Dually, consider the set $\mathrm{Hom}_{\mathbb{R}}(U, rV)$ of all $\mathbb{R}$-linear maps $f_{\flat} \colon U \to rV$. The commutative diagram (A.2) states that the isomorphism

$$
\sharp \colon f_{\flat} \to f^{\sharp} = \varepsilon_V \circ cf_{\flat} \in \mathrm{Hom}_{\mathbb{C}}(cU, V)
$$

is one-to-one. In particular, we have $\varepsilon_V = \mathrm{id}_{rV}^{\sharp}$ and $(\varepsilon_V)_{\flat} = \mathrm{id}_{rV}$. The isomorphisms $\flat$ and $\sharp$ are each other's inverses.

Indeed, by definition, the image $f_{\flat}$ of $f^{\sharp} \colon cU \to V$ under the map $\flat$ is the composite map $U \xrightarrow{\ \eta_U\ } rcU \xrightarrow{\ rf^{\sharp}\ } rV$. It follows that the map $(\sharp \circ \flat)(f^{\sharp})$ is the composition of the upper horizontal maps and the right-hand vertical map in the diagram

$$
\begin{array}{ccccc}
cU & \xrightarrow{\ c\eta_U\ } & crcU & \xrightarrow{\ crf^{\sharp}\ } & crV \\
 & {\scriptstyle \mathrm{id}_{cU}}\searrow & \big\downarrow {\scriptstyle \varepsilon_{cU}} & & \big\downarrow {\scriptstyle \varepsilon_V} \\
 & & cU & \xrightarrow{\ f^{\sharp}\ } & V.
\end{array}
$$

The left-hand triangle commutes due to the triangle identities, the right-hand square commutes because ε is a natural transformation.

Dually, the image $f^{\sharp}$ of $f_{\flat} \colon U \to rV$ under the map $\sharp$ is the composite map $cU \xrightarrow{\ cf_{\flat}\ } crV \xrightarrow{\ \varepsilon_V\ } V$. The map $(\flat \circ \sharp)(f_{\flat})$ is the composition of the left-hand vertical map and lower horizontal maps in the diagram

$$U \xrightarrow{f_\flat} rV$$

with η_U, η_{rV}, id_{rV}, $rcf_\flat$, $r\varepsilon_V$ maps; $rcU \xrightarrow{rcf_\flat} rcrV \xrightarrow{r\varepsilon_V} rV.$

The left-hand square commutes because η is a natural transformation, the right-hand triangle commutes due to the triangle identities.

This example leads to the following definition. Let C and D be two categories.

Definition 128. An *adjunction* from C to D is a triple (L, R, φ), where L (resp., R) is a covariant functor from C to D (resp., from D to C), and the function φ assigns a one-to-one correspondence

$$\varphi_{X,Y} \colon \mathrm{Hom}_{\mathsf{D}}(LX, Y) \to \mathrm{Hom}_{\mathsf{C}}(X, RY)$$

to each pair of objects X of C and Y of D in such a way that for any objects X' of C and Y' of D, and all morphisms $h \colon X \to X'$, $k \colon Y \to Y'$, the diagrams

$$
\begin{array}{ccc}
\mathrm{Hom}_{\mathsf{D}}(LX, Y) & \xrightarrow{\flat} & \mathrm{Hom}_{\mathsf{C}}(X, RY) \\
{\scriptstyle \circ Lh} \downarrow & & \downarrow {\scriptstyle \circ h} \\
\mathrm{Hom}_{\mathsf{D}}(LX', Y) & \xrightarrow{\flat} & \mathrm{Hom}_{\mathsf{C}}(X', RY)
\end{array}
$$

$$
\begin{array}{ccc}
\mathrm{Hom}_{\mathsf{D}}(LX, Y) & \xrightarrow{\flat} & \mathrm{Hom}_{\mathsf{C}}(X, RY) \\
{\scriptstyle k\circ} \downarrow & & \downarrow {\scriptstyle Rk\circ} \\
\mathrm{Hom}_{\mathsf{D}}(LX, Y') & \xrightarrow{\flat} & \mathrm{Hom}_{\mathsf{C}}(X, RY')
\end{array}
$$

commute. The functor L is *left-adjoint* to R, the functor R is *right-adjoint* to L.

Having an adjunction, for any object X of C, we can construct the comma category $(X \to R)$ with an initial object (LX, η_X), where $\eta_X = (\mathrm{id}_{LX})^\sharp$. The pair (LX, η_X) is universal for (X, R). Moreover, the function $X \mapsto \eta_X$ is a natural transformation $\mathrm{id}_X \to RL$. The map $\varphi_{X,Y}$ maps $f^\sharp \in \mathrm{Hom}_{\mathsf{D}}(LX, Y)$ to

$$f_\flat = Rf^\sharp \circ \eta_X \in \mathrm{Hom}_{\mathsf{C}}(X, RY). \tag{A.4}$$

The composite natural transformation $L \xrightarrow{L\eta} LRL \xrightarrow{\varepsilon L} L$ is the identity of L.

Dually, for any object Y od D, we can construct the comma category $(Y \leftarrow L)$ with a terminal object (RY, ε_Y), where $\varepsilon_Y = \mathrm{id}_{RY}^\sharp$. The pair (RY, ε_Y)

is couniversal for (Y, L). Moreover, the function $\varepsilon_Y \mapsto Y$ is a natural transformation $LR \to \mathrm{id}_Y$. The map $\varphi_{X,Y}^{-1}$ maps $f_\flat \in \mathrm{Hom}_\mathsf{C}(X, RY)$ to

$$f^\sharp = \varepsilon_Y \circ L f_\flat \in \mathrm{Hom}_\mathsf{D}(LX, Y). \tag{A.5}$$

The composite natural transformation $R \xrightarrow{\ \eta R\ } RLR \xrightarrow{\ R\varepsilon\ } R$ is the identity of R.

Conversely, if the natural transformations $\eta \colon \mathrm{id}_L \to RL$ called the *unit* and $\varepsilon \colon LT \to \mathrm{id}_R$ called the *counit* satisfy the *triangular identities*, that is, for any objects X of C and Y of D the diagrams (A.3) commute, then Equations (A.4) and (A.5) define an adjunction between L and R.

Example 110. The functor q is left-adjoint to c', while c' is right-adjoint to q.

Example 111. Let G be a compact Lie group. Consider the category $\mathsf{Rep}_\mathbb{K}(G)$ whose objects are continuous representations of G in Hilbert spaces over $\mathbb{K}$ that respect the inner product, and whose morphisms are intertwining operators. Let H be a closed subgroup of G. Define a functor Ind that maps a representation L of the group H to the representation of G induced by L. Another functor, Res, restricts a representation R of the group G to H.

We will sketch a proof of the Frobenius reciprocity, which says that Res is left-adjoint to Ind and Ind is right-adjoint to Res. To perform that, we construct a unit and a counit.

Let L be a representation of H and R a representation of G. For each $\mathbf{y} \in R$, define the square-integrable section $\eta_R(\mathbf{y})$ of the homogeneous fibre bundle $\xi[L] = (E, \pi_L, G/H)$ over G/H with fibre L and associated principal bundle $(G, \pi, G/H, H)$ by $\eta_R(\mathbf{y})x = g^{-1} \cdot \mathbf{y}$, where $x \in G/H$, and an arbitrary element $g \in \pi_L^{-1}(x)$ acts on $\mathbf{y}$ by the representation R.

Dually, for each square-integrable section $\mathbf{s}(x) \in L \uparrow G$, define the element $\varepsilon_L(\mathbf{s}) \in L$ by $\varepsilon_L(\mathbf{s}) = \mathbf{s}(e)$.

Consider the isomorphism $\flat$ between the set $\mathrm{Hom}_H(\mathrm{Res}\, R, L)$ of all H-intertwining operators $f^\sharp \colon \mathrm{Res}\, R \to L$ and $\mathrm{Hom}_G(R, \mathrm{Ind}\, L)$ given by

$$\flat \colon f^\sharp \to f_\flat = \mathrm{Ind}\, f^\sharp \circ \eta_R \in \mathrm{Hom}_G(R, \mathrm{Ind}\, L).$$

Dually, the map

$$\sharp \colon f^\flat \to f_\sharp = \varepsilon_L \circ \mathrm{Res} \in \mathrm{Hom}_H(\mathrm{Res}\, R, L)$$

is an isomorphism. Moreover, we have $\eta_R = (\mathrm{id}_{\mathrm{Res}\, R})_\flat$, $\eta_R^\sharp = \mathrm{id}_{\mathrm{Res}\, R}$, $\varepsilon_L = \mathrm{id}_{\mathrm{Ind}\, L}^\sharp$ and $(\varepsilon_L)_\flat = \mathrm{id}_{\mathrm{Ind}\, L}$. The isomorphisms $\flat$ and $\sharp$ are each other's inverses.

Proofs can be left to the reader. We only mention that the equality $(\sharp \circ \flat)(f^\sharp) = f^\sharp$ is proved with the help of the commutative diagram

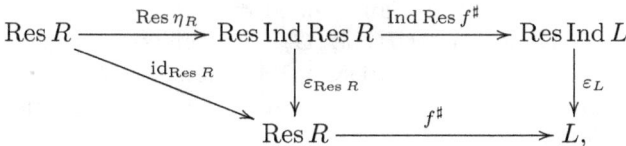

$$
\begin{array}{ccccc}
\mathrm{Res}\, R & \xrightarrow{\ \mathrm{Res}\, \eta_R\ } & \mathrm{Res}\, \mathrm{Ind}\, \mathrm{Res}\, R & \xrightarrow{\ \mathrm{Ind}\, \mathrm{Res}\, f^\sharp\ } & \mathrm{Res}\, \mathrm{Ind}\, L \\[2pt]
& \searrow{\scriptstyle \mathrm{id}_{\mathrm{Res}\, R}} & \downarrow{\scriptstyle \varepsilon_{\mathrm{Res}\, R}} & & \downarrow{\scriptstyle \varepsilon_L} \\[2pt]
& & \mathrm{Res}\, R & \xrightarrow[\ f^\sharp\ ]{} & L,
\end{array}
$$

while the equality $(\flat \circ \sharp)(f_\flat) = f_\flat$ with the help of the commutative diagram

$$
\begin{array}{ccc}
R & \xrightarrow{\ f_\flat\ } & \operatorname{Ind} L \\
\downarrow{\scriptstyle \eta_R} & & \downarrow{\scriptstyle \eta_{\operatorname{Ind} L}} \quad\searrow{\scriptstyle \operatorname{id}_{\operatorname{Ind} L}} \\
\operatorname{Ind}\operatorname{Res} R & \xrightarrow{\operatorname{Ind}\operatorname{Res} f_\flat} & \operatorname{Ind}\operatorname{Res}\operatorname{Ind} L \xrightarrow{\ \operatorname{Ind}\varepsilon_L\ } \operatorname{Ind} L.
\end{array}
$$

A.4 Spin Manifolds

Let (TM, π, M) be the tangent bundle of a smooth Riemannian manifold M of dimension n. There is an open covering $\{\mathcal{U}_\alpha : \alpha \in A\}$ of M and a cocycle $f_{\alpha\beta}$ taking values in $O(n)$. Our first question is to establish a condition under which there is a cocycle with values in $SO(n)$.

Let q be a nonnegative integer, G be a Lie group with identity e, and consider the set of all smooth maps

$$\{\, f_{\alpha_1 \cdots \alpha_{q+1}} : \mathcal{U}_{\alpha_1} \cap \mathcal{U}_{\alpha_2} \cap \cdots \cap \mathcal{U}_{q+1} \to G \,\}.$$

Call this set a *Čech q-cochain*. We observe that it is a group with respect to point-wise multiplication.

Let the *coboundary map* d_0 maps a 0-cochain to the 1-cochain by

$$(d_0 f)_{\alpha\beta}(p) = f_\beta(p) f_\alpha^{-1}(p), \qquad p \in \mathcal{U}_\alpha \cap \mathcal{U}_\beta.$$

Denote by $\check{H}_0(M; G)$ the set of all 0-*cocycles*: smooth maps $f_\alpha : \mathcal{U}_\alpha \to G$ with $d_0 f = e$. We see that f_α is a 0-cocycle if and only if for all sets $\mathcal{U}_\alpha$ and $\mathcal{U}_\beta$ with $\mathcal{U}_\alpha \cap \mathcal{U}_\beta \neq \varnothing$, and for all $p \in \mathcal{U}_\alpha \cap \mathcal{U}_\beta$, we have $f_\alpha(p) = f_\beta(p)$, that is, all the maps f_α perfectly patch together to form a smooth function $f : M \to G$. Therefore, we have $\check{H}_0(M; G) = C^\infty(M; G)$, the set of all smooth G-valued functions on M. Call $\check{H}_0(M; G)$ the *non-abelian Čech cohomology group*.

The coboundary map d_1 maps a 1-cochain to the 2-cochain by

$$(d_1 h)_{\alpha\beta\gamma}(p) = h_{\alpha\beta}(p) h_{\beta\gamma}(p) h_{\gamma\alpha}(p), \qquad p \in \mathcal{U}_\alpha \cap \mathcal{U}_\beta \cap \mathcal{U}_\gamma.$$

A 1-*cocycle* is a 1-cochain h with $(d_1 h)_{\alpha\beta\gamma}(p) = e$ for all α, β, and γ *in* A.

The non-abelian Čech cohomology group acts on the set of 1- cocycles by

$$(f \cdot h)_{\alpha\beta}(p) = f_\beta(p) h_{\alpha\beta}(p) f_\alpha^{-1}(p). \tag{A.6}$$

Denote by $\check{H}_1(M; G)$ the set of orbits for this action. It is possible to prove the following. There exists such an open covering $\{\mathcal{U}_\alpha : \alpha \in A\}$ of M that its refining does not change the set $\check{H}_1(M; G)$. Two different constructions of such a covering are given in (Scorpan 2005, p. 190). Call $\check{H}_1(M; G)$ the *non-abelian Čech cohomology set*. This set has a distinguished element: the orbit of the trivial 1-cochain $h_{\alpha\beta}(p) = e$.

The non-abelian Čech cohomology set has a simple geometric meaning. Indeed, an orbit of the action (A.6) is determined by a 1-cocycle, say h. Two 1-cocycles, h_1 and h_2, belong to the same orbit if and only if they *differ by a coboundary*: there is a 0-cochain f_α with $(h_1)_{\alpha\beta}(p) = f_\alpha(p)(h_2)_{\alpha\beta}(p)f_\beta^{-1}(p)$, $p \in \mathcal{U}_\alpha \cap \mathcal{U}_\beta$. We recognise Definition 55 of equivalent principal bundles with common base M and common fibres G and see that the non-abelian Čech cohomology set is the set of equivalence classes of principal bundles, and its distinguished element is the trivial bundle of Example 34.

If G is a discrete abelian group, then we have $\check{H}_q(M; G) = H_q(M; G)$, $q = 0, 1$, where the right hand side is the group of singular cohomologies of M. For a proof, see (p. 190–191), and for missing definitions see the references given in Section A.5.

Let G_1 and G_2 be two Lie groups, and let $\varphi \colon G_1 \to G_2$ be a smooth group homomorphism. Define the map $\varphi_* \colon \check{H}_q(M; G_1) \to \check{H}_q(M; G_2)$ as follows: if $q = 0$ then $(\varphi_* f)(p) = \varphi(f(p))$, while if $q = 1$ and a class $h \in \check{H}_q(M; G_1)$ is given by the maps $h_{\alpha\beta} \colon \mathcal{U}_\alpha \cap \mathcal{U}_\beta \to G_1$, then put $(\varphi_* h)_{\alpha\beta}(p) = \varphi(h_{\alpha\beta}(p))$ for all $p \in \mathcal{U}_\alpha \cap \mathcal{U}_\beta$.

When $q = 1$ and a class $h \in \check{H}_q(M; G)$ corresponds to a principal bundle $\xi = (P, \pi, M, G_1)$, then the principal bundle $\varphi_* \xi$ is a fibre bundle over M with fibre G_2 associated with ξ with the help of the right action of G_1 on the Cartesian product $P \times G_2$ by $(x, g) \cdot g_1 = (x \cdot g_1, \varphi(g_1^{-1})g)$, $(x, g) \in P \times G_2$, $g_1 \in G_1$. That is, $\varphi_* \xi = \xi[H]$.

Now we need a definition.

Definition 129. A sequence $G_0 \xrightarrow{f_1} G_1 \xrightarrow{f_2} G_2 \xrightarrow{f_3} \cdots \xrightarrow{f_n} G_n$ of groups and group homomorphisms (resp., of sets with distinguished elements and maps) is called *exact at* G_i if the image of f_i is equal to the kernel (resp., to the inverse image of the distinguished element) of f_{i+1}. The above sequence is called *exact* if it is exact at each G_i, $1 \le i \le n-1$, *short exact* if it is exact and $n = 4$, and *long exact* if it is exact and $n \ge 5$.

Consider a short exact sequence of groups

$$\{e\} \longrightarrow G_1 \xrightarrow{i} G \xrightarrow{\pi} G_2 \longrightarrow \{e\} \,,$$

where the first term is the identity of G_1, while the last one is the identity of G_2. The second map is denoted by i (inclusion) because the sequence is exact at G_1 if and only if G_1 is isomorphic to $i(G_1)$. Similarly, the third map is denoted by π (projection) because the sequence is exact at G_2 if and only if π is a group covering.

The above short exact sequence of groups generates the following long exact sequence in cohomology:

$$\{e\} \longrightarrow \check{H}_0(M;G_1) \xrightarrow{i_*} \check{H}_0(M;G) \xrightarrow{\pi_*} \check{H}_0(M;G_2)$$

$$\xrightarrow{\delta} \check{H}_1(M;G_1) \xrightarrow{i_*} \check{H}_1(M;G) \xrightarrow{\pi_*} \check{H}_1(M;G_2).$$

Only the map δ should be constructed. The tuple (G, π, G_2, G_1) is a principal bundle over G_2 with fibre G_1. Let h be the class of that bundle in $\check{H}_1(G_2; G_1)$. Let $\varphi \in \check{H}_0(M; G_2) = C^\infty(M, G_2)$. Define the map $\varphi^* \colon \check{H}_1(G_2; G_1) \to \check{H}_1(M; G_1)$ by $(\varphi^*(k))_{\alpha\beta}(p) = k_{\alpha\beta}(\varphi(p))$, $k \in \check{H}_1(G_2; G_1)$. The map δ has the form $\delta(\varphi) = \varphi^*(h)$.

Put $G_1 = \mathrm{SO}(n)$, $G = \mathrm{O}(n)$, $G_2 = \mathrm{O}(1)$. The last four terms of the corresponding long exact sequence in cohomology have the form

$$\check{H}_0(M;\mathrm{O}(1)) \xrightarrow{\delta} \check{H}_1(M;\mathrm{SO}(n)) \xrightarrow{i_*} \check{H}_1(M;\mathrm{O}(n)) \xrightarrow{\pi_*} H_1(M;\mathrm{O}(1)),$$
$$(\mathrm{A.7})$$

where we omit the check sign in the last term because the group $\mathrm{O}(1)$ is discrete.

Recall that the second (resp., the third) term of the above sequence is the set of equivalence classes of oriented (resp., all) vector bundles of dimension n over M. The map i_* forgets the orientation. Our task is to give an interpretation to the map π_*. We denote a vector bundle $(E, \pi, M) \in \check{H}_1(M; \mathrm{O}(n))$ just by E.

Theorem 28. *There exists unique maps* $w_q \colon \check{H}_1(M; \mathrm{O}(n)) \to H_q(M; \mathrm{O}(1))$, $q \geq 0$, *that satisfy the following conditions.*

- *If $q > n$, then $w_q(E) = \{e\}$.*

- *For any smooth map $\varphi \colon M' \to M$ we have $w_q(\varphi^*(E)) = \varphi^*(w_q(E))$.*

- $w_q(E_1 \oplus E_2) = \sum_{i=0}^{q} w_i(E_1) \smile w_{q-i}(E_2)$.

- *The map w_1 is not equal to identity on the Möbius bundle.*

The maps w_q are called the *Stiefel–Whitney characteristic classes*. The symbol $\smile$ denotes the cup product, see Section A.5. We say that the Stiefel–Whitney characteristic class w_q *vanishes* if $w_q(E) = \{e\}$.

It is possible to prove that the map π_* of the exact sequence (A.7) satisfies the conditions of Theorem 28, that is, $\pi_* = w_1$.

Assume that the tangent bundle (TM, π, M) has a $\mathrm{SO}(n)$-valued cocycle, that is, M is orientable. Its image in $\check{H}_1(M; \mathrm{O}(n))$ belongs to the kernel of w_1, because the sequence (A.7) is exact. This means that the first Stiefel–Whitney characteristic class of (TM, π, M) vanishes.

Conversely, assume that $w_1(TM) = \{e\}$. Then $TM \in \check{H}_1(M; \mathrm{SO}(n))$ by exactness, so (TM, π, M) has a $\mathrm{SO}(n)$-valued cocycle and is orientable. The two orientations of M correspond to the two elements of the group $\check{H}_0(M; \mathrm{O}(1)) = \mathrm{O}(1)$.

We investigate the second question: under which conditions the $\mathrm{SO}(n)$-valued cocycle of the oriented tangent bundle can be lifted to a $\mathrm{Spin}(n)$-valued one?

Put $G_1 = \mathrm{O}(1)$, $G = \mathrm{Spin}(n)$, and $G_2 = \mathrm{SO}(n)$. Observe that G_1 is abelian and $i(G_1)$ is the centre of G. We may add one more term to the exact sequence (A.7) as follows:

$$H_1(M; \mathrm{O}(1)) \xrightarrow{\ i_*\ } \check{H}_1(M; \mathrm{Spin}(n)) \xrightarrow{\ \pi_*\ } H_1(M; \mathrm{SO}(n))$$

$$\downarrow{\scriptstyle \delta}$$

$$H^2(M; \mathrm{O}(1)),$$

where the last term exists because $\mathrm{O}(1)$ is abelian. Again, it is possible to prove that the map δ satisfies all conditions of Theorem 28, so $\delta = w_2$. The same arguments as before show that M is spin if and only if the second Stiefel–Whitney characteristic class of TM vanishes, and different spin structures are parameterized by the elements of the group $H_1(M; \mathrm{O}(1))$ of $\mathrm{O}(1)$-valued singular cohomologies of M.

A.5   Bibliographical Remarks

Commutative diagrams are part of *homological algebra* developed in 1940s. See (Adámek, Herrlich, and Strecker 1990; Bourbaki 2007; Cartan and Eilenberg 1999; Hilton and Stammbach 1997; Rotman 2009).

Categories appeared at the same time. A vast literature on categories includes books (Leinster 2014; Riehl 2016) and a paper Baez and Lauda (2011), among others.

For characteristic classes and the cup product $\smile$, see Milnor and Stasheff (1974). Spin structures on manifolds are discussed in Milnor (1963).

Bibliography

Abbott, B.P. et al. 2016. 'Observation of Gravitational Waves from a Binary Black Hole Merger'. *Phys. Rev. Lett.* 116 (6): 061102. https://link.aps.org/doi/10.1103/PhysRevLett.116.061102.

———. 2018. 'Search for Tensor, Vector, and Scalar Polarizations in the Stochastic Gravitational-Wave Background'. *Phys. Rev. Lett.* 120 (20): 201102. https://link.aps.org/doi/10.1103/PhysRevLett.120.201102.

Abłamowicz, Rafał, William E. Baylis, Thomas Branson, Pertti Lounesto, Ian Porteous, John Ryan, J. M. Selig, and Garret Sobczyk. 2004. *Lectures on Clifford (geometric) algebras and applications.* xviii+221. Edited by Abłamowicz and Sobczyk. Birkhäuser Boston, Boston, MA. ISBN: 0-8176-3257-3. https://doi.org/10.1007/978-0-8176-8190-6.

Abłamowicz, Rafał, and Bertfried Fauser, eds. 2000. *Clifford algebras and their applications in mathematical physics. Vol. 1.* Vol. 18. Progress in Physics. Algebra and physics, Papers from the 5th International Conference held in Ixtapa-Zihuatanejo, June 27–July 4, 1999. Birkhäuser Boston, Boston, MA. ISBN: 0-8176-4182-3. https://doi.org/10.1007/978-1-4612-1368-0.

Abłamowicz, Rafał, and Pertti Lounesto, eds. 1995. *Clifford algebras and spinor structures.* Vol. 321. Mathematics and its Applications. A special volume dedicated to the memory of Albert Crumeyrolle (1919–1992). Kluwer Academic Publishers, Dordrecht. ISBN: 0-7923-3366-7. https://doi.org/10.1007/978-94-015-8422-7.

Acrami, Y. et al. 2020. 'Planck 2018 results - IX. Constraints on primordial non-Gaussianity'. *A&A* 641:A9. https://doi.org/10.1051/0004-6361/201935891.

Adámek, Jiři, Horst Herrlich, and George E. Strecker. 1990. *Abstract and concrete categories. The joy of cats.* xiv+482. Pure and Applied Mathematics (New York). A Wiley-Interscience Publication. John Wiley & Sons, New York. ISBN: 0-471-60922-6.

Adams, J. Frank. 1969. *Lectures on Lie groups.* xii+182. W. A. Benjamin, New York–Amsterdam.

Adler, Ronald J. 2021. *General relativity and cosmology—a first encounter.* xiv+313. Graduate Texts in Physics. Springer, Cham. ISBN: 978-3-030-61573-4. https://doi.org/10.1007/978-3-030-61574-1. https://doi.org/10.1007/978-3-030-61574-1.

Afzal, Adeela, Waqas Ahmed, Mansoor Ur Rehman, and Qaisar Shafi. 2022. 'μ-hybrid inflation, gravitino dark matter, and stochastic gravitational wave background from cosmic strings'. *Phys. Rev. D* 105 (10): 103539. https://link.aps.org/doi/10.1103/PhysRevD.105.103539.

Agazie, Gabriella et al. 2023. 'The NANOGrav 15 yr Data Set: Evidence for a Gravitational-wave Background'. *The Astrophysical Journal Letters* 951, no. 1 (June): L8. https://doi.org/10.3847/2041-8213/acdac6.

Aghanim, N. et al. 2020a. 'Planck 2018 results - V. CMB power spectra and likelihoods'. *A&A* 641:A5. https://doi.org/10.1051/0004-6361/201936386.

———. 2020b. 'Planck 2018 results - VI. Cosmological parameters'. *A&A* 641:A6. https://doi.org/10.1051/0004-6361/201833910.

Altmann, Simon L. 1986. *Rotations, quaternions, and double groups.* xiv+317. Oxford Science Publications. The Clarendon Press, Oxford University Press, New York. ISBN: 0-19-855372-2.

Alvey, James, Miguel Escudero, Nashwan Sabti, and Thomas Schwetz. 2022. 'Cosmic neutrino background detection in large-neutrino-mass cosmologies'. *Phys. Rev. D* 105 (6): 063501. https://link.aps.org/doi/10.1103/PhysRevD.105.063501.

Anglès, Pierre. 2008. *Conformal groups in geometry and spin structures.* 50:xxviii+283. Progress in Mathematical Physics. With forewords by Jaime Keller and José Bertin. Birkhäuser Boston, Boston, MA. ISBN: 978-0-8176-3512-1. https://doi.org/10.1007/978-0-8176-4643-1.

Artz, Ray E. 1981. 'Classical mechanics in Galilean space-time'. *Found. Phys.* 11 (9-10): 679–697. ISSN: 0015-9018. https://doi.org/10.1007/BF00726944.

Athar, M. Sajjad, and S. K. Singh. 2020. *Physics of Neutrino Interactions.* Cambridge University Press.

Atkinson, Kendall, and Weimin Han. 2012. *Spherical harmonics and approximations on the unit sphere: an introduction.* 2044:x+244. Lecture Notes in Mathematics. Springer, Heidelberg. ISBN: 978-3-642-25982-1. https://doi.org/10.1007/978-3-642-25983-8.

Auclair, Pierre. 2020. 'Impact of the small-scale structure on the Stochastic Background of Gravitational Waves from cosmic strings'. *Journal of Cosmology and Astroparticle Physics* 2020, no. 11 (November): 050. https://dx.doi.org/10.1088/1475-7516/2020/11/050.

Auclair, Pierre, Jose J. Blanco-Pillado, Daniel G. Figueroa, Alexander C. Jenkins, Marek Lewicki, Mairi Sakellariadou, Sotiris Sanidas, et al. 2020. 'Probing the gravitational wave background from cosmic strings with LISA'. *Journal of Cosmology and Astroparticle Physics* 2020, no. 04 (April): 034. https://dx.doi.org/10.1088/1475-7516/2020/04/034.

Audren, Benjamin, Emilio Bellini, Antonio J. Cuesta, Satya Gontcho A. Gontcho, Julien Lesgourgues, Viviana Niro, Marcos Pellejero-Ibanez, et al. 2015. 'Robustness of cosmic neutrino background detection in the cosmic microwave background'. *Journal of Cosmology and Astroparticle Physics* 2015, no. 03 (March): 036. https://dx.doi.org/10.1088/1475-7516/2015/03/036.

Audretsch, Jürgen, and Wolfgang Graf. 1970. 'Neutrino radiation in gravitational fields'. *Comm. Math. Phys.* 19:315–326. ISSN: 0010-3616. http://projecteuclid.org./euclid.cmp/1103842744.

Auger, Gerard, and Eric Plagnol. 2017. *Overview of Gravitational Waves. Theory, Sources, and Detection.* World Scientific, Singapore.

Baez, John Carlos. 2012. 'Division algebras and quantum theory'. *Found. Phys.* 42 (7): 819–855. ISSN: 0015-9018. https://doi.org/10.1007/s10701-011-9566-z.

Baez, John Carlos, and John Huerta. 2010. 'The algebra of grand unified theories'. *Bull. Amer. Math. Soc. (N.S.)* 47 (3): 483–552. ISSN: 0273-0979. https://doi.org/10.1090/S0273-0979-10-01294-2.

Baez, John Carlos, and Aaron D. Lauda. 2011. 'A prehistory of n-categorical physics'. In *Deep beauty. Understanding the quantum world through mathematical innovation,* edited by Hans Halvorson, 13–128. Cambridge Univ. Press, Cambridge.

Baldi, Paolo, and Maurizia Rossi. 2014. 'Representation of Gaussian isotropic spin random fields'. *Stochastic Process. Appl.* 124 (5): 1910–1941. ISSN: 0304-4149. https://doi.org/10.1016/j.spa.2014.01.007.

Bambi, Cosimo. 2018. *Introduction to general relativity. A course for undergraduate students of physics.* xvi+335. Undergraduate Lecture Notes in Physics. Springer, Singapore. ISBN: 978-981-13-1089-8. https://doi.org/10.1007/978-981-13-1090-4.

Başkal, Sibel, Young Suh Kim, and Marilyn E. Noz. 2019. *Mathematical Devices for Optical Sciences.* 2053-2563. IOP Publishing. ISBN: 978-0-7503-1614-9. https://doi.org/10.1088/2053-2563/aafe78.

Bauer, Martin, and Jack D. Shergold. 2023. 'Limits on the cosmic neutrino background'. *Journal of Cosmology and Astroparticle Physics* 2023, no. 01 (January): 003. https://dx.doi.org/10.1088/1475-7516/2023/01/003.

Baulieu, Laurent, John Iliopoulos, and Roland Sénéor. 2017. *From classical to quantum fields.* xvii+931. Oxford University Press, Oxford. ISBN: 978-0-19-878840-9. https://doi.org/10.1093/oso/9780198788393.001.0001.

Baumann, Daniel. 2022. *Cosmology.* Cambridge University Press. https://doi.org/10.1017/9781108937092.

Baumgarte, Thomas W., and Stuart L. Shapiro. 2010. *Numerical relativity. Solving Einstein's equations on the computer.* Cambridge University Press.

Beck, Dominic, Giulio Fabbian, and Josquin Errard. 2018. 'Lensing reconstruction in post-Born cosmic microwave background weak lensing'. *Phys. Rev. D* 98 (4): 043512, 20. ISSN: 2470-0010. https://doi.org/10.1103/physrevd.98.043512.

Benn, Ian M., and Robin W. Tucker. 1989. *An introduction to spinors and geometry with applications in physics.* x+358. Reprint of the 1987 original. Bristol: Adam Hilger. ISBN: 0-85274-261-4.

Berger, Marcel. 2003. *A panoramic view of Riemannian geometry.* xxiv+824. Springer-Verlag, Berlin. ISBN: 3-540-65317-1. https://doi.org/10.1007/978-3-642-18245-7.

———. 2009. *Geometry I.* xiv+428. Universitext. Translated from the 1977 French original by M. Cole and S. Levy, Fourth printing of the 1987 English translation. Berlin: Springer-Verlag. ISBN: 978-3-540-11658-5.

Bergmann, Peter G., and Venzo Sabbata, eds. 1980. *Cosmology and gravitation. Spin, torsion, rotation, and supergravity.* Vol. 58. NATO advanced study institute series. B. Physics. Plenum Press, New York.

Bernal, Antonio N., and Miguel Sánchez. 2003. 'Leibnizian, Galilean and Newtonian structures of space-time'. *J. Math. Phys.* 44 (3): 1129–1149. ISSN: 0022-2488. https://doi.org/10.1063/1.1541120.

Bernal, José Luis, Andrea Caputo, Francisco Villaescusa-Navarro, and Marc Kamionkowski. 2021. 'Searching for the Radiative Decay of the Cosmic Neutrino Background with Line-Intensity Mapping'. *Phys. Rev. Lett.* 127 (13): 131102. https://link.aps.org/doi/10.1103/PhysRevLett.127.131102.

Bernardis, Francesco De, Luca Pagano, Paolo Serra, Alessandro Melchiorri, and Asantha Cooray. 2008. 'Anisotropies in the cosmic neutrino background after Wilkinson Microwave Anisotropy Probe five-year data'. *Journal of Cosmology and Astroparticle Physics* 2008, no. 06 (June): 013. https://doi.org/10.1088/1475-7516/2008/06/013.

Bernardo, Reginald Christian, and Kin-Wang Ng. 2022. 'Pulsar and cosmic variances of pulsar timing-array correlation measurements of the stochastic gravitational wave background'. *Journal of Cosmology and Astroparticle Physics* 2022, no. 11 (November): 046. https://dx.doi.org/10.1088/1475-7516/2022/11/046.

Besse, Arthur L. 2008. *Einstein manifolds.* xii+516. Classics in Mathematics. Reprint of the 1987 edition. Springer-Verlag, Berlin. ISBN: 978-3-540-74120-6.

Bethke, Laura Bianca. 2015. *Exploring the early universe with gravitational waves.* xiii+139. Springer Theses. Doctoral thesis accepted by Imperial College London, UK. Springer, Cham. ISBN: 978-3-319-17448-8. https://doi.org/10.1007/978-3-319-17449-5.

Bian, Ligong, Jing Shu, Bo Wang, Qiang Yuan, and Junchao Zong. 2022. 'Searching for cosmic string induced stochastic gravitational wave background with the Parkes Pulsar Timing Array'. *Phys. Rev. D* 106 (10): L101301. https://link.aps.org/doi/10.1103/PhysRevD.106.L101301.

Bishop, Nigel T. 2005. 'Linearized solutions of the Einstein equations within a Bondi–Sachs framework, and implications for boundary conditions in numerical simulations'. *Classical and Quantum Gravity* 22, no. 12 (May): 2393. https://doi.org/10.1088/0264-9381/22/12/006.

Bishop, Nigel T., and Luciano Rezzolla. 2016. 'Extraction of gravitational waves in numerical relativity'. *Living Rev. Relativ.* 19 (2): 2393. https://doi.org/10.1007/s41114-016-0001-9.

Blanco, Miguel A., M. Flórez, and M. Bermejo. 1997. 'Evaluation of the rotation matrices in the basis of real spherical harmonics'. *Journal of Molecular Structure: THEOCHEM* 419 (1): 19–27. ISSN: 0166-1280. https://www.sciencedirect.com/science/article/pii/S0166128097001851.

Blanco-Pillado, Jose J., and Ken D. Olum. 2017. 'Stochastic gravitational wave background from smoothed cosmic string loops'. *Phys. Rev. D* 96 (10): 104046. https://link.aps.org/doi/10.1103/PhysRevD.96.104046.

Blas, Diego, Julien Lesgourgues, and Thomas Tram. 2011. 'The Cosmic Linear Anisotropy Solving System (CLASS). Part II: Approximation schemes'. *Journal of Cosmology and Astroparticle Physics* 2011, no. 7 (July): 034. https://doi.org/10.1088/1475-7516/2011/07/034.

Blattner, Robert J. 1961. 'On induced representations'. *Amer. J. Math.* 83:79–98. ISSN: 0002-9327. https://doi.org/10.2307/2372722.

Bohr, Aage, and Ben R. Mottelson. 1998. *Nuclear Structure.* World Scientific, Singapore. https://doi.org/10.1142/3530.

Boileau, Guillaume, Alexander C. Jenkins, Mairi Sakellariadou, Renate Meyer, and Nelson Christensen. 2022. 'Ability of LISA to detect a gravitational-wave background of cosmological origin: The cosmic string case'. *Phys. Rev. D* 105 (2): 023510. https://link.aps.org/doi/10.1103/PhysRevD.105.023510.

Boskoff, Wladimir-Georges, and Salvatore Capozziello. 2020. *A mathematical journey to relativity. Deriving special and general relativity with basic mathematics.* xxii+397. Unitext for Physics. Springer, Cham. ISBN: 978-3-030-47894-0. https://doi.org/10.1007/978-3-030-47894-0.

Bourbaki, Nicolas. 2003. *Algebra II. Chapters 4–7.* viii+461. Elements of Mathematics (Berlin). Translated from the 1981 French edition by P. M. Cohn and J. Howie, Reprint of the 1990 English edition [Springer, Berlin]. Springer-Verlag, Berlin. ISBN: 3-540-00706-7. https://doi.org/10.1007/978-3-642-61698-3.

———. 2007. *Éléments de mathématique. Algèbre. Chapitre 10. Algèbre homologique.* viii+216. Reprint of the 1980 original [Masson, Paris; MR0610795]. Springer-Verlag, Berlin. ISBN: 978-3-540-34492-6.

Bourguignon, Jean-Pierre, Oussama Hijazi, Jean-Louis Milhorat, Andrei Moroianu, and Sergiu Moroianu. 2015. *A spinorial approach to Riemannian and conformal geometry.* ix+452. EMS Monographs in Mathematics. European Mathematical Society (EMS), Zürich. ISBN: 978-3-03719-136-1. https://doi.org/10.4171/136.

Boyle, McBlaine Michael. 2016. 'How should spin-weighted spherical functions be defined?' *J. Mathematical Phys.* 57 (9): 092504. https://doi.org/10.1063/1.4962723.

Bredon, Glen E. 1972. *Introduction to compact transformation groups.* 46:xiii+459. Pure and Applied Mathematics. Academic Press, New York–London.

Brémaud, Pierre. 2020. *Probability theory and stochastic processes.* xvii+713. Universitext. Springer, Cham. ISBN: 978-3-030-40183-2. https://doi.org/10.1007/978-3-030-40183-2.

Breuer, Reinhard A., J. Tiomno, and C. V. Vishveshwara. 1975. 'Polarization of gravitational geodesic synchrotron radiation'. *Nuovo Cimento B (11)* 25:851–870. ISSN: 0369-3554. https://doi.org/10.1007/BF02724757.

Brink, David M., and George Raymond Satchler. 1993. *Angular momentum.* Third ed. Oxford Library of Physical Science. Clarendon Press, Oxford.

Bröcker, Theodor, and Tammo tom Dieck. 1995. *Representations of compact Lie groups.* 98:x+313. Graduate Texts in Mathematics. Translated from the German manuscript, Corrected reprint of the 1985 translation. Springer-Verlag, New York. ISBN: 0-387-13678-9.

Bucher, Martin. 2015. 'Physics of the cosmic microwave background aniso-tropy'. *Internat. J. Modern Phys. D* 24 (2): 1530004, 107. ISSN: 0218-2718. https://doi.org/10.1142/S0218271815300049.

Buchmüller, Wilfried, Valerie Domcke, and Kai Schmitz. 2021. 'Stochastic gravitational-wave background from metastable cosmic strings'. *Journal of Cosmology and Astroparticle Physics* 2021, no. 12 (December): 006. https://dx.doi.org/10.1088/1475-7516/2021/12/006.

Budinich, Marco. 2019. 'On complex representations of Clifford algebra'. *Adv. Appl. Clifford Algebr.* 29 (1): Paper No. 18, 15. ISSN: 0188-7009. https://doi.org/10.1007/s00006-018-0930-3.

Budinich, Paolo, and Andrzej Trautman. 1988. *The spinorial chessboard.* viii+128. Trieste Notes in Physics. Springer-Verlag, Berlin. ISBN: 3-540-19078-3. https://doi.org/10.1007/978-3-642-83407-3.

Burgess, Clifford P., and Guy D. Moore. 2007. *The standard model: a primer.* xvi+542. Cambridge University Press, New York. ISBN: 978-0-521-86036-9.

Byerly, William Elwood. 1959. *An elementary treatise on Fourier's series and spherical, cylindrical, and ellipsoidal harmonics, with applications to problems in mathematical physics.* ix+287. Dover Publications, New York.

Cabella, Paolo, and Domenico Marinucci. 2009. 'Statistical challenges in the analysis of cosmic microwave background radiation'. *Ann. Appl. Stat.* 3 (1): 61–95. ISSN: 1932-6157. https://doi.org/10.1214/08-AOAS190.

Cai, Hongbo, Yilun Guan, Toshiya Namikawa, and Arthur Kosowsky. 2023. 'Impact of anisotropic birefringence on measuring cosmic microwave background lensing'. *Phys. Rev. D* 107 (4): Paper No. 043513, 13. ISSN: 2470-0010. https://doi.org/10.1103/physrevd.107.043513.

Calabrese, Matteo, Carmelita Carbone, Giulio Fabbian, Marco Baldi, and Carlo Baccigalupi. 2015. 'Multiple lensing of the cosmic microwave background anisotropies'. *J. Cosmol. Astropart. Phys.*, no. 3, 049, front matter+29. https://doi.org/10.1088/1475-7516/2015/03/049.

Campbell, S. J., and John Wainwright. 1977. 'Algebraic computing and the Newman-Penrose formalism in general relativity'. *Gen. Relativity Gravitation* 8 (12): 987–1001. ISSN: 0001-7701. https://doi.org/10.1007/bf00759742.

Caprini, Chiara, and Daniel G Figueroa. 2018. 'Cosmological backgrounds of gravitational waves'. *Classical and Quantum Gravity* 35, no. 16 (July): 163001. https://dx.doi.org/10.1088/1361-6382/aac608.

Cardoso, J. G. 2005. 'The Infeld-van der Waerden formalisms for general relativity'. *Czechoslovak J. Phys.* 55 (4): 401–462. ISSN: 0011-4626. https://doi.org/10.1007/s10582-005-0051-9.

Carlip, Steven Jonathan. 2019. *General Relativity: A Concise Introduction.* Oxford University Press. ISBN: 9780198822158. https://doi.org/10.1093/oso/9780198822158.001.0001.

Carmeli, Moshe, and Shimon Malin. 2000. *Theory of spinors. An introduction.* xiv+212. World Scientific Publishing Co., River Edge, NJ. ISBN: 981-02-4261-1. https://doi.org/10.1142/4380.

Carroll, Sean. 2004. *Spacetime and geometry. An introduction to general relativity.* xiv+513. Addison Wesley, San Francisco, CA. ISBN: 0-8053-8732-3.

Carrón Duque, Javier, and Domenico Marinucci. 2023. *Geometric Methods for Spherical Data, with Applications to Cosmology.* arXiv: 2303.15278 [astro-ph.CO].

Cartan, Élie. 1894. *Sur la structure des groupes de transformations finis et continus.* Vol. 826. Thèses présentées a la Faculté des Sciences de Paris pour obtenir le grade de docteur ès sciences mathématiques. Paris: Nony.

————. 1913. 'Les groupes projectifs qui ne laissent invariante aucune multiplicité plane'. *Bull. Soc. Math. France* 41:53–96. ISSN: 0037-9484. http://www.numdam.org/item?id=BSMF_1913__41__53_1.

————. 1981. *The theory of spinors.* v+157. Dover Books on Advanced Mathematics. With a foreword by Raymond Streater, A reprint of the 1966 English translation. New York: Dover Publications. ISBN: 0-486-64070-1.

Cartan, Henri, and Samuel Eilenberg. 1999. *Homological algebra.* xvi+390. Princeton Landmarks in Mathematics. With an appendix by David A. Buchsbaum, Reprint of the 1956 original. Princeton University Press, Princeton, NJ. ISBN: 0-691-04991-2.

Cayuso, Juan, Richard Bloch, Selim C. Hotinli, Matthew Craig Johnson, and Fiona McCarthy. 2023. 'Velocity reconstruction with the cosmic microwave background and galaxy surveys'. *J. Cosmol. Astropart. Phys.,* no. 2, Paper No. 051, 86.

Cenzer, Douglas, Jean Larson, Christopher Porter, and Jindrich Zapletal. 2020. *Set theory and foundations of mathematics – an introduction to mathematical logic. Vol. 1. Set theory.* xi+209. World Scientific, Hackensack, NJ. ISBN: 978-981-120-193-6.

Cervantes-Cota, Jorge L., Salvador Galindo-Uribarri, and George F. Smoot. 2016. 'A Brief History of Gravitational Waves'. *Universe* 2 (3). ISSN: 2218-1997. https://www.mdpi.com/2218-1997/2/3/22.

Challinor, Anthony. 2000a. 'Microwave background anisotropies from gravitational waves: the $1 + 3$ covariant approach'. *Classical and Quantum Gravity* 17, no. 4 (February): 871. https://doi.org/10.1088/0264-9381/17/4/309.

———. 2000b. 'Microwave background polarization in cosmological models'. *Phys. Rev. D* 62 (4): 043004. https://doi.org/10.1103/PhysRevD.62.043004.

Chandra, Debabrata, and Supratik Pal. 2018. 'A proposal for constraining initial vacuum by cosmic microwave background'. *Classical Quantum Gravity* 35 (1): 015008, 21. ISSN: 0264-9381. https://doi.org/10.1088/1361-6382/aa95f7.

Chandrasekhar, Subrahmanyan. 1960. *Radiative transfer.* xiv+393. Dover Publications, New York.

Chang, Chia-Feng, and Yanou Cui. 2020. 'Stochastic gravitational wave background from global cosmic strings'. *Physics of the Dark Universe* 29:100604. ISSN: 2212-6864. https://www.sciencedirect.com/science/article/pii/S2212686420301072.

Chen, Zu-Cheng, Yu-Mei Wu, and Qing-Guo Huang. 2022. 'Search for the Gravitational-wave Background from Cosmic Strings with the Parkes Pulsar Timing Array Second Data Release'. *The Astrophysical Journal* 936, no. 1 (August): 20. https://dx.doi.org/10.3847/1538-4357/ac86cb.

Chen, Joe Zhiyu, Amol Upadhye, and Yvonne Y.Y. Wong. 2021. 'The cosmic neutrino background as a collection of fluids in large-scale structure simulations'. *Journal of Cosmology and Astroparticle Physics* 2021, no. 03 (March): 065. https://dx.doi.org/10.1088/1475-7516/2021/03/065.

Cheng, Ta-Pei. 2010. *Relativity, gravitation and cosmology. A basic introduction,* Second ed. 11:xiv+435. Oxford Master Series in Particle Physics, Astrophysics, and Cosmology. Oxford University Press, Oxford. ISBN: 978-0-19-957364-6.

Chern, Shiing-shen. 1995. *Complex manifolds without potential theory (with an appendix on the geometry of characteristic classes).* Second ed. vi+160. Universitext. Springer-Verlag, New York. ISBN: 0-387-90422-0.

Chevalley, Claude. 1997. *The algebraic theory of spinors and Clifford algebras.* xiv+214. Collected works. Vol. 2, Edited and with a foreword by Pierre Cartier and Catherine Chevalley, With a postface by J.-P. Bourguignon. Berlin: Springer-Verlag. ISBN: 3-540-57063-2.

Choi, Cheol Ho, Joseph Ivanic, Mark S. Gordon, and Klaus Ruedenberg. 1999. 'Rapid and stable determination of rotation matrices between spherical harmonics by direct recursion'. *J. Chem. Phys.* 111 (19): 8825–8831.

Choquet-Bruhat, Yvonne. 1968. 'Construction de solutions radiatives approchées des équations d'Einstein'. *Atti Accad. Naz. Lincei Rend. Cl. Sci. Fis. Mat. Nat. (8)* 44:649–652. ISSN: 0392-7881.

Chruściel, Piotr T. 2019. *Elements of general relativity*. ix+283. Compact Textbooks in Mathematics. Birkhäuser/Springer, Cham. ISBN: 978-3-030-28416-9. https://doi.org/10.1007/978-3-030-28416-9.

Clarke, Thomas J., Edmund J. Copeland, and Adam Moss. 2020. 'Constraints on primordial gravitational waves from the cosmic microwave background'. *J. Cosmol. Astropart. Phys.*, no. 10, 002, 31. https://doi.org/10.1088/1475-7516/2020/10/002.

Clifford, William Kingdon. 1873. 'Preliminary Sketch of Biquaternions'. *Proc. Lond. Math. Soc.* 4:381–395. ISSN: 0024-6115. https://doi.org/10.1112/plms/s1-4.1.381.

Clifton, Timothy, Pedro G. Ferreira, Antonio Padilla, and Constantinos Skordis. 2012. 'Modified gravity and cosmology'. *Physics Reports* 513 (1): 1–189. ISSN: 0370-1573. https://doi.org/10.1016/j.physrep.2012.01.001.

Conneely, Ciarán, Andrew H. Jaffe, and Chiara M. F. Mingarelli. 2019. 'On the amplitude and Stokes parameters of a stochastic gravitational-wave background'. *Monthly Notices of the Royal Astronomical Society* 487, no. 1 (April): 562–579. ISSN: 0035-8711. https://doi.org/10.1093/mnras/stz1022.

Corinaldesi, Ernesto, and Franco Strocchi. 1963. *Relativistic wave mechanics.* x+310. North-Holland Publishing Co., Amsterdam; Interscience Publishers John Wiley & Sons, Inc., New York.

Curtis, Charles W. 1999. *Pioneers of representation theory: Frobenius, Burnside, Schur, and Brauer.* 15:xvi+287. History of Mathematics. American Mathematical Society, Providence, RI; London Mathematical Society, London. ISBN: 0-8218-9002-6. https://doi.org/10.1090/hmath/015.

Curtis, W. Dan, and David E. Lerner. 1978. 'Complex line bundles in relativity'. *J. Mathematical Phys.* 19 (4): 874–877. ISSN: 0022-2488. https://doi.org/10.1063/1.523750.

Cusin, Giulia, Ruth Durrer, and Pedro G. Ferreira. 2019. 'Polarization of a stochastic gravitational wave background through diffusion by massive structures'. *Phys. Rev. D* 99 (2): 023534. https://link.aps.org/doi/10.1103/PhysRevD.99.023534.

Dąbrowski, Ludwik. 1988. *Group actions on spinors.* 9:viii+119. Monographs and Textbooks in Physical Science. Lecture Notes. Bibliopolis, Naples. ISBN: 88-7088-205-5.

Das, Anirban, Yuber F. Perez-Gonzalez, and Manibrata Sen. 2022. 'Neutrino secret self-interactions: A booster shot for the cosmic neutrino background'. *Phys. Rev. D* 106 (9): 095042. https://link.aps.org/doi/10.1103/PhysRevD.106.095042.

Das, Santanu, and Anh Phan. 2020. 'Cosmic microwave background anisotropy numerical solution (CMBAns). Part I. An introduction to C_l calculation'. *J. Cosmol. Astropart. Phys.*, no. 5, 00649. https://doi.org/10.1088/1475-7516/2020/05/006.

Das, Santanu, and Tarun Souradeep. 2014. 'SCoPE: an efficient method of Cosmological Parameter Estimation'. *Journal of Cosmology and Astroparticle Physics* 2014, no. 07 (July): 018. https://doi.org/10.1088/1475-7516/2014/07/018.

Davis, Tamara M., and Charles H. Lineweaver. 2004. 'Expanding Confusion: Common Misconceptions of Cosmological Horizons and the Superluminal Expansion of the Universe'. *Publications of the Astronomical Society of Australia* 21 (1): 97–109. https://doi.org/10.1071/AS03040.

Davydov, A. S. 1976. *Quantum mechanics.* 1:xiii+636. International Series in Natural Philosophy. Translated from the second Russian edition, edited and with additions by D. ter Haar. Pergamon Press, Oxford–New York–Toronto.

Degl'Innocenti, Egidio Landi. 2013. *Atomic Spectroscopy and Radiative Processes.* UNITEXT for physics. Springer.

Deruelle, Nathalie, and Jean-Philippe Uzan. 2018. *Relativity in modern physics.* xi+691. Oxford Graduate Texts. With a foreword by David Langlois. Oxford University Press, Oxford. ISBN: 978-0-19-878639-9. https://doi.org/10.1093/oso/9780198786399.001.0001.

Dhurandhar, Sanjeev, and Sanjit Mitra. 2022. *General Relativity and Gravitational Waves. Essentials of Theory and Practice.* Springer.

Diek, Adel, and R. Kantowski. 1995. 'Some Clifford algebra history'. In *Clifford algebras and spinor structures,* 321:3–12. Math. Appl. Kluwer Acad. Publ., Dordrecht. https://doi.org/10.1007/978-94-015-8422-7_1.

Dodelson, Scott, and Fabian Schmidt. 2021. *Modern Cosmology.* Second ed. Academic Press. ISBN: 978-0-12-815948-4. https://www.elsevier.com/books/modern-cosmology/dodelson/978-0-12-815948-4.

Dodelson, Scott, and Mika Vesterinen. 2009. 'Cosmic Neutrino Last Scattering Surface'. *Phys. Rev. Lett.* 103 (17): 171301. https://link.aps.org/doi/10.1103/PhysRevLett.103.171301.

Dolan, Sam R. 2018. 'Geometrical optics for scalar, electromagnetic and gravitational waves on curved spacetime'. *International Journal of Modern Physics D* 27 (11): 1843010. https : / / doi . org / 10 . 1142 / S0218271818430101.

Doran, Michael. 2005. 'CMBEASY: an object oriented code for the cosmic microwave background'. *Journal of Cosmology and Astroparticle Physics* 2005, no. 10 (October): 011. https://doi.org/10.1088/1475-7516/2005/10/011.

Doran, Michael, and Christian M. Müller. 2004. 'Analyse this! A cosmological constraint package for CMBEASY'. *Journal of Cosmology and Astroparticle Physics* 2004, no. 09 (September): 003. https://doi.org/10.1088/1475-7516/2004/09/003.

Dray, Tevian. 1985. 'The relationship between monopole harmonics and spin-weighted spherical harmonics'. *J. Math. Phys.* 26 (5): 1030–1033. ISSN: 0022-2488. https://doi.org/10.1063/1.526533.

———. 1986. 'A unified treatment of Wigner $\mathcal{D}$ functions, spin-weighted spherical harmonics, and monopole harmonics'. *J. Math. Phys.* 27 (3): 781–792. ISSN: 0022-2488. https://doi.org/10.1063/1.527183.

Duda, Gintaras, Graciela Gelmini, and Shmuel Nussinov. 2001. 'Expected signals in relic neutrino detectors'. *Phys. Rev. D* 64 (12): 122001. https://link.aps.org/doi/10.1103/PhysRevD.64.122001.

Duistermaat, Johannes Jisse, and Johan Antoon Casper Kolk. 2000. *Lie groups.* viii+344. Universitext. Springer-Verlag, Berlin. ISBN: 3-540-15293-8. https://doi.org/10.1007/978-3-642-56936-4.

Durastanti, Claudio. 2017. 'Tail behavior of Mexican needlets'. *J. Math. Anal. Appl.* 447 (2): 716–735. ISSN: 0022-247X. https://doi.org/10.1016/j.jmaa.2016.10.046.

Durrer, Ruth. 2020. *The Cosmic Microwave Background.* Second ed. Cambridge University Press, Cambridge.

Durrett, Rick. 2019. *Probability—theory and examples.* 49:xii+419. Cambridge Series in Statistical and Probabilistic Mathematics. Fifth edition. Cambridge University Press, Cambridge. ISBN: 978-1-108-47368-2. https://doi.org/10.1017/9781108591034.

Dyson, Freeman. 2011. *Advanced quantum mechanics.* Second ed. xxvi+289. Translated and transcribed from the French by David Derbes, With a foreword by David Kaiser. World Scientific, Hackensack, NJ. ISBN: 978-981-4383-41-7. https://doi.org/10.1142/8356.

Eardley, Douglas M., David L. Lee, and Alan P. Lightman. 1973. 'Gravitational-Wave Observations as a Tool for Testing Relativistic Gravity'. *Phys. Rev. D* 8 (10): 3308–3321. https://doi.org/10.1103/PhysRevD. 8.3308. https://link.aps.org/doi/10.1103/PhysRevD.8.3308.

Earman, John. 1989. *World enough and space-time. Absolute versus relational theories of space and time.* xvi+233. A Bradford Book. MIT Press, Cambridge, MA. ISBN: 0-262-05040-4.

Earman, John, Clark Glymour, and John Stachel, eds. 1977. *Foundations of space-time theories.* Vol. VIII. Minnesota studies in the philosophy of science. University of Minnesota Press, Minneapolis.

Eastwood, Michael, and Paul Tod. 1982. 'Edth—a differential operator on the sphere'. *Math. Proc. Cambridge Philos. Soc.* 92 (2): 317–330. ISSN: 0305-0041. https://doi.org/10.1017/S0305004100059971.

Edmonds, A. R. 1957. *Angular momentum in quantum mechanics.* 4:viii+146. Investigations in Physics. Princeton University Press, Princeton, N.J.

Efthimiou, Costas, and Christopher Frye. 2014. *Spherical harmonics in p dimensions.* xii+143. World Scientific, Hackensack, NJ. ISBN: 978-981-4596-69-5. https://doi.org/10.1142/9134.

Eguchi, Tohru, Peter B. Gilkey, and Andrew J. Hanson. 1980. 'Gravitation, gauge theories and differential geometry'. *Phys. Rep.* 66 (6): 213–393. ISSN: 0370-1573. https://doi.org/10.1016/0370-1573(80)90130-1.

Einstein, Albert. 1916. 'Näherungsweise Integration der Feldgleichungen der Gravitation'. *Sitzungsber. Preuss. Akad. Wiss. Berlin (Math. Phys.)* (January): 688–696. https://ui.adsabs.harvard.edu/abs/1916SPAW. ......688E.

———. 1917. 'Kosmologische Betrachtungen zur allgemeinen Relativitätstheorie'. *Sitzungsber. Preuss. Akad. Wiss. Berlin (Math. Phys.)* (January): 142–152.

———. 1931. 'Zum kosmologischen Problem der allgemeinen Relativitätstheorie'. *Sitzungsber. Preuss. Akad. Wiss. Berlin (Math. Phys.),* 235–237.

Ereditato, Antonio, ed. 2018. *State of the Art of Neutrino Physics: A Tutorial for Graduate Students and Young Researchers.* Vol. 28. Advanced series on directions in high energy physics. World Scientific, Singapore.

Esposito, Giampiero. 1995. *Complex general relativity.* 69:xii+201. Fundamental Theories of Physics. Kluwer Academic Publishers Group, Dordrecht. ISBN: 0-7923-3340-3.

Exton, Albert R., Ezra Ted Newman, and Roger Penrose. 1969. 'Conserved quantities in the Einstein–Maxwell theory'. *J. Mathematical Phys.* 10:1566–1570. ISSN: 0022-2488. https://doi.org/10.1063/1.1665006.

Fano, Ugo, and Giulio Racah. 1959. *Irreducible tensorial sets.* 4:vii+171. Pure and Applied Physics. Academic Press, New York.

Faria, Edson de, and Welington de Melo. 2010. *Mathematical aspects of quantum field theory.* 127:xiv+298. Cambridge Studies in Advanced Mathematics. With a foreword by Dennis Sullivan. Cambridge University Press, Cambridge. ISBN: 978-0-521-11577-3. https://doi.org/10.1017/CBO9780511760532.

Felice, Fernando de, and Chris J. S. Clarke. 1990. *Relativity on curved manifolds.* xii+448. Cambridge Monographs on Mathematical Physics. Cambridge University Press, Cambridge. ISBN: 0-521-26639-4.

Ferrari, Valeria, Leonardo Gualtieri, and Paolo Pani. 2020. *General Relativity.* Taylor & Francis Group.

Ferrers, Norman Macleod. 1877. *An elementary treatise on spherical harmonics and subjects connected with them.* Cambridge University Press, Cambridge.

Floerchinger, Stefan. 2021. 'Real Clifford Algebras and Their Spinors for Relativistic Fermions'. *Universe* 7 (6). ISSN: 2218-1997. https://doi.org/10.3390/universe7060168.

Folland, Gerald B. 2008. *Quantum field theory. A tourist guide for mathematicians.* 149:xii+325. Mathematical Surveys and Monographs. American Mathematical Society, Providence, RI. ISBN: 978-0-8218-4705-3. https://doi.org/10.1090/surv/149.

———. 2016. *A course in abstract harmonic analysis.* Second ed. xiii+305 pp.+loose errata. Textbooks in Mathematics. CRC Press, Boca Raton, FL. ISBN: 978-1-4987-2713-6.

Follin, Brent, Lloyd Knox, Marius Millea, and Zhen Pan. 2015. 'First Detection of the Acoustic Oscillation Phase Shift Expected from the Cosmic Neutrino Background'. *Phys. Rev. Lett.* 115 (9): 091301. https://link.aps.org/doi/10.1103/PhysRevLett.115.091301.

Foster, James, and J. David Nightingale. 2006. *A short course in general relativity.* Third ed. x+292. Springer, New York. ISBN: 978-0387-26078-5. https://doi.org/10.1007/978-0-387-27583-3.

Francis, Matthew R., and Arthur Kosowsky. 2005. 'The construction of spinors in geometric algebra'. *Annals of Physics* 317 (2): 383–409. ISSN: 0003-4916. https://www.sciencedirect.com/science/article/pii/S0003491604002209.

Frankel, Theodore. 1979. *Gravitational curvature. An introduction to Einstein's theory.* xviii+172. W. H. Freeman, San Francisco, Calif. ISBN: 0-7167-1062-5.

Freeden, Willi, Volker Michel, and Frederik J. Simons. 2018. 'Spherical harmonics based special function systems and constructive approximation methods'. In *Handbook of mathematical geodesy. Functional analytic and potential theoretic methods,* edited by Willi Freeden and M. Zuhair Nashed, 753–819. Geosyst. Math. Birkhäuser/Springer, Cham.

Friedman, Michael. 1983. *Foundations of space-time theories.* xvi+386. Relativistic physics and philosophy of science. Princeton University Press, Princeton, NJ. ISBN: 0-691-02039-6.

Friedrich, Thomas, and Andrzej Trautman. 2000. 'Spin spaces, Lipschitz groups, and spinor bundles', 18:221–240. 3-4. Special issue in memory of Alfred Gray (1939–1998). https://doi.org/10.1023/A:1006713405277.

Frobenius, Ferdinand Georg. 1897. 'Über die Darstellung der endlichen Gruppen durch lineare Substitutionen'. *Sitz. Preuss. Akad. Wiss.,* 944–1015.

———. 1898. 'Über Relationen zwischen den Charakteren einer Gruppe unddenen ihrer Untergruppen'. *Sitz. Preuss. Akad. Wiss.,* 501–515.

Frobenius, Ferdinand Georg, and Issai Schur. 1906. 'Über die reellen Darstellungen der endlichen Gruppen'. *Sitz. Preuss. Akad. Wiss.,* 186–208.

Fuente, Daniel de la, José A. S. Pelegrín, and Rafael María Rubio. 2021. 'On the geometry of stationary Galilean spacetimes'. *Gen. Relativity Gravitation* 53 (1): Paper No. 8, 15. ISSN: 0001-7701. https://doi.org/10.1007/s10714-020-02772-1.

———. 2022. 'Completeness of uniformly accelerated observers in Galilean spacetimes'. *Lett. Math. Phys.* 112 (6): Paper No. 111, 10. ISSN: 0377-9017. https://doi.org/10.1007/s11005-022-01575-6.

Fukugita, Masataka, and Tsutomu Yanagida. 2003. *Physics of Neutrinos and Application to Astrophysics.* Texts and monographs in physics. Springer.

Fulton, William, and Joe Harris. 1991. *Representation theory. A first course.* 129:xvi+551. Graduate Texts in Mathematics. Springer-Verlag, New York. ISBN: 0-387-97527-6. https://doi.org/10.1007/978-1-4612-0979-9.

Gallier, Jean, and Jocelyn Quaintance. 2020a. *Differential geometry and Lie groups—a computational perspective.* 12:xv+777. Geometry and Computing. Springer, Cham. ISBN: 978-3-030-46039-6.

———. 2020b. *Differential geometry and Lie groups—a second course.* 13:xiv+620. Geometry and Computing. Springer, Cham. ISBN: 978-3-030-46046-4.

Garling, D. J. H. 2011. *Clifford algebras: an introduction.* 78:viii+200. London Mathematical Society Student Texts. Cambridge University Press, Cambridge. ISBN: 978-1-107-42219-3. https : / / doi . org / 10 . 1017 / CBO9780511972997.

Gel'fand, I. M., and Z. Ya. Šapiro. 1952. 'Representations of the group of rotations in three-dimensional space and their applications'. *Uspehi Matem. Nauk (N.S.)* 7 (1(47)): 3–117. ISSN: 0042-1316.

Geller, Daryl, and Domenico Marinucci. 2010. 'Spin wavelets on the sphere'. *J. Fourier Anal. Appl.* 16 (6): 840–884. ISSN: 1069-5869. https://doi.org/10.1007/s00041-010-9128-3.

———. 2011. 'Mixed needlets'. *J. Math. Anal. Appl.* 375 (2): 610–630. ISSN: 0022-247X. https://doi.org/10.1016/j.jmaa.2010.09.046.

Gentili, Stefano. 2020. *Measure, integration and a primer on probability theory.* Vol. 1. 125:xi+463. Unitext. Translated from the Italian original by Simon G. Chiossi, La Matematica per il 3+2. Springer, Cham. ISBN: 978-3-030-54939-8. https://doi.org/10.1007/978-3-030-54940-4.

Geroch, Robert P., Alan Held, and Roger Penrose. 1973. 'A space-time calculus based on pairs of null directions'. *J. Mathematical Phys.* 14:874–881. ISSN: 0022-2488. https://doi.org/10.1063/1.1666410.

Giaquinta, Mariano, and Stefan Hildebrandt. 1996a. *Calculus of variations. I. The Lagrangian formalism.* 310:xxx+474. Grundlehren der mathematischen Wissenschaften [Fundamental Principles of Mathematical Sciences]. Springer-Verlag, Berlin. ISBN: 3-540-50625-X.

———. 1996b. *Calculus of variations. II. The Hamiltonian formalism.* 311:xxx+652. Grundlehren der mathematischen Wissenschaften [Fundamental Principles of Mathematical Sciences]. Springer-Verlag, Berlin. ISBN: 3-540-57961-3.

Gibson, W. M., and B. R. Pollard. 1976. *Symmetry principles in elementary particle physics.* xii+380. Cambridge Monographs on Physics. Cambridge University Press, Cambridge–New York–Melbourne.

Gil Pérez, José J., and Razvigor Ossikovski. 2016. *Polarized light and the Mueller matrix approach.* xxi+383. Series in Optics and Optoelectronics. CRC Press, Boca Raton, FL. ISBN: 978-1-4822-5155-5.

Gilmore, Robert. 2012. 'Relations among low-dimensional simple Lie groups'. *J. Geom. Symmetry Phys.* 28:1–45. ISSN: 1312-5192.

Gimbutas, Zydrunas, and L. Greengard. 2009. 'A fast and stable method for rotating spherical harmonic expansions'. *J. Comp. Phys.* 228 (16): 5621–5627. ISSN: 0021-9991.

Giovannini, Massimo. 2018. 'Probing large-scale magnetism with the cosmic microwave background'. *Classical Quantum Gravity* 35 (8): 084003, 108. ISSN: 0264-9381. https://doi.org/10.1088/1361-6382/aab17d.

Gluscevic, Vera, Marc Kamionkowski, and Asantha Cooray. 2009. 'Derotation of the cosmic microwave background polarization: Full-sky formalism'. *Phys. Rev. D* 80 (2): 023510. https://link.aps.org/doi/10.1103/PhysRev D.80.023510.

Godement, Roger. 2017. *Introduction to the theory of Lie groups.* ix+293. Universitext. Translated from the 2004 French edition by Urmie Ray. Springer, Cham. ISBN: 978-3-319-54373-4. https://doi.org/10.1007/978-3-319-54375-8.

Goldberg, Joshua N., Alan J. Macfarlane, Ezra Ted Newman, Fritz Rohrlich, and Ennackel Chandy George Sudarshan. 1967. 'Spin-s spherical harmonics and ð'. *J. Mathematical Phys.* 8:2155–2161. ISSN: 0022-2488. https://doi.org/10.1063/1.1705135.

Gómez, Roberto, Luis R. Lehner, Philippos Papadopoulos, and Jeffrey Winicour. 1997. 'The eth formalism in numerical relativity'. *Classical Quantum Gravity* 14 (4): 977–990. ISSN: 0264-9381. https://doi.org/10.1088/0264-9381/14/4/013.

Gourgoulhon, Éric. 2012. *3+1 formalism in general relativity. Bases of numerical relativity.* 846:xviii+294. Lecture Notes in Physics. Springer, Heidelberg. ISBN: 978-3-642-24524-4. https://doi.org/10.1007/978-3-642-24525-1.

Grant, Ian P. 2007. *Relativistic quantum theory of atoms and molecules. Theory and computation.* 40:xxiv+797. Springer Series on Atomic, Optical, and Plasma Physics. Springer, New York. ISBN: 978-0-387-34671-7. https://doi.org/10.1007/978-0-387-35069-1.

Greub, Werner, Stephen Halperin, and Ray Vanstone. 1972. *Connections, curvature, and cohomology. Vol. I: De Rham cohomology of manifolds and vector bundles.* 47:xix+443. Pure and Applied Mathematics. Academic Press, New York–London.

———. 1973. *Connections, curvature, and cohomology. Vol. II: Lie groups, principal bundles, and characteristic classes.* 47-II:xxi+541. Pure and Applied Mathematics. Academic Press [Harcourt Brace Jovanovich, Publishers], New York–London.

———. 1976. *Connections, curvature, and cohomology. Cohomology of principal bundles and homogeneous spaces.* 47-III:xxi+593. Pure and Applied Mathematics. Academic Press [Harcourt Brace Jovanovich, Publishers], New York–London.

Griffiths, Jerry B. 1991. *Colliding plane waves in general relativity.* xiv+232. Oxford Mathematical Monographs. Oxford Science Publications. The Clarendon Press, Oxford University Press, New York. ISBN: 0-19-853209-1.

Griffiths, Jerry B., and R. A. Newing. 1971. 'A note on neutrino radiation fields with positive energy density'. *J. Phys. A: Gen. Phys.* 4, no. 5 (September): L81. https://doi.org/10.1088/0305-4470/4/5/016.

Griffiths, Jerry B., and Jiří Podolský. 2009. *Exact space-times in Einstein's general relativity.* xviii+525. Cambridge Monographs on Mathematical Physics. Cambridge University Press, Cambridge. ISBN: 978-0-521-88927-8. https://doi.org/10.1017/CBO9780511635397.

Grimmett, Geoffrey R., and David R. Stirzaker. 2020. *Probability and random processes.* xii+669. Fourth edition. Oxford University Press, Oxford. ISBN: 978-0-19-884759-5.

Grimus, Walter, and Helmuth K. Urbantke. 1997. 'A note on the reality of tensor products'. In *Geometry and physics,* vol. 14, A165–A170. 1A. A special issue in honor of Andrzej Trautman on the occasion of his 64th birthday, Classical Quantum Gravity 14 (1997), no. 1A. https://doi.org/10.1088/0264-9381/14/1A/014.

Grøn, Øyvind, and Sigbjørn Hervik. 2007. *Einstein's general theory of relativity. With modern applications in cosmology.* xx+538. Springer, New York. ISBN: 978-0-387-69199-2. https://doi.org/10.1007/978-0-387-69200-5.

Guan, Yilun, and Arthur Kosowsky. 2022. 'Distinguishing primordial magnetic fields from inflationary tensor perturbations in the cosmic microwave background'. *Phys. Rev. D* 106 (6): Paper No. 063505, 18. ISSN: 2470-0010. https://doi.org/10.1103/physrevd.106.063505.

Guedes, G. S. F., P. P. Avelino, and L. Sousa. 2018. 'Signature of inflation in the stochastic gravitational wave background generated by cosmic string networks'. *Phys. Rev. D* 98 (12): 123505. https://link.aps.org/doi/10.1103/PhysRevD.98.123505.

Guidry, Mike. 2019. *Modern General Relativity: Black Holes, Gravitational Waves, and Cosmology.* Cambridge University Press, Cambridge. https://doi.org/10.1017/9781108181938.

Hall, Brian. 2015. *Lie groups, Lie algebras, and representations.* 2nd ed. 222:xiv+449. Graduate Texts in Mathematics. An elementary introduction. Springer, Cham. ISBN: 978-3-319-13466-6. https://doi.org/10.1007/978-3-319-13467-3.

Hamann, Jan, Quôc Thông Lê Gia, Ian H. Sloan, Yu Guang Wang, and Robert S. Womersley. 2021. 'A new probe of Gaussianity and isotropy with application to cosmic microwave background maps'. *Internat. J. Modern Phys. C* 32 (6): Paper No. 2150084, 26. ISSN: 0129-1831. https://doi.org/10.1142/S0129183121500844.

Hamilton, Mark J. D. 2017. *Mathematical gauge theory. With applications to the standard model of particle physics.* xviii+657. Universitext. Springer, Cham. ISBN: 978-3-319-68438-3. https://doi.org/10.1007/978-3-319-68439-0.

Hamilton, William Rowan. 1844. 'LXXVIII. On quaternions; or on a new system of imaginaries in Algebra'. *The London, Edinburgh, and Dublin Philosophical Magazine and Journal of Science* 25 (169): 489–495. eprint: https://doi.org/10.1080/14786444408645047.

———. 1854. 'LXXVII. On some extensions of quaternions'. *The London, Edinburgh, and Dublin Philosophical Magazine and Journal of Science* 7 (48): 492–499. https://doi.org/10.1080/14786445408651871.

Hannestad, Steen. 2010. 'Neutrino physics from precision cosmology'. *Progress in Particle and Nuclear Physics* 65 (2): 185–208. ISSN: 0146-6410. https://www.sciencedirect.com/science/article/pii/S0146641010000499.

Hannestad, Steen, and Jacob Brandbyge. 2010. 'The Cosmic Neutrino Background anisotropy—linear theory'. *Journal of Cosmology and Astroparticle Physics* 2010, no. 03 (March): 020. https://doi.org/10.1088/1475-7516/2010/03/020.

Harish-Chandra. 1946. 'A note on the σ-symbols'. *Proc. Indian Acad. Sci., Sect. A.* 23:152–163.

Hartle, James B. 2013. *Gravity: An Introduction to Einstein's General Relativity.* Pearson Education.

Harvey, Frank Reese. 1990. *Spinors and calibrations.* 9:xiv+323. Perspectives in Mathematics. Boston, MA: Academic Press. ISBN: 0-12-329650-1.

Hawking, Stephen W., and George Francis Rayner Ellis. 2023. *The large scale structure of space-time. 50th anniversary edition.* xi+391. Cambridge Monographs on Mathematical Physics, No. 1. Cambridge University Press, London–New York.

Hill, E. L. 1954. 'The theory of vector spherical harmonics'. *Amer. J. Phys.* 22:211–214. ISSN: 0002-9505. https://doi.org/10.1119/1.1933682.

Hilton, Peter John, and Urs Stammbach. 1997. *A course in homological algebra.* Second ed. 4:xii+364. Graduate Texts in Mathematics. Springer-Verlag, New York. ISBN: 0-387-94823-6. https://doi.org/10.1007/978-1-4419-8566-8.

Hobson, E. W. 1955. *The theory of spherical and ellipsoidal harmonics.* xi+500. Chelsea Publishing Co., New York.

Hobson, Michael P., George P. Efstathiou, and Anthony N. Lasenby. 2006. *General relativity. An introduction for physicists.* xviii+572 pp.+errata. Cambridge University Press, Cambridge. ISBN: 978-0-521-82951-9. https://doi.org/10.1017/CBO9780511790904.

Hofmann, Karl Heinrich, and Sidney A. Morris. 2020. *The structure of compact groups—a primer for the student—a handbook for the expert.* Fourth ed. 25:xxvii+1006. De Gruyter Studies in Mathematics. De Gruyter, Berlin. ISBN: 978-3-11-069599-1. https://doi.org/10.1515/9783110695991.

Hsiang, Wu-Yi. 2017. *Lectures on Lie groups.* Second. 9:viii+152. Series on University Mathematics. World Scientific Publishing Co., Hackensack, NJ. ISBN: 978-981-4740-71-5. https://doi.org/10.1142/9912.

Hu, Wayne, Douglas Scott, Naoshi Sugiyama, and Martin White. 1995. 'Effect of physical assumptions on the calculation of microwave background anisotropies'. *Phys. Rev. D* 52 (10): 5498–5515. https://doi.org/10.1103/PhysRevD.52.5498.

Husemoller, Dale. 1994. *Fibre bundles.* Third ed. 20:xx+353. Graduate Texts in Mathematics. Springer-Verlag, New York. ISBN: 0-387-94087-1. https://doi.org/10.1007/978-1-4757-2261-1.

Husemöller, Dale, Michael Joachim, Branislav Jurčo, and Martin Schottenloher. 2008. *Basic bundle theory and K-cohomology invariants.* 726:xvi+340. Lecture Notes in Physics. With contributions by Siegfried Echterhoff, Stefan Fredenhagen and Bernhard Krötz. Springer, Berlin. ISBN: 978-3-540-74955-4.

Huybrechts, Daniel. 2005. *Complex geometry. An introduction.* xii+309. Universitext. Springer-Verlag, Berlin. ISBN: 3-540-21290-6.

Iglewska-Nowak, Ilona. 2019. 'Spin weighted spherical wavelets derived from approximate identities'. *J. Math. Anal. Appl.* 479 (1): 242–259. ISSN: 0022-247X. https://doi.org/10.1016/j.jmaa.2019.06.025.

———. 2022. 'Spin weighted wavelets on the sphere–Frames'. *J. Comput. Appl. Math.* 407:Paper No. 114078, 11. ISSN: 0377-0427. https://doi.org/10.1016/j.cam.2021.114078.

Inayoshi, Kohei, Kazumi Kashiyama, Eli Visbal, and Zoltán Haiman. 2021. 'Gravitational Wave Backgrounds from Coalescing Black Hole Binaries at Cosmic Dawn: An Upper Bound'. *The Astrophysical Journal* 919, no. 1 (September): 41. https://dx.doi.org/10.3847/1538-4357/ac106d.

Infeld, Leopold, and Bartel Leendert van der Waerden. 1933. 'Die Wellengleichung des Elektrons in der allgemeinen Relativitätstheorie'. *Sitz. Preuss. Akad. Wiss.*, 380–401.

Inomata, Keisuke, and Marc Kamionkowski. 2019. 'Circular polarization of the cosmic microwave background from vector and tensor perturbations'. *Phys. Rev. D* 99 (4): 043501, 15. ISSN: 2470-0010. https://doi.org/10.1103/physrevd.99.043501.

Isaacson, Richard A. 1968. 'Gravitational Radiation in the Limit of High Frequency. II. Nonlinear Terms and the Effective Stress Tensor'. *Phys. Rev.* 166 (5): 1272–1280. https://link.aps.org/doi/10.1103/PhysRev.166.1272.

Itzkowitz, Gerald, Sheldon Rothman, and Helen Strassberg. 1991. 'A note on the real representations of SU(2, **C**)'. *J. Pure Appl. Algebra* 69 (3): 285–294. ISSN: 0022-4049. https://doi.org/10.1016/0022-4049(91)90023-U.

Itzykson, Claude, and Jean Bernard Zuber. 1980. *Quantum field theory.* xxii+705. International Series in Pure and Applied Physics. McGraw-Hill International Book Co., New York. ISBN: 0-07-032071-3.

Ivanic, Joseph, and Klaus Ruedenberg. 1996. 'Rotation Matrices for Real Spherical Harmonics. Direct Determination by Recursion'. *J. Phys. Chem. A* 100 (15): 6342–6347. ISSN: 0122-3654. https : / / www . sciencedirect.com/science/article/pii/S0166128097001851.

———. 1998. 'Correction: Rotation matrices for real spherical harmonics. Direct determination by recursion'. *J. Phys. Chem. A* 102 (45): 9099–9100.

Jackson, John David. 1975. *Classical electrodynamics.* Second ed. xxii+848. John Wiley & Sons, Inc., New York–London–Sydney.

Jeevanjee, Nadir. 2015. *An introduction to tensors and group theory for physicists.* Second ed. xvi+305. Springer, Cham. ISBN: 978-3-319-14793-2. https://doi.org/10.1007/978-3-319-14794-9.

Jenkins, Alexander C., and Mairi Sakellariadou. 2018. 'Anisotropies in the stochastic gravitational-wave background: Formalism and the cosmic string case'. *Phys. Rev. D* 98 (6): 063509. https://link.aps.org/doi/10.1103/PhysRevD.98.063509.

Jetzer, Philippe. 2022. *Applications of General Relativity. With Problems.* UNITEXT for physics. Springer.

Kachelriess, Michael. 2018. *Quantum fields. From the Hubble to the Planck scale.* xv+528. Oxford Graduate Texts. Oxford University Press, Oxford. ISBN: 978-0-19-880287-7. https://doi.org/10.1093/oso/9780198802877.001.0001.

Kallenberg, Olav. 2021. *Foundations of modern probability.* 99:xii+946. Probability Theory and Stochastic Modelling. Third edition. Springer, Cham. ISBN: 978-3-030-61871-1. https://doi.org/10.1007/978-3-030-61871-1.

Kamionkowski, Marc, Arthur Kosowsky, and Albert Stebbins. 1997. 'Statistics of cosmic microwave background polarization'. *Phys. Rev. D* 55 (12): 7368–7388. https://doi.org/10.1103/PhysRevD.55.7368.

Kamionkowski, Marc, and Ely D. Kovetz. 2016. 'The Quest for B Modes from Inflationary Gravitational Waves'. *Annual Review of Astronomy and Astrophysics* 54 (1): 227–269. https://doi.org/10.1146/annurev-astro-081915-023433.

Kaniuth, Eberhard, and Keith F. Taylor. 2013. *Induced representations of locally compact groups.* 197:xiv+343. Cambridge Tracts in Mathematics. Cambridge University Press, Cambridge. ISBN: 978-0-521-76226-7.

Kato, Ryo, and Jiro Soda. 2016. 'Probing circular polarization in stochastic gravitational wave background with pulsar timing arrays'. *Phys. Rev. D* 93 (6): 062003. https://link.aps.org/doi/10.1103/PhysRevD.93.062003.

Killing, Wilhelm. 1888a. 'Die Zusammensetzung der stetigen endlichen Transformations-gruppen'. *Math. Ann.* 31 (2): 252–290. ISSN: 0025-5831. https://doi.org/10.1007/BF01211904.

———. 1888b. 'Die Zusammensetzung der stetigen endlichen Transformationsgruppen'. *Math. Ann.* 33 (1): 1–48. ISSN: 0025-5831. https://doi.org/10.1007/BF01444109.

———. 1889a. 'Die Zusammensetzung der stetigen endlichen Transformationsgruppen'. *Math. Ann.* 34 (1): 57–122. ISSN: 0025-5831. https://doi.org/10.1007/BF01446792.

———. 1889b. 'Erweiterung des Begriffes der Invarianten von Transformationsgruppen'. *Math. Ann.* 35 (3): 423–432. ISSN: 0025-5831. https://doi.org/10.1007/BF01443863.

———. 1890. 'Die Zusammensetzung der stetigen endlichen Transformationsgruppen'. *Math. Ann.* 36 (2): 161–189. ISSN: 0025-5831. https://doi.org/10.1007/BF01207837.

Kim, Young Suh, and Marilyn E. Noz. 1986. *Theory and applications of the Poincaré group.* xvi+331. Fundamental Theories of Physics. D. Reidel Publishing Co., Dordrecht. ISBN: 90-277-2141-6. https://doi.org/10.1007/978-94-009-4558-6.

Klenke, Achim. 2020. *Probability theory—a comprehensive course.* xiv+716. Universitext. Third edition. Springer, Cham. ISBN: 978-3-030-56402-5. https://doi.org/10.1007/978-3-030-56402-5.

Ko, M. K. W., Ezra Ted Newman, and Roger Penrose. 1977. 'The Kähler structure of asymptotic twistor space'. *J. Mathematical Phys.* 18 (1): 58–64. ISSN: 0022-2488. https://doi.org/10.1063/1.523151.

Kobayashi, Shoshichi. 1995. *Transformation groups in differential geometry.* viii+182. Classics in Mathematics. Reprint of the 1972 edition. Springer-Verlag, Berlin. ISBN: 3-540-58659-8.

———. 2014. *Differential geometry of complex vector bundles.* xi + 304. Princeton Legacy Library. Reprint of the 1987 edition. Princeton University Press, Princeton, NJ. ISBN: 978-0-691-60329-2.

Kobayashi, Shoshichi, and Katsumi Nomizu. 1996a. *Foundations of differential geometry. Vol. I.* xii+329. Wiley Classics Library. Reprint of the 1963 original, A Wiley-Interscience Publication. John Wiley & Sons, New York. ISBN: 0-471-15733-3.

———. 1996b. *Foundations of differential geometry. Vol. II.* xvi+468. Wiley Classics Library. Reprint of the 1969 original, A Wiley-Interscience Publication. John Wiley & Sons, New York. ISBN: 0-471-15732-5.

Komatsu, Eiichiro. 2010. 'Hunting for primordial non-Gaussianity in the cosmic microwave background'. *Classical Quantum Gravity* 27 (12): 124010, 26. ISSN: 0264-9381. https://doi.org/10.1088/0264-9381/27/12/124010.

Kopczyński, Wojciech, and Andrzej Trautman. 1992. *Spacetime and gravitation.* vi+170. A Wiley-Interscience Publication. Translated and revised from the 1984 Polish original, Translated by Jerzy Bałdyga and Antoni Pol. John Wiley & Sons, Ltd., Chichester; PWN—Polish Scientific Publishers, Warsaw. ISBN: 0-471-92186-6.

Kuchowicz, Bronisław. 1974. 'Neutrinos in general relativity: four(?) levels of approach'. *Gen. Relativity Gravitation* 5 (2): 201–234. ISSN: 0001-7701. https://doi.org/10.1016/0003-4916(73)90103-6.

Künzle, Hans P. 1972. 'Galilei and Lorentz structures on space-time: comparison of the corresponding geometry and physics'. *Ann. Inst. H. Poincaré Sect. A (N.S.)* 17:337–362. ISSN: 0246-0211.

———. 1976. 'Covariant Newtonian limit of Lorentz space-times'. *Gen. Relativity Gravitation* 7 (5): 445–457. ISSN: 0001-7701. https://doi.org/10.1007/bf00766139.

Lambiase, G., G. Papini, R. Punzi, and G. Scarpetta. 2005. 'Neutrino optics and oscillations in gravitational fields'. *Phys. Rev. D* 71 (7): 073011. https://link.aps.org/doi/10.1103/PhysRevD.71.073011.

Lambourne, Robert J. 2010. *Relativity, gravitation and cosmology.* 312. Cambridge University Press. ISBN: 9780521761192.

Landau, Lev, and Evgenyi Lifshitz. 1975. *Course of theoretical physics, Vol. 2. The classical theory of fields.* Fourth ed. xiv+402. Translated from the Russian by Morton Hamermesh. Pergamon Press, Oxford–New York–Toronto.

Lang, Serge. 2002. *Algebra.* Third ed. 211:xvi+914. Graduate Texts in Mathematics. New York: Springer-Verlag. ISBN: 0-387-95385-X. https://doi.org/10.1007/978-1-4613-0041-0.

Lawson, H. Blaine, Jr., and Marie-Louise Michelsohn. 1989. *Spin geometry.* 38:xii+427. Princeton Mathematical Series. Princeton, NJ: Princeton University Press. ISBN: 0-691-08542-0.

Le Gall, Jean-François. 2022. *Measure theory, probability, and stochastic processes.* 295:xiv+406. Graduate Texts in Mathematics. Springer, Cham. ISBN: 978-3-031-14204-8. https://doi.org/10.1007/978-3-031-14205-5.

Lê Gia, Quôc Thông, Ian H. Sloan, Yu Guang Wang, and Robert S. Womersley. 2017. 'Needlet approximation for isotropic random fields on the sphere'. *J. Approx. Theory* 216:86–116. ISSN: 0021-9045. https://doi.org/10.1016/j.jat.2017.01.001.

Lee, Taeyoung. 2022. 'Real Harmonic Analysis on the Special Orthogonal Group'. *Int. J. Anal. Appl.* 20 (April): 21. https://doi.org/10.28924/2291-8639-20-2022-21.

Leeuwen, Marc van. 2011. *On 'complexifying' vector spaces.* Mathematics Stack Exchange. URL:https://math.stackexchange.com/q/85758 (version: 2011-11-27). https://math.stackexchange.com/users/18880/marc-van-leeuwen, November.

Leinster, Tom. 2014. *Basic category theory.* 143:viii+183. Cambridge Studies in Advanced Mathematics. Cambridge University Press, Cambridge. ISBN: 978-1-107-04424-1. https://doi.org/10.1017/CBO9781107360068.

Leonenko, Mykola, and Lyudmyla Sakhno. 2012. 'On Spectral Representations of Tensor Random Fields on the Sphere'. *Stochastic Analysis and Applications* 30 (1): 44–66. https://doi.org/10.1080/07362994.2012.628912.

Lerario, Antonio, Domenico Marinucci, Maurizia Rossi, and Michele Stecconi. 2022. *Geometry and topology of spin random fields.* arXiv: 2207.08413 [math.PR].

Lesgourgues, Julien, Gianpiero Mangano, Gennaro Miele, and Sergio Pastor. 2013. *Neutrino Cosmology.* Cambridge University Press.

Lesgourgues, Julien, and Sergio Pastor. 2006. 'Massive neutrinos and cosmology'. *Physics Reports* 429 (6): 307–379. ISSN: 0370-1573. https://www.sciencedirect.com/science/article/pii/S0370157306001359.

Lewis, Antony. 2013. 'Efficient sampling of fast and slow cosmological parameters'. *Phys. Rev. D* 87 (10): 103529. https://doi.org/10.1103/PhysRevD.87.103529.

Lewis, Antony, and Sarah Bridle. 2002. 'Cosmological parameters from CMB and other data: A Monte Carlo approach'. *Phys. Rev. D* 66 (10): 103511. https://doi.org/10.1103/PhysRevD.66.103511.

Li, Jun, and Guang-Hai Guo. 2022. 'Measuring the primordial gravitational waves from cosmic microwave background and stochastic gravitational wave background observations'. *Modern Physics Letters A* 37 (10): 2250066. https://doi.org/10.1142/S0217732322500663.

Liddle, Andrew. 2015. *An Introduction to Modern Cosmology.* Wiley. ISBN: 1118502140.

Liddle, Andrew R., and David H. Lyth. 2000. *Cosmological inflation and large-scale structure.* xiv+400. Cambridge University Press, Cambridge. ISBN: 0-521-66022-X. https://doi.org/10.1017/CBO9781139175180.

Lie, Sophus Marius. 2015. *Theory of transformation groups. I.* xvi+643. General properties of continuous transformation groups, A contemporary approach and translation, With the collaboration of Friedrich Engel, Edited and translated from the German and with a foreword by Joël Merker. Heidelberg: Springer. ISBN: 978-3-662-46210-2.

Liebscher, Dierck-Ekkehard. 2005. *Cosmology.* Vol. 210. Springer tracts in modern physics. Springer.

Lin, Joshua Yao-Yu, and Gilbert Holder. 2020. 'Gravitational lensing of the cosmic neutrino background'. *Journal of Cosmology and Astroparticle Physics* 2020, no. 04 (April): 054. https://dx.doi.org/10.1088/1475-7516/2020/04/054.

Lisanti, Mariangela, Benjamin R. Safdi, and Christopher G. Tully. 2014. 'Measuring anisotropies in the cosmic neutrino background'. *Phys. Rev. D* 90 (7): 073006. https://link.aps.org/doi/10.1103/PhysRevD.90.073006.

Liu, Jing, Rong-Gen Cai, and Zong-Kuan Guo. 2021. 'Large Anisotropies of the Stochastic Gravitational Wave Background from Cosmic Domain Walls'. *Phys. Rev. Lett.* 126 (14): 141303. https://link.aps.org/doi/10.1103/PhysRevLett.126.141303.

Lledó, Fernando. 2004. 'Massless relativistic wave equations and quantum field theory'. *Ann. Henri Poincaré* 5 (4): 607–670. ISSN: 1424-0637. https://doi.org/10.1007/s00023-004-0179-3.

Long, Andrew J., Cecilia Lunardini, and Eray Sabancilar. 2014. 'Detecting non-relativistic cosmic neutrinos by capture on tritium: phenomenology and physics potential'. *Journal of Cosmology and Astroparticle Physics* 2014, no. 08 (August): 038. https://dx.doi.org/10.1088/1475-7516/2014/08/038.

Lounesto, Pertti. 2001. *Clifford algebras and spinors*. 2nd ed. 286:x+338. London Mathematical Society Lecture Note Series. Cambridge: Cambridge University Press. ISBN: 0-521-00551-5. https://doi.org/10.1017/CBO9780511526022.

Lovelock, David. 1969. 'The uniqueness of the Einstein field equations in a four-dimensional space'. *Arch. Rational Mech. Anal.* 33 (1): 54–70. ISSN: 0003-9527. https://doi.org/10.1007/BF00248156.

MacCallum, Malcolm A. H., and Abraham Haskel Taub. 1973. 'The averaged Lagrangian and high-frequency gravitational waves'. *Comm. Math. Phys.* 30:153–169. ISSN: 0010-3616. http://projecteuclid.org/euclid.cmp/11038 58809.

Mackey, George Whitelaw. 1949. 'Imprimitivity for representations of locally compact groups. I'. *Proc. Nat. Acad. Sci. U.S.A.* 35:537–545. ISSN: 0027-8424. https://doi.org/10.1073/pnas.35.9.537.

———. 1951. 'On induced representations of groups'. *Amer. J. Math.* 73:576–592. ISSN: 0002-9327. https://doi.org/10.2307/2372309.

———. 1952. 'Induced representations of locally compact groups. I'. *Ann. of Math. (2)* 55:101–139. ISSN: 0003-486X. https://doi.org/10.2307/1969423.

———. 1953. 'Induced representations of locally compact groups. II. The Frobenius reciprocity theorem'. *Ann. of Math. (2)* 58:193–221. ISSN: 0003-486X. https://doi.org/10.2307/1969786.

MacRobert, Thomas Murray. 1947. *Spherical Harmonics. An Elementary Treatise on Harmonic Functions with Applications*. xv+372. 2d ed. Methuen & Co., London.

Madison, Dustin R. 2021. 'Framework for describing perturbations to the cosmic microwave background from a gravitational wave burst with memory'. *Phys. Rev. D* 103 (8): 083515. https://link.aps.org/doi/10.1103/PhysRevD.103.083515.

Madore, J. 1972. 'The dispersion of gravitational waves'. *Comm. Math. Phys.* 27 (4): 291–302. ISSN: 0010-3616. https://doi.org/10.1007/BF01645516.

———. 1973. 'The absorption of gravitational radiation by a dissipative fluid'. *Comm. Math. Phys.* 30 (4): 335–340. ISSN: 0010-3616. https://doi.org/10.1007/BF01645508.

Maggiore, Michele. 2005. *A modern introduction to quantum field theory*. Oxford master series in statistical, computational, and theoretical physics. Oxford University Press, Oxford.

————. 2007. *Gravitational Waves: Volume 1: Theory and Experiments.* Oxford University Press, Oxford, October. ISBN: 9780198570745. https://doi.org/10.1093/acprof:oso/9780198570745.001.0001.

————. 2018. *Gravitational Waves: Volume 2: Astrophysics and Cosmology.* Oxford University Press, Oxford, March. ISBN: 9780198570899. https://doi.org/10.1093/oso/9780198570899.001.0001.

Malament, David B. 2012. *Topics in the foundations of general relativity and Newtonian gravitation theory.* xii+349. Chicago Lectures in Physics. University of Chicago Press, Chicago, IL. ISBN: 978-0-226-50245-8. https://doi.org/10.7208/chicago/9780226502472.001.0001.

Malyarenko, Anatoliy. 2011. 'Invariant random fields in vector bundles and application to cosmology'. *Ann. Inst. Henri Poincaré Probab. Stat.* 47 (4): 1068–1095. ISSN: 0246-0203. https://doi.org/10.1214/10-AIHP409.

————. 2013. *Invariant random fields on spaces with a group action.* xviii+261. Probability and its Applications (New York). With a foreword by Nikolai Leonenko. Springer, Heidelberg. ISBN: 978-3-642-33405-4. https://doi.org/10.1007/978-3-642-33406-1.

Malyarenko, Anatoliy, and Martin Ostoja-Starzewski. 2023. 'Tensor- and spinor-valued random fields with applications to continuum physics and cosmology'. *Probab. Surv.* 20:1–86. https://doi.org/10.1214/22-ps12.

Man, Chi-Sing. 2022. 'Crystallographic texture and group representations'. *J. Elasticity* 149 (1-2): 3–445. ISSN: 0374-3535. https://doi.org/10.1007/s10659-022-09882-8.

Marathe, Kishore. 2010. *Topics in physical mathematics.* xxii+442. Springer-Verlag, London. ISBN: 978-1-84882-938-1. https://doi.org/10.1007/978-1-84882-939-8.

Marinucci, Domenico. 2004. 'Testing for non-Gaussianity on cosmic microwave background radiation: a review'. *Statist. Sci.* 19 (2): 294–307. ISSN: 0883-4237. https://doi.org/10.1214/088342304000000783.

————. 2023. 'Some recent developments on the geometry of random spherical eigenfunctions'. In *Proceedings of the 8th Congress (8ECM) held in Portorož, 20–26 June 2021,* edited by Ademir Hujdurović, Klavdija Kutnar, Dragan Marušič, Štefko Miklavič, Tomaž Pisanski, and Primož Šparl, 337–365. EMS Press, Berlin.

Marinucci, Domenico, and Giovanni Peccati. 2011. *Random fields on the sphere. Representation, limit theorems and cosmological applications.* 389:xii+341. London Mathematical Society Lecture Note Series. Cambridge University Press, Cambridge. ISBN: 978-0-521-17561-6. https://doi.org/10.1017/CBO9780511751677.

Marsden, Jerrold E., and Tudor S. Ratiu. 1999. *Introduction to mechanics and symmetry. A basic exposition of classical mechanical systems.* Second ed. 17:xviii+582. Texts in Applied Mathematics. Springer-Verlag, New York. ISBN: 0-387-98643-X. https://doi.org/10.1007/978-0-387-21792-5.

Martel, Karl, and Eric Poisson. 2005. 'Gravitational perturbations of the Schwarzschild spacetime: A practical covariant and gauge-invariant formalism'. *Phys. Rev. D* 71 (10): 104003. https://doi.org/10.1103/PhysRevD.71.104003.

Mathews, Jon. 1962. 'Gravitational multipole radiation'. *J. Soc. Indust. Appl. Math.* 10:768–780. ISSN: 0368-4245.

Matsui, Yuka, and Sachiko Kuroyanagi. 2019. 'Gravitational-wave background from kink-kink collisions on infinite cosmic strings'. *Phys. Rev. D* 100 (12): 123515. https://link.aps.org/doi/10.1103/PhysRevD.100.123515.

McEwen, Jason D., Claudio Durastanti, and Yves Wiaux. 2018. 'Localisation of directional scale-discretised wavelets on the sphere'. *Appl. Comput. Harmon. Anal.* 44 (1): 59–88. ISSN: 1063-5203. https://doi.org/10.1016/j.acha.2016.03.009.

Messiah, Albert. 1959. *Mécanique quantique.* Vol. 1: xv+pp. 1–430, Vol. 2: xv+pp. 431–974. 2 vols. Dunod, Paris.

Michel, Volker, Alain Plattner, and K. Seibert. 2022. 'A unified approach to scalar, vector, and tensor Slepian functions on the sphere and their construction by a commuting operator'. *Anal. Appl. (Singap.)* 20 (5): 947–988. ISSN: 0219-5305. https://doi.org/10.1142/S0219530521500317.

Michney, R. J., and R. R. Caldwell. 2007. 'Anisotropy of the cosmic neutrino background'. *Journal of Cosmology and Astroparticle Physics* 2007, no. 01 (January): 014. https://doi.org/10.1088/1475-7516/2007/01/014.

Michor, Peter W. 2008. *Topics in differential geometry.* 93:xii+494. Graduate Studies in Mathematics. American Mathematical Society, Providence, RI. ISBN: 978-0-8218-2003-2. https://doi.org/10.1090/gsm/093.

Milnor, John Willard. 1963. 'Spin structures on manifolds'. *Enseign. Math. (2)* 9:198–203. ISSN: 0013-8584.

Milnor, John Willard, and James Dillon Stasheff. 1974. *Characteristic classes.* vii+331. Annals of Mathematics Studies, No. 76. Princeton University Press, Princeton, N. J.; University of Tokyo Press, Tokyo.

Misner, Charles W., Kip S. Thorne, and John Archibald Wheeler. 2017. *Gravitation.* Princeton University Press, Princeton.

Mohapatra, Rabindra N., and Pal B. Palash. 2004. *Massive neutrinos in physics and astrophysics.* Third ed. Vol. 72. World Scientific Lecture Notes in Physics. World Scientific, Singapore.

Morrey, Charles B., Jr. 2008. *Multiple integrals in the calculus of variations.* x+506. Classics in Mathematics. Reprint of the 1966 edition. Springer-Verlag, Berlin. ISBN: 978-3-540-69915-6. https://doi.org/10.1007/978-3-540-69952-1.

Morrow, James, and Kunihiko Kodaira. 2006. *Complex manifolds.* x+194. Reprint of the 1971 edition with errata. AMS Chelsea Publishing, Providence, RI. ISBN: 0-8218-4055-X. https://doi.org/10.1090/chel/355.

Müller, Claus. 1966. *Spherical harmonics.* 17:iv+45. Lecture Notes in Mathematics. Springer-Verlag, Berlin–New York.

Mulliken, Robert Sanderson. 1955. 'Report on Notation for the Spectra of Polyatomic Molecules'. *The Journal of Chemical Physics* 23 (11): 1997–2011. https://doi.org/10.1063/1.1740655.

———. 1956. 'Erratum : Report on Notation for the Spectra of Polyatomic Molecules'. *The Journal of Chemical Physics* 24 (5): 1118–1118. https://doi.org/10.1063/1.1742716.

Munteanu, Gheorghe. 2004. *Complex spaces in Finsler, Lagrange and Hamilton geometries.* 141:xii+221. Fundamental Theories of Physics. Kluwer Academic Publishers, Dordrecht. ISBN: 1-4020-2205-0. https://doi.org/10.1007/978-1-4020-2206-7.

Naber, Gregory L. 1992. *The geometry of Minkowski spacetime. An introduction to the mathematics of the special theory of relativity.* 92:xvi+257. Applied Mathematical Sciences. Springer-Verlag, New York. ISBN: 0-387-97848-8. https://doi.org/10.1007/978-1-4757-4326-5.

———. 2011. *Topology, geometry and gauge fields. Interactions.* Second ed. 141:xii+419. Applied Mathematical Sciences. Springer, New York. ISBN: 978-1-4419-7894-3.

Nakahara, Mikio. 2003. *Geometry, topology and physics.* Second ed. xxii+573. Graduate Student Series in Physics. Institute of Physics, Bristol. ISBN: 0-7503-0606-8. https://doi.org/10.1201/9781420056945.

Narlikar, Jayant V. 2010. *An Introduction to Relativity.* Cambridge University Press, Cambridge. https://doi.org/10.1017/CBO9780511801341.

Newman, Ezra Ted, and Roger Penrose. 1962. 'An approach to gravitational radiation by a method of spin coefficients'. *J. Mathematical Phys.* 3:566–578. ISSN: 0022-2488. https://doi.org/10.1063/1.1724257.

———. 1966. 'Note on the Bondi–Metzner–Sachs group'. *J. Mathematical Phys.* 7:863–870. ISSN: 0022-2488. https://doi.org/10.1063/1.1931221.

Newton, Roger G. 1966. *Scattering theory of waves and particles.* xviii+681. McGraw-Hill Book Co., New York–Toronto–London.

Niederer, Ulrich H., and Lochlainn O'Raifeartaigh. 1974a. 'Realizations of the unitary representations of the inhomogeneous space-time groups. I. General structure'. *Fortschr. Physik* 22:111–129. ISSN: 0015-8208. https://doi.org/10.1002/prop.19740220302.

———. 1974b. 'Realizations of the unitary representations of the inhomogeneous space-time groups. II. Covariant realizations of the Poincaré group'. *Fortschr. Physik* 22:131–157. ISSN: 0015-8208. https://doi.org/10.1002/prop.19740220303.

Nussinov, Shmuel, and Zohar Nussinov. 2022. 'Quantum induced broadening: A challenge for cosmic neutrino background discovery'. *Phys. Rev. D* 105 (4): 043502. https://link.aps.org/doi/10.1103/PhysRevD.105.043502.

O'Donnell, Peter. 2003. *Introduction to 2-spinors in general relativity.* xii+191. World Scientific Publishing Co., River Edge, NJ. ISBN: 981-238-307-7. https://doi.org/10.1142/9789812795311.

O'Raifeartaigh, Cormac, Michael O'Keeffe, Werner Nahm, and Simon Mitton. 2017. 'Einstein's 1917 static model of the universe: a centennial review'. *The European Physical Journal H* 42, no. 3 (July): 431–474. https://doi.org/10.1140/epjh/e2017-80002-5.

Ohanian, Hans C., and Remo Ruffini. 2013. *Gravitation and Spacetime.* Third ed. Cambridge University Press, Cambridge. https://doi.org/10.1017/CBO9781139003391.

Ohnuki, Yoshio. 1988. *Unitary representations of the Poincaré group and relativistic wave equations.* xiv+213. Translated from the Japanese by S. Kitakado and T. Sugiyama. World Scientific Publishing Co., Teaneck, NJ. ISBN: 9971-50-250-X. https://doi.org/10.1142/0537.

Okonek, Christian, Michael Schneider, and Heinz Spindler. 2011. *Vector bundles on complex projective spaces.* viii+239. Modern Birkhäuser Classics. Corrected reprint of the 1988 edition, With an appendix by S. I. Gelfand. Birkhäuser/Springer Basel AG, Basel. ISBN: 978-3-0348-0150-8.

Olver, Frank William John, Daniel W. Lozier, Ronald F. Boisvert, and Charles W. Clark, eds. 2010. *NIST handbook of mathematical functions.* With 1 CD-ROM (Windows, Macintosh and UNIX). U.S. Department of Commerce, National Institute of Standards / Technology, Washington, DC; Cambridge University Press, Cambridge. ISBN: 978-0-521-14063-8.

Ostrogradsky, Mykhailo Vasyliovych. 1850. 'Mémoires sur les équations différentielles, relatives au problème des isopérimètres'. *Mem. Acad. St. Petersbourg* 6 (4): 385–517.

Ota, Atsuhisa. 2022. 'Cosmic microwave background spectral distortions from Rayleigh scattering at second order'. *Phys. Rev. D* 106 (10): Paper No. 103521, 13. ISSN: 2470-0010. https://doi.org/10.1103/physrevd.106.103521.

Pal, Palash B. 2011. 'Dirac, Majorana, and Weyl fermions'. *American Journal of Physics* 79 (5): 485–498. https://doi.org/10.1119/1.3549729.

Paykari, Paniez, Jean-Luc Starck, and Jalal M. Fadili. 2013. 'Theoretical power spectrum estimation from cosmic microwave background data'. In *Statistical challenges in modern astronomy V,* edited by Eric D. Feigelson, 209:539–542. Lect. Notes Stat. Springer, New York.

Penrose, Roger. 1960. 'A spinor approach to general relativity'. *Ann. Physics* 10:171–201. ISSN: 0003-4916. https://doi.org/10.1016/0003-4916(60)90021-X.

———. 1967. 'Twistor algebra'. *J. Mathematical Phys.* 8:345–366. ISSN: 0022-2488. https://doi.org/10.1063/1.1705200.

———. 1968. 'Structure of space-time'. In *Battelle Rencontres. 1967 lectures in mathematics and physics,* edited by Cécile M. DeWitt-Morette and John Archibald Wheeler, 121–235. New York–Amsterdam: W. A. Benjamin.

———. 1969. 'Solutions of the Zero-Rest-Mass Equations'. *Journal of Mathematical Physics* 10 (1): 38–39. https://doi.org/10.1063/1.1664756.

Penrose, Roger, and Wolfgang Rindler. 1987. *Spinors and space-time. Vol. 1. Two-spinor calculus and relativistic fields.* x+458. Cambridge Monographs on Mathematical Physics. Cambridge University Press, Cambridge. ISBN: 0-521-33707-0.

———. 1988. *Spinors and space-time. Vol. 2. Spinor and twistor methods in space-time geometry.* Second ed. x+501. Cambridge Monographs on Mathematical Physics. Cambridge University Press, Cambridge. ISBN: 0-521-34786-6.

Penzias, Arno Alan, and Robert Woodrow Wilson. 1965. 'A Measurement of Excess Antenna Temperature at 4080 Mc/s'. *Astrophys. J.* 142 (July): 419–421. https://doi.org/10.1086/148307.

Perlick, Volker. 2000. *Ray optics, Fermat's principle, and applications to general relativity.* 61:x+220. Lecture Notes in Physics. Monographs. Springer-Verlag, Berlin. ISBN: 3-540-66898-5.

Peter, Patrick, and Jean-Philippe Uzan. 2009. *Primordial cosmology.* Oxford University Press.

Philippoz, Lionel Antoine. 2018. 'On the polarization of gravitational waves'. PhD diss., University of Zurich. https://doi.org/10.5167/uzh-169555.

Piattella, Oliver. 2018. *Lecture notes in cosmology.* xviii+418. Unitext for Physics. Springer, Cham. ISBN: 978-3-319-95569-8. https://doi.org/10. 1007/978-3-319-95570-4.

Pinchon, Didier, and Philip E. Hoggan. 2007. 'Rotation matrices for real spherical harmonics: general rotations of atomic orbitals in space-fixed axes'. *J. Phys. A: Math. and Theoret.* 40 (7): 1597.

Poincaré, Henri. 2010. *Papers on topology.* 37:xx+228. History of Mathematics. ıt Analysis situs and its five supplements, Translated and with an introduction by John Stillwell. American Mathematical Society, Providence, RI; London Mathematical Society, London. ISBN: 978-0-8218-5234-7. https://doi.org/10.1090/hmath/037.

Poisson, Eric, and Clifford M. Will. 2014. *Gravity. Newtonian, Post-Newtonian, Relativistic.* Cambridge Unversity Press.

Porteous, Ian R. 1995. *Clifford algebras and the classical groups.* 50:x+295. Cambridge Studies in Advanced Mathematics. Cambridge University Press, Cambridge. ISBN: 0-521-55177-3. https://doi.org/10.1017/CBO9780511470912.

Pound, Adam, and Barry Wardell. 2020. 'Black Hole Perturbation Theory and Gravitational Self-Force'. In *Handbook of Gravitational Wave Astronomy,* edited by Cosimo Bambi, Stavros Katsanevas, and Konstantinos D. Kokkotas, 1–119. Singapore: Springer. ISBN: 978-981-15-4702-7. https://doi.org/10.1007/978-981-15-4702-7_38-1.

Preti, Giovanni, Fernando de Felice, and Luca Masiero. 2009. 'On the Galilean non-invariance of classical electromagnetism'. *European J. Phys.* 30 (2): 381–391. ISSN: 0143-0807. https://doi.org/10.1088/0143-0807/30/2/017.

Prince, Heather, Kavilan Moodley, Jethro Ridl, and Martin Bucher. 2018. 'Real space lensing reconstruction using cosmic microwave background polarization'. *J. Cosmol. Astropart. Phys.,* no. 1, 034, front matter+33. https://doi.org/10.1088/1475-7516/2018/01/034.

Puget, Jean-Loup. 2021. 'The Planck mission and the cosmic microwave background'. In *The universe,* edited by Bertrand Duplantier and Vincent Rivasseau, 76:73–92. Prog. Math. Phys. Poincaré Seminar 2015. Birkhäuser/Springer, Cham.

Rahaman, Farook. 2021. *General Theory of Relativity. A Mathematical Approach.* University of Cambridge ESOL Examinations.

Ravenni, Andrea, Michele Liguori, Nicola Bartolo, and Maresuke Shiraishi. 2017. 'Primordial non-Gaussianity with μ-type and y-type spectral distortions: exploiting cosmic microwave background polarization and dealing with secondary sources'. *J. Cosmol. Astropart. Phys.,* no. 9, 042, front matter+24. https://doi.org/10.1088/1475-7516/2017/09/042.

Regge, Tullio, and John Archibald Wheeler. 1957. 'Stability of a Schwarzschild Singularity'. *Phys. Rev.* 108 (4): 1063–1069. https://doi.org/10.1103/PhysRev.108.1063.

Ricciardone, A., L. Valbusa Dall'Armi, N. Bartolo, D. Bertacca, M. Liguori, and S. Matarrese. 2021. 'Cross-Correlating Astrophysical and Cosmological Gravitational Wave Backgrounds with the Cosmic Microwave Background'. *Phys. Rev. Lett.* 127 (27): 271301. https://link.aps.org/doi/10.1103/PhysRevLett.127.271301.

Riehl, Emily. 2016. *Category Theory in Context.* 272. Aurora: Dover modern math originals. Dover.

Riemann, Georg Friedrich Bernhard. 2016. 'On the hypotheses which lie at the bases of geometry', x+172. Classic Texts in the Sciences. Edited and with commentary by Jürgen Jost, Expanded English translation of the German original. Cham: Birkhäuser/Springer. ISBN: 978-3-319-26040-2. https://doi.org/10.1007/978-3-319-26042-6.

Rindler, Filip. 2018. *Calculus of variations.* xii+444. Universitext. Springer, Cham. ISBN: 978-3-319-77636-1. https://doi.org/10.1007/978-3-319-77637-8.

Ringwald, Andreas, and Yvonne Y. Y. Wong. 2004. 'Gravitational clustering of relic neutrinos and implications for their detection'. *Journal of Cosmology and Astroparticle Physics* 2004, no. 12 (December): 005. https://dx.doi.org/10.1088/1475-7516/2004/12/005.

Robbin, Joel W., and Dietmar A. Salamon. 2022. *Introduction to differential geometry.* xiii+418. Springer Studium Mathematik—Master. Springer Spektrum, Wiesbaden. ISBN: 978-3-662-64339-6. https://doi.org/10.1007/978-3-662-64340-2.

Roche, Cian, Amir Babak Aazami, and Carla Cederbaum. 2023. 'Exact parallel waves in general relativity'. *Gen. Relativity Gravitation* 55 (2): Paper No. 40. ISSN: 0001-7701. https://doi.org/10.1007/s10714-023-03083-x.

Rose, Morris Edgar. 1955. *Multipole fields.* viii+99. John Wiley & Sons, New York; Chapman & Hall, London.

———. 1957. *Elementary theory of angular momentum.* x+248. John Wiley & Sons, New York; Chapman & Hall, London.

Rotman, Joseph J. 2009. *An introduction to homological algebra.* 2nd ed. xiv+709. Universitext. Springer, New York. ISBN: 978-0-387-24527-0. https://doi.org/10.1007/b98977.

Rowe, E. G. Peter. 2001. *Geometrical physics in Minkowski spacetime.* xvi+248. Springer Monographs in Mathematics. With a foreword by Wojtek J. Zakrzewski. Springer-Verlag London, London. ISBN: 1-85233-366-9. https://doi.org/10.1007/978-1-4471-3893-8.

Rudolph, Gerd, and Matthias Schmidt. 2013. *Differential geometry and mathematical physics. Part I. Manifolds, Lie groups and Hamiltonian systems.* xiv+759. Theoretical and Mathematical Physics. Springer, Dordrecht. ISBN: 978-94-007-5344-0. https://doi.org/10.1007/978-94-007-5345-7.

———. 2017. *Differential geometry and mathematical physics. Part II. Fibre bundles, topology and gauge fields.* xv+830. Theoretical and Mathematical Physics. Springer, Dordrecht. ISBN: 978-94-024-0958-1. https://doi.org/10.1007/978-94-024-0959-8.

Ryder, Lewis H. 1996. *Quantum field theory.* Second ed. xx+487. Cambridge University Press, Cambridge. ISBN: 0-521-47814-6. https://doi.org/10.1017/CBO9780511813900.

———. 2009. *Introduction to general relativity.* xvi+441. Cambridge University Press, Cambridge. ISBN: 978-0-521-84563-2. https://doi.org/10.1017/CBO9780511809033.

Sachs, Rainer Kurt, and A. M. Wolfe. 2007. 'Republication of: "Perturbations of a cosmological model and angular variations of the microwave background" [Astrophys. J. 147 (1967), 73–90]'. Edited and with commentary by George Ellis and with notes on Sachs-Wolfe integration by Jürgen Ehlers, With brief autobiographies by Sachs and Wolfe, *Gen. Relativity Gravitation* 39 (11): 1929–1961. ISSN: 0001-7701. https://doi.org/10.1007/s10714-007-0448-9.

Sachs, Rainer Kurt, and Hung Hsi Wu. 1977a. 'General relativity and cosmology'. *Bull. Amer. Math. Soc.* 83 (6): 1101–1164. ISSN: 0002-9904. https://doi.org/10.1090/S0002-9904-1977-14394-2.

———. 1977b. *General relativity for mathematicians.* 48:xii+291. Graduate Texts in Mathematics. Springer-Verlag, New York–Heidelberg. ISBN: 0-387-90218-X.

Safdi, Benjamin R., Mariangela Lisanti, Joshua Spitz, and Joseph A. Formaggio. 2014. 'Annual modulation of cosmic relic neutrinos'. *Phys. Rev. D* 90 (4): 043001. https://link.aps.org/doi/10.1103/PhysRevD.90.043001.

Scholz, Erhard. 1980. *Geschichte des Mannigfaltigkeitsbegriffs von Riemann bis Poincaré.* 430. Boston, Mass.: Birkhäuser. ISBN: 3-7643-3023-6.

———. 1999. 'The Concept of Manifold, 1850–1950'. Chap. 2 in *History of Topology,* edited by I. M. James, 25–64. Amsterdam: North-Holland. ISBN: 978-0-444-82375-5. https://doi.org/10.1016/B978-044482375-5/50003-1.

Schur, Issai. 1906. 'Neue Begründung der Theorie der Gruppencharaktere'. *Sitz. Preuss. Akad. Wiss.*, 406–432.

Schwartz, Matthew D. 2014. *Quantum field theory and the standard model.* xviii+850. Cambridge University Press, Cambridge. ISBN: 978-1-107-03473-0.

Scorpan, Alexandru. 2005. *The wild world of 4-manifolds.* xx+609. American Mathematical Society, Providence, RI. ISBN: 0-8218-3749-4.

Seljak, Uros, and Matias Zaldarriaga. 1996. 'A Line-of-Sight Integration Approach to Cosmic Microwave Background Anisotropies'. *Astrophys. J.* 469 (October): 437. https://doi.org/10.1086/177793.

Seto, Naoki, and Atsushi Taruya. 2008. 'Polarization analysis of gravitational-wave backgrounds from the correlation signals of ground-based interferometers: Measuring a circular-polarization mode'. *Phys. Rev. D* 77 (10): 103001. https://link.aps.org/doi/10.1103/PhysRevD.77.103001.

Sexl, Roman U., and Helmuth K. Urbantke. 2001. *Relativity, groups, particles. Special relativity and relativistic symmetry in field and particle physics.* xii+388. Springer Physics. Revised and translated from the third German (1992) edition by Urbantke. Springer-Verlag, Vienna. ISBN: 3-211-83443-5. https://doi.org/10.1007/978-3-7091-6234-7.

de-Shalit, Amos, and Igal Talmi. 1963. *Nuclear shell theory.* 14:x+573. Pure and Applied Physics. Academic Press, New York–London.

Sharpe, Richard W. 1997. *Differential geometry. Cartan's generalization of Klein's Erlangen program.* 166:xx+421. Graduate Texts in Mathematics. With a foreword by S. S. Chern. Springer-Verlag, New York. ISBN: 0-387-94732-9.

Shibata, Masaru. 2015. *Numerical relativity.* Vol. 1. 100 years of general relativity. World Scientific, Singapore.

Soffel, Michael H., and Wen-Biao Han. 2019. *Applied general relativity—theory and applications in astronomy, celestial mechanics and metrology.* xx+538. Astronomy and Astrophysics Library. Springer, Cham. ISBN: 978-3-030-19672-1. https://doi.org/10.1007/978-3-030-19673-8.

Song, Iickho, So Ryoung Park, and Seokho Yoon. 2022. *Probability and random variables: theory and applications.* Korean. xii+496. Springer, Cham. ISBN: 978-3-030-97678-1. https://doi.org/10.1007/978-3-030-97679-8.

Sousa, L., P. P. Avelino, and G. S. F. Guedes. 2020. 'Full analytical approximation to the stochastic gravitational wave background generated by cosmic string networks'. *Phys. Rev. D* 101 (10): 103508. https://link.aps.org/doi/10.1103/PhysRevD.101.103508.

Spivak, Michael David. 1979. *A comprehensive introduction to differential geometry. Vol. I.* 2nd ed. xiv+668. Wilmington, Del.: Publish or Perish. ISBN: 0-914098-83-7.

Srednicki, Mark. 2010. *Quantum field theory.* xxii+641. Corrected 4th printing of the 2007 original. Cambridge University Press, Cambridge. ISBN: 978-0-521-86449-7.

Steane, Andrew M. 2017. 'Matter-wave coherence limit owing to cosmic gravitational wave background'. *Physics Letters A* 381 (47): 3905–908. ISSN: 0375-9601. https://www.sciencedirect.com/science/article/pii/S0375960117308010.

Stecconi, Michele. 2022. 'Isotropic random spin weighted functions on S^2 vs isotropic random fields on S^3'. *Theory Probab. Math. Statist.*, no. 107, 77–109. ISSN: 0094-9000. https://doi.org/10.1090/tpms/1177.

Stephani, Hans. 1982. *General relativity. An introduction to the theory of the gravitational field.* xvi+298. Translated from the German by Martin Pollock and John Stewart, Edited and with a preface by Stewart. Cambridge University Press, Cambridge–New York. ISBN: 0-521-24008-5.

Stephani, Hans, Dietrich Kramer, Malcolm MacCallum, Cornelius Hoenselaers, and Eduard Herlt. 2003. *Exact solutions of Einstein's field equations.* Second ed. xxx+701. Cambridge Monographs on Mathematical Physics. Cambridge University Press, Cambridge. ISBN: 0-521-46136-7. https://doi.org/10.1017/CBO9780511535185.

Sternberg, Wolfgang J., and Turner L. Smith. 1944. *The Theory of Potential and Spherical Harmonics.* xii+312. Mathematical Expositions, No. 3. University of Toronto Press, Toronto, Ont.

Stewart, John. 1990. *Advanced general relativity.* viii+228. Cambridge Monographs on Mathematical Physics. Cambridge University Press, Cambridge. ISBN: 0-521-32319-3.

Stokes, George Gabriel. 1851. 'On the Composition and Resolution of Streams of Polarized Light from different Sources'. *Transactions of the Cambridge Philosophical Society* 9 (January): 399–416.

Straumann, Norbert. 1974. 'Minimal assumptions leading to a Robertson–Walker model of the universe'. *Helv. Phys. Acta* 47:379–383. ISSN: 0018-0238.

―――. 2006. 'From primordial quantum fluctuations to the anisotropies of the cosmic microwave background radiation'. *Ann. Phys.* 15 (10-11): 701–847. ISSN: 0003-3804. https://doi.org/10.1002/andp.200610212.

————. 2013. *General relativity*. Second ed. xx+735. Graduate Texts in Physics. Springer, Dordrecht. ISBN: 978-94-007-5409-6. https://doi.org/10.1007/978-94-007-5410-2.

Streater, Raymond F., and Arthur Strong Wightman. 2000. *PCT, spin and statistics, and all that*. x+207. Princeton Landmarks in Physics. Corrected third printing of the 1978 edition. Princeton University Press, Princeton, NJ. ISBN: 0-691-07062-8.

Switzer, Eric R., and Duncan J. Watts. 2016. 'Robust likelihoods for inflationary gravitational waves from maps of cosmic microwave background polarization'. *Phys. Rev. D* 94 (6): 063526, 11. ISSN: 2470-0010. https://doi.org/10.1103/physrevd.94.063526.

Tapp, Kristopher. 2016. *Matrix groups for undergraduates*. Second ed. 79:viii+239. Student Mathematical Library. American Mathematical Society, Providence, RI. ISBN: 978-1-4704-2722-1. https://doi.org/10.1090/stml/079.

Taubes, Clifford Henry. 2011. *Differential geometry. Bundles, connections, metrics and curvature*. 23:xiv+298. Oxford Graduate Texts in Mathematics. Oxford University Press, Oxford. ISBN: 978-0-19-960588-0. https://doi.org/10.1093/acprof:oso/9780199605880.001.0001.

Taylor, Stephen R. 2021. *Nanohertz Gravitational Wave Astronomy*. Taylor & Francis. ISBN: 9781032147062.

Thompson, Jonathan E., Barry Wardell, and Bernard F. Whiting. 2019. 'Gravitational self-force regularization in the Regge–Wheeler and easy gauges'. *Phys. Rev. D* 99 (12): 124046. https://doi.org/10.1103/PhysRevD.99.124046.

Thomson, William Lord Kelvin, and Peter Guthrie Tait. 1867. *Treatise on Natural Philosophy*. Clarendon Press.

Thorne, Kip S. 1980. 'Multipole expansions of gravitational radiation'. *Rev. Modern Phys.* 52 (2, part 1): 299–339. ISSN: 0034-6861. https://doi.org/10.1103/RevModPhys.52.299.

Thorne, Kip S., and Roger D. Blandford. 2017. *Modern classical physics. Optics, fluids, plasmas, elasticity, relativity, and statistical physics*. xl+1511. Princeton University Press, Princeton, NJ. ISBN: 978-0-691-15902-7.

Tiec, Alexandre Le, and Jérôme Novak. 2017. 'Theory of Gravitational Waves'. Chap. 1 in *An Overview of Gravitational Waves*, 1–41. World Scientific. https://www.worldscientific.com/doi/abs/10.1142/9789813141766_0001.

Tinkham, Michael. 1964. *Group theory and quantum mechanics*. xii+340. McGraw-Hill Book Co., New York–Toronto–London.

Todorov, Ivan T. 2011. 'Clifford algebras and spinors'. *Bulg. J. Phys.* 38 (1): 3–28. ISSN: 1310-0157.

Torre, C. G. 2014. 'The spacetime geometry of a null electromagnetic field'. *Classical and Quantum Gravity* 31, no. 4 (February): 045022. https://dx.doi.org/10.1088/0264-9381/31/4/045022.

Torres del Castillo, Gerardo Francisko. 2003. *3-D spinors, spin-weighted functions and their applications.* 32:x+246. Progress in Mathematical Physics. Birkhäuser Boston, Boston, MA. ISBN: 0-8176-3249-2. https://doi.org/10.1007/978-0-8176-8146-3.

Trautman, Andrzej. 1965. 'Foundations and Current Problems of General Relativity: Notes by Graham Dixon, Petros Florides and Gerald Lemmer'. In *Lectures on General Relativity*, 1:1–248. Brandeis summer institute in theoretical physics. Englewood Cliffs, N.J.: Prentice-Hall.

———. 1984. *Differential geometry for physicists.* 2:v+145. Monographs and Textbooks in Physical Science. Stony Brook lectures. Bibliopolis, Naples. ISBN: 88-7088-087-7.

———. 1997. 'Clifford and the "square root" ideas'. In *Geometry and nature (Madeira, 1995),* edited by Hanna Nencka and Jean-Pierre Bourguignon, 203:3–24. Contemp. Math. In memory of W. K. Clifford, Papers from the Conference on New Trends in Geometrical and Topological Methods held in Madeira, July 30–August 5, 1995. Amer. Math. Soc., Providence, RI. https://doi.org/10.1090/conm/203/02577.

———. 1998. 'Reflections and spinors on manifolds'. In *Particles, fields, and gravitation (Łódź, 1998),* edited by Jakub Rembieliński, 453:518–527. AIP Conf. Proc. Papers from the conference dedicated to the memory of Ryszard Rączka held in Łódź, April 15–19, 1998. Amer. Inst. Phys., Woodbury, NY.

———. 2008. 'Connections and the Dirac operator on spinor bundles'. *J. Geom. Phys.* 58 (2): 238–252. ISSN: 0393-0440. https://doi.org/10.1016/j.geomphys.2007.11.001.

Trotta, Roberto, and Alessandro Melchiorri. 2005. 'Indication for Primordial Anisotropies in the Neutrino Background from the Wilkinson Microwave Anisotropy Probe and the Sloan Digital Sky Survey'. *Phys. Rev. Lett.* 95 (1): 011305. https://doi.org/10.1103/PhysRevLett.95.011305.

Tully, Christopher G., and Gemma Zhang. 2021. 'Multi-messenger astrophysics with the cosmic neutrino background'. *Journal of Cosmology and Astroparticle Physics* 2021, no. 06 (June): 053. https://dx.doi.org/10.1088/1475-7516/2021/06/053.

Vakhania, Nikoloz, Vazha Tarieladze, and Sergei Chobanyan. 1987. *Probability distributions on Banach spaces.* 14:xxvi+482. Mathematics and its Applications (Soviet Series). Translated from the Russian and with a preface by Wojbor A. Woyczynski. D. Reidel Publishing Co., Dordrecht. ISBN: 90-277-2496-2. https://doi.org/10.1007/978-94-009-3873-1.

Vakhaniya, Nikoloz. 1998. 'Random vectors with values in quaternion Hilbert spaces'. *Teor. Veroyatnost. i Primenen.* 43 (1): 18–40. ISSN: 0040-361X. https://doi.org/10.1137/S0040585X97976696.

Vakhaniya, Nikoloz, and Nodar Kandelaki. 1996. 'Random vectors with values in complex Hilbert spaces'. *Teor. Veroyatnost. i Primenen.* 41 (1): 31–52. ISSN: 0040-361X. https://doi.org/10.1137/TPRBAU00004100000100011 6000001.

Valle, José W. F., and Jorge C. Romão. 2014. *Neutrinos In High Energy And Astroparticle Physics.* Wiley–VCH.

Vaz, Jayme, Jr., and Roldão da Rocha Jr. 2019. *An introduction to Clifford algebras and spinors.* xiv+242. Oxford: Oxford University Press. ISBN: 978-0-19-883628-5.

Vecchiato, Alberto. 2017. *Variational approach to gravity field theories. From Newton to Einstein and beyond.* xii+361. Undergraduate Lecture Notes in Physics. Springer, Cham. ISBN: 978-3-319-51209-9.

Velo, Giorgio, and Arthur Strong Wightman. 1978. *Invariant Wave Equations.* Vol. 73. Lecture Notes in Physics. Springer.

Vilenkin, Naum Yakovlevich, and Anatoliy Ulianovich Klimyk. 1991. *Representation of Lie groups and special functions. Vol. 1. Simplest Lie groups, special functions and integral transforms.* 72:xxiv+608. Mathematics and its Applications (Soviet Series). Translated from the Russian by V. A. Groza and A. A. Groza. Kluwer Academic Publishers Group, Dordrecht. ISBN: 0-7923-1466-2. https://doi.org/10.1007/978-94-011-3538-2.

———. 1992. *Representation of Lie groups and special functions. Vol. 3. Classical and quantum groups and special functions.* 75:xx+634. Mathematics and its Applications (Soviet Series). Translated from the Russian by V. A. Groza and A. A. Groza. Kluwer Academic Publishers Group, Dordrecht. ISBN: 0-7923-1493-X. https://doi.org/10.1007/978-94-017-2881-2.

———. 1993. *Representation of Lie groups and special functions. Vol. 2. Class I representations, special functions, and integral transforms.* 74:xviii+607. Mathematics and its Applications (Soviet Series). Translated from the Russian by V. A. Groza and A. A. Groza. Kluwer Academic Publishers Group, Dordrecht. ISBN: 0-7923-1492-1. https://doi.org/10.1007/978-94-017-2883-6.

Vilenkin, Naum Yakovlevich, and Anatoliy Ulianovich Klimyk. 1995. *Representation of Lie groups and special functions. Recent advances.* 316:xvi+497. Mathematics and its Applications. Translated from the Russian manuscript by V. A. Groza and A. A. Groza. Kluwer Academic Publishers Group, Dordrecht. ISBN: 0-7923-3210-5. https://doi.org/10.1007/978-94-017-2885-0.

Vogel, Petr. 2015. 'How difficult it would be to detect cosmic neutrino background?' *AIP Conference Proceedings* 1666 (1): 140003. https://aip.scitation.org/doi/abs/10.1063/1.4915587.

Voigt, Woldemar. 1898. *Die fundamentalen physikalischen Eigenschaften der Krystalle in elementarer Darstellung.* Leipzig: Von Veit.

Waerden, Bartel Leendert van der. 1991a. *Algebra. Vol. I.* xiv+265. Based in part on lectures by E. Artin and E. Noether, Translated from the seventh German edition by Fred Blum and John R. Schulenberger. New York: Springer-Verlag. ISBN: 0-387-97424-5. https://doi.org/10.1007/978-1-4612-4420-2.

———. 1991b. *Algebra. Vol. II.* xii+284. Based in part on lectures by E. Artin and E. Noether, Translated from the fifth German edition by John R. Schulenberger. New York: Springer-Verlag. ISBN: 0-387-97425-3.

Wald, Robert M. 1984. *General relativity.* xiii+491. Chicago, IL: University of Chicago Press. ISBN: 0-226-87032-4. https://doi.org/10.7208/chicago/9780226870373.001.0001.

———. 1994. *Quantum field theory in curved spacetime and black hole thermodynamics.* xiv+205. Chicago Lectures in Physics. University of Chicago Press, Chicago, IL. ISBN: 0-226-87025-1.

Wallach, Nolan R. 1973. *Harmonic analysis on homogeneous spaces.* xv+361. Pure and Applied Mathematics, No. 19. Marcel Dekker, New York.

Wardell, Barry, and Niels Warburton. 2015. 'Applying the effective-source approach to frequency-domain self-force calculations: Lorenz-gauge gravitational perturbations'. *Phys. Rev. D* 92 (8): 084019. https://doi.org/10.1103/PhysRevD.92.084019.

Wawrzyńczyk, Antoni. 1984. *Group representations and special functions.* xvi+637. Mathematics and its Applications (East European Series). Examples and problems prepared by Aleksander Strasburger, Translated from the Polish by Bogdan Ziemian. D. Reidel Publishing Co., Dordrecht; PWN—Polish Scientific Publishers, Warsaw. ISBN: 90-277-1269-7. https://doi.org/10.1007/978-94-009-6531-7.

Weinberg, Steven. 1972. *Gravitation and cosmology: principles and applications of the general theory of relativity.* Wiley.

————. 2005a. *The quantum theory of fields. Vol. I. Foundations.* xxvi+609. Cambridge University Press, Cambridge. ISBN: 0-521-55001-7.

————. 2005b. *The quantum theory of fields. Vol. II. Modern applications.* xxii+489. Cambridge University Press, Cambridge. ISBN: 0-521-55002-5.

————. 2005c. *The quantum theory of fields. Vol. III. Supersymmetry.* xxii+419. Cambridge University Press, Cambridge. ISBN: 0-521-66000-9.

————. 2008. *Cosmology.* xviii+593. Oxford University Press, Oxford. ISBN: 978-0-19-852682-7.

Wells, Raymond O., Jr. 2008. *Differential analysis on complex manifolds.* 3rd ed. 65:xiv+299. Graduate Texts in Mathematics. With a new appendix by Oscar Garcia-Prada. Springer, New York. ISBN: 978-0-387-73891-8. https://doi.org/10.1007/978-0-387-73892-5.

Weyl, Hermann. 1997. *The classical groups. Their invariants and representations.* xiv+320. Princeton Landmarks in Mathematics. Fifteenth printing, Princeton Paperbacks. Princeton University Press, Princeton, NJ. ISBN: 0-691-05756-7.

Wheeler, John Archibald. 1998. *Geons, black holes, and quantum foam: A life in physics, With Kenneth Ford.* 380. New York: W. W. Norton & Co. ISBN: 0-393-04642-7.

Whitney, Hassler. 1936. 'Differentiable manifolds'. *Ann. of Math. (2)* 37 (3): 645–680. ISSN: 0003-486X. https://doi.org/10.2307/1968482.

Wibig, Tadeusz, and Arnold W. Wolfendale. 2016. 'Cosmic ray contributions to the WMAP polarization data on the cosmic microwave background'. *Internat. J. Modern Phys. D* 25 (3): 1650029, 11. ISSN: 0218-2718. https://doi.org/10.1142/S0218271816500292.

Wigmans, Richard. 2000. 'On Big Bang relics, the neutrino mass and the spectrum of cosmic rays'. *Nuclear Physics B—Proceedings Supplements* 85 (1): 305–310. ISSN: 0920-5632. https://doi.org/10.1016/S0920-5632(00)00522-3.

Wigner, Eugene Paul. 1927. 'Einige Folgerungen aus der Schrödingerschen Theorie für die Termstrukturen'. *Z. Physik* 43 (9): 624–652.

————. 1959. *Group theory and its application to the quantum mechanics of atomic spectra.* 5:xi+372. Pure and Applied Physics. Expanded and improved ed. Translated from the German by J. J. Griffin. Academic Press, New York–London.

Workman, R. L. et al. 2022. 'Review of Particle Physics'. *Progress of Theoretical and Experimental Physics* 2022, no. 8 (August): 083C01. ISSN: 2050-3911. https://doi.org/10.1093/ptep/ptac097.

Xing, Zhi-Zhong, and Shun Zhou. 2011. *Neutrinos in Particle Physics, Astronomy and Cosmology.* Advanced topics in science and technology in China. Springer, Berlin–Heidelberg.

Yadrenko, M. Ĭ. 1959. 'Isotropic Gauss random fields of the Markov type on a sphere'. *Dopovidi Akad. Nauk Ukrain. RSR* 1959:231–236.

Yano, Kentaro. 1965. *Differential geometry on complex and almost complex spaces.* 49:xii+326. International Series of Monographs in Pure and Applied Mathematics. The Macmillan Company, New York.

Yokota, Ichiro, and Toshikazu Miyashita. 1990. 'Global isomorphisms of lower-dimensional Lie groups'. *J. Fac. Sci. Shinshu Univ.* 25 (2): 59–63. ISSN: 0583-063X.

Yutsis, A. P., J. Levinsonas, and V. V. Vanagas. 1962. *Mathematical apparatus of the theory of angular momentum.* vi+158 pp. (1 fold–out). Translated from the Russian by A. Sen and R. N. Sen. Israel Program for Scientific Translations, Jerusalem.

Zaldarriaga, Matias, and Uroš Seljak. 1997. 'All-sky analysis of polarization in the microwave background'. *Phys. Rev. D* 55 (4): 1830–1840. https://doi.org/10.1103/PhysRevD.55.1830.

Zaldarriaga, Matias, Uroš Seljak, and Edmund Bertschinger. 1998. 'Integral Solution for the Microwave Background Anisotropies in Nonflat Universes'. *The Astrophysical Journal* 494, no. 2 (February): 491. https://doi.org/10.1086/305223.

Zee, Anthony. 2010. *Quantum field theory in a nutshell.* Second ed. xxviii+576. Princeton University Press, Princeton, NJ. ISBN: 978-0-691-14034-6.

Zerilli, Frank J. 1970a. 'Gravitational Field of a Particle Falling in a Schwarzschild Geometry Analyzed in Tensor Harmonics'. *Phys. Rev. D* 2 (10): 2141–2160. https://doi.org/10.1103/PhysRevD.2.2141. https://link.aps.org/doi/10.1103/PhysRevD.2.2141.

———. 1970b. 'Tensor harmonics in canonical form for gravitational radiation and other applications'. *J. Mathematical Phys.* 11:2203–2208. ISSN: 0022-2488. https://doi.org/10.1063/1.1665380.

Zhou, Yi. 2022. 'Methods of Discovering New Physics from the Cosmic Neutrino Background'. *Journal of Physics: Conference Series* 2381, no. 1 (December): 012079. https://dx.doi.org/10.1088/1742-6596/2381/1/012079.

Zlochower, Yosef, Roberto Gómez, Sascha Husa, Luis Lehner, and Jeffrey Winicour. 2003. 'Mode coupling in the nonlinear response of black holes'. *Phys. Rev. D* 68 (8): 084014. https://doi.org/10.1103/PhysRevD.68.084014.

Zuber, Kai. 2020. *Neutrino Physics.* Third ed. Series in high energy physics, cosmology, and gravitation. Taylor & Francis.

Index

For Product Safety Concerns and Information please contact our EU
representative GPSR@taylorandfrancis.com
Taylor & Francis Verlag GmbH, Kaufingerstraße 24, 80331 München, Germany

www.ingramcontent.com/pod-product-compliance
Lightning Source LLC
Chambersburg PA
CBHW060345220326
41598CB00023B/2809